SOLID OXIDE FUEL CELLS

Materials Properties and Performance

T0136380

Green Chemistry and Chemical Engineering

SOLID OXIDE FUEL CELLS

Materials Properties and Performance

EDITED BY

Jeffrey W. Fergus
Rob Hui
Xianguo Li
David P. Wilkinson
Jiujun Zhang

CRC Press
Taylor & Francis Group
Boca Raton London New York

CRC Press is an imprint of the
Taylor & Francis Group, an **informa** business

CRC Press
Taylor & Francis Group
6000 Broken Sound Parkway NW, Suite 300
Boca Raton, FL 33487-2742

First issued in paperback 2019

© 2009 by Taylor & Francis Group, LLC
CRC Press is an imprint of Taylor & Francis Group, an Informa business

No claim to original U.S. Government works

ISBN-13: 978-1-4200-8883-0 (hbk)
ISBN-13: 978-0-367-38643-6 (pbk)

Library of Congress Cataloging-in-Publication Data

Solid oxide fuel cells : materials properties and performance / editors, Jeffrey
  Fergus ... [et al.].
       p. cm. -- (Green chemistry and chemical engineering series)
    Includes bibliographical references and index.
    ISBN 978-1-4200-8883-0 (alk. paper)
    1. Solid oxide fuel cells--Materials. I. Fergus, Jeffrey. II. Title: SOFCs materials
properties and performance. III. Series.

TK2931.S63 2009
621.31'2429--dc22   2008041366

**Visit the Taylor & Francis Web site at
http://www.taylorandfrancis.com**

**and the CRC Press Web site at
http://www.crcpress.com**

Contents

Green Chemistry and Chemical Engineering
A Book Series by CRC Press/Taylor & Francis

The subjects and disciplines of chemistry and chemical engineering have encountered a new landmark in the way of thinking about, developing, and designing chemical products and processes. This revolutionary philosophy, termed **green chemistry and chemical engineering**, focuses on the designs of products and processes which are conducive to reducing or eliminating the use and/or generation of hazardous substances. In dealing with hazardous or potentially hazardous substances, there may be some overlaps and interrelationships between environmental chemistry and green chemistry. While environmental chemistry is the chemistry of the natural environment and the pollutant chemicals in nature, green chemistry proactively aims to reduce and prevent pollution at its very source. In essence, the philosophies of green chemistry and chemical engineering tend to focus more on industrial application and practice rather than academic principles and phenomenological science. However, as both a chemistry and chemical engineering philosophy, green chemistry and chemical engineering derives from and builds upon organic chemistry, inorganic chemistry, polymer chemistry, fuel chemistry, biochemistry, analytical chemistry, physical chemistry, environmental chemistry, thermodynamics, chemical reaction engineering, transport phenomena, chemical process design, separation technology, automatic process control, and more. In short, green chemistry and chemical engineering is the rigorous use of chemistry and chemical engineering for pollution prevention and environmental protection.

The Pollution Prevention Act of 1990 in the United States established a national policy to prevent or reduce pollution at its source whenever feasible. And adhering to the spirit of this policy, the Environmental Protection Agency (EPA) launched its Green Chemistry Program, in order to promote innovative chemical technologies which reduce or eliminate the use or generation of hazardous substances in the design, manufacture, and use of chemical products. The global efforts in green chemistry and chemical engineering have recently gained a substantial amount of support from the international community of science, engineering, academia, industry, and governments in all phases and aspects.

Some of the successful examples and key technological developments include the use of supercritical carbon dioxide as green solvent in separation technologies, application of supercritical water oxidation for destruction of harmful substances, process integration with carbon dioxide sequestration steps, solvent-free synthesis of chemicals and polymeric materials, exploitation of biologically degradable materials, use of aqueous hydrogen peroxide for efficient oxidation, development of hydrogen proton exchange membrane (PEM) fuel cells for a variety of power generation needs, advanced biofuel productions, devulcanization of spent tire rubber, avoidance of the use of chemicals and processes causing generation of volatile organic compounds (VOCs), replacement of traditional petrochemical processes by microorganism-based bioengineering processes, replacement of chlorofluoroacrbons (CFCs) with nonhazardous alternatives, advances in design of energy efficient processes, use of clean, alternative and renewable energy sources in manufacturing, and much more. This list, even though it is only a partial compilation, is undoubtedly growing exponentially.

This book series on Green Chemistry and Chemical Engineering by CRC Press/Taylor & Francis is designed to meet the new challenges of the 21st century in the chemistry and chemical engineering disciplines by publishing books and monographs based upon cutting-edge research and development to the effect of reducing adverse impacts upon the environment by chemical enterprise. And in achieving this, the series will detail the development of alternative sustainable technologies which will minimize the hazard and maximize the efficiency of any chemical choice. The series aims at delivering the readers in academia and industry with an authoritative information source in the field of green chemistry and chemical engineering. The Publisher and its series editor are fully aware of the rapidly evolving nature of the subject and its long-lasting impact upon the quality of human life in both the present and future. As such, the team is committed to making this series the most comprehensive and accurate literary source in the field of green chemistry and chemical engineering.

Sunggyu Lee

Preface

A fuel cell is an electrochemical device that provides efficient and clean power generation. Due to their unique operation and performance characteristics, solid oxide fuel cells (SOFCs) are well suited for distributed on-site cogeneration of heat and power. The SOFC is an important enabling technology for future sustainable energy systems. SOFCs, in particular, can be used with a wide variety of fuels and are affected less by impurities than polymer electrolyte membrane (PEM) fuel cells. This fuel tolerance is a result of the high operating temperature, which not only accelerates the fuel reactions, but also accelerates unwanted chemical reactions that can lead to degradation of materials. Therefore, progress in SOFCs relies heavily on materials development, and, although significant progress has been made over the past several decades, many technical challenges remain to be overcome before SOFCs can be widely commercialized.

The objective of this book is therefore to provide a summary of the progress that has been made so far in the field, the current state-of-the-art materials and future materials development, as well as the significant technical challenges that remain to be resolved, such that this book can serve as a valuable resource on the materials-related issues in the development of SOFCs for beginners as well as for experienced researchers and developers of SOFCs.

This book reviews the critical issues and recent progress on materials for SOFCs. The first three chapters focus on materials for the electrochemical cell, starting with the electrolyte, which is the component that converts the chemical energy to electrical energy. Recent efforts to reduce the operating temperature of SOFCs require electrolytes with higher conductivities than the most commonly used yttria-stabilized zirconia (YSZ). Because of their high operating temperatures, SOFCs do not require platinum catalysts for the electrodes like PEM fuel cells. However, the difference between the high-temperature gas compositions at the anode and cathode requires that different materials be used for the two electrodes, so the anode and cathode materials are covered in two separate chapters. The most common anode material is a two-phase mixture of nickel and YSZ, but improvements in the microstructural stability and performance in the presence of non-hydrogen fuels, such as sulfur-containing and hydrocarbon gases, are needed. The most common cathode material is lanthanum strontium manganate (LSM), but alternative cathode materials with improved electrochemical properties and resistance to chromium poisoning are being developed for improved performance.

The next two chapters discuss two supporting components of the fuel cell stack —specifically, interconnects and sealants. The interconnect conducts the electrical current between the two electrodes through the external circuit and is thus simultaneously exposed to both high oxygen partial pressure (air) and low oxygen partial pressure (fuel), which places stringent requirements on the materials stability. Ceramic interconnects have been used, but metallic interconnects offer promise for the production of lower cost SOFCs. Sealants are needed to separate the two gases. The most common approach is to use a glass or glass-ceramic, which can provide a gas-tight seal, but is susceptible to thermal and mechanical stresses. Another approach is to use compressive seals, which are more forgiving of differences in thermal expansion, but require application of a load during operation.

The final chapter addresses the cross-cutting issue of materials processing for SOFC applications. The challenges in developing materials that satisfy the stringent materials property requirements is further complicated by the need to fabricate the materials in the desired shapes and with the desired microstructures. The production of a cost-effective SOFC requires compromises between materials properties and processing methods to produce materials with adequate properties at an acceptable cost.

About the Editors

Jeffrey W. Fergus, Ph.D., received his B.S. in metallurgical engineering from the University of Illinois, Urbana-Champaign, in 1985, and his Ph.D. in materials science and engineering from the University of Pennsylvania, Philadelphia, in 1990. He was a postdoctoral research associate in the Center for Sensor Materials at the University of Notre Dame, Notre Dame, IN, and, in 1992, joined the materials engineering program at Auburn University, Auburn, AL, where he is currently an associate professor. His research interests are generally in high-temperature and solid-state chemistry of materials, including electrochemical devices (e.g., chemical sensors and fuel cells) and the chemical stability of materials (e.g., high-temperature oxidation). Dr. Fergus is an active member of The Electrochemical Society, the Metals, Minerals and Materials Society, the American Ceramics Society, the Materials Research Society, and the American Society for Engineering Education.

Rob Hui, Ph.D., received his Ph.D. in materials science and engineering at McMaster University, Hamilton, Ontario, Canada, in 2000. Dr. Hui is senior research officer and acting group leader for high-temperature fuel cells at NRC Institute for Fuel Cell Innovation in Vancouver, British Columbia. He is an adjunct professor at the University of British Columbia in Canada and three major universities in China. He has conducted research and development for materials, processing, and characterization for more than 20 years. Dr. Hui has worked on various projects including chemical sensors, solid oxide fuel cells, magnetic materials, gas separation membranes, nano-structured materials, thin film fabrication, and protective coatings for metals. He has led or has been involved in more than 30 national and international projects funded by governments or industries in China, the United States, and Canada. Dr. Hui is an active member of The Electrochemical Society, and has authored more than 60 publications.

Xianguo Li, Ph.D., received his B.Sc. from Tianjin University, China, in 1982, M.Sc. in 1986, and Ph.D. in 1989 from Northwestern University, Evanston, IL. Dr. Li's academic career formally began in 1992 when he was appointed as an assistant professor in the Department of Mechanical Engineering, University of Victoria. In 1997, he joined the University of Waterloo, and was promoted to the rank of full professor in 2000. Dr. Li's current research involves both experimental and theoretical analyses in the areas of fuel cells, green energy systems, liquid atomization, and sprays. He has published extensively, including journal and conference articles, confidential contract reports, and invited seminars and presentations. Dr. Li is also active in the professional community, serving as the editor-in-chief for the *International Journal of Green Energy* and on the editorial board for a number of international journals and an encyclopedia on energy engineering and technology; he has also served as guest editor for a number of international journals.

David P. Wilkinson, Ph.D., received his B.A.Sc. in chemical engineering from the University of British Columbia in 1978 and his Ph.D. in chemistry from the University of Ottawa in 1987. He has more than 20 years of industrial experience in the areas of fuel cells and advanced rechargeable lithium batteries. In 2004 Dr. Wilkinson was awarded a Tier 1 Canada Research Chair in Clean Energy and Fuel Cells in the Department of Chemical and Biological Engineering at the University of British Columbia. He presently maintains a joint appointment with the university and the Canadian National Research Council Institute for Fuel Cell Innovation (NRC-IFCI). Prior to this appointment, Dr. Wilkinson was the director, and then vice president of research and development at Ballard Power Systems Inc., involved with the research, development, and application of fuel cell technology for transportation, stationary power, and portable applications. Until 2003, Dr. Wilkinson was the leading all-time fuel cell inventor by issued U.S. patents. Dr. Wilkinson's main research interest is in electrochemical power sources and processes to create clean and sustainable energy.

Jiujun Zhang, Ph.D., received his B.S. and M.Sc. in electrochemistry from Peking University in 1982 and 1985, and his Ph.D. in electrochemistry from Wuhan University in 1988. Dr. Zhang is a senior research officer and PEM catalysis core competency leader at the National Research Council of Canada Institute for Fuel Cell Innovation (NRC-IFCI) in Vancouver, British Columbia. After completing his Ph.D., he was an associate professor at the Huazhong Normal University, Wuhan, China, for two years. Starting in 1990, he carried out three terms of postdoctoral research at the California Institute of Technology, York University, Toronto, Ontario, Canada, and the University of British Columbia. Dr. Zhang has more than 26 years of research and development (R&D) experience in theoretical and applied electrochemistry, including more than 12 years of fuel cell R&D, among these, 6 years at Ballard Power Systems, 4 years at NRC-IFCI, and 3 years of electrochemical sensor experience. Dr. Zhang holds seven adjunct professorships, including one at the University of Waterloo and one at the University of British Columbia. His research is based on low/non-Pt cathode catalyst development with long-term stability for catalyst cost reduction, preparation of novel material-supported Pt catalysts through ultrasonic spray pyrolysis, catalyst layer/cathode structure, fundamental understanding through first principles theoretical modeling, catalyst layer characterization and electrochemical evaluation, and preparation of cost-effective Membrane Electrolyte Assemblies (MEAs) for fuel cell testing and evaluation. Dr. Zhang has coauthored more than 140 research papers published in refereed journals and holds more than 10 U.S. patents. He has also produced in excess of 70 industrial technical reports. Dr. Zhang is an active member of The Electrochemical Society, the International Society of Electrochemistry, and the American Chemical Society.

Contributors

Zhe Cheng
School of Materials Science
and Engineering
Georgia Institute of Technology
Atlanta, Georgia, USA

S. Elangovan
Ceramatec, Inc.
Salt Lake City, Utah, USA

J. Hartvigsen
Ceramatec, Inc.
Salt Lake City, Utah, USA

San Ping Jiang
School of Mechanical and Aerospace
Engineering
Nanyang Technological University
Singapore

Olivera Kesler
Department of Mechanical
and Industrial Engineering
University of Toronto
Toronto, Ontario, Canada

P.A. Lessing
Idaho National Laboratory
Idaho Falls, Idaho, USA

Jian Li
School of Materials Science
and Engineering
Huazhong University of Science
and Technology
Wuhan, People's Republic of China

Meilin Liu
School of Materials Science
and Engineering
Georgia Institute of Technology
Atlanta, Georgia, USA

Paolo Marcazzan
Department of Mechanical Engineering
University of British Columbia
Vancouver, British Columbia, Canada

Jeng-Han Wang
School of Materials Science
and Engineering
Georgia Institute of Technology
Atlanta, Georgia, USA

Changrong Xia
Department of Materials Science
and Engineering
University of Science and Technology
of China
Hefei, People's Republic of China

Zhenguo (Gary) Yang
Pacific Northwest National Laboratory
(PNNL)
Richland, Washington, USA

1 Electrolytes

Changrong Xia

CONTENTS

1.1 SOLID ELECTROLYTES FOR SOFCS

The electrolytes for solid oxide fuel cells (SOFCs) are generally oxygen ion conductors, in which current flow occurs by the movement of oxygen ions through the crystal lattice. This movement is a result of thermally activated hopping of the oxygen ion, moving from one crystal lattice site to its neighbor site. To achieve the movement, the crystal must contain unoccupied sites equivalent to those occupied by the lattice oxygen ions, and the energy involved in the process of migration from one site to the unoccupied equivalent site must be small, certainly less than about 1 eV. This small barrier to migration would seem, at first glance, difficult to attain since the oxygen ions are the largest components of the lattice, with an ionic radius of 0.14 nm. Intuitively, it would be expected that the smaller metal ions would be more likely to have an appreciable mobility in the lattice and, hence, carry the current. However, in certain crystal structures, oxygen defects are predominant, so oxygen ions migrate in the electric field. Examples of these crystal structures with partially occupied oxygen sites are ZrO_2-, CeO_2-, and Bi_2O_3-based oxides with fluorite structure, $LaGaO_3$-based perovskites, derivatives of $Bi_4V_2O_{11}$ and $La_2Mo_2O_9$, perovskite- and brownmillerite-like phases (e.g., derived from $Ba_2In_2O_5$), pyrochlores, and $(Gd, Ca)_2Ti_2O_{7-\delta}$[1].

Shown in Figure 1.1 is the oxygen ion conductivity of selected oxides. Among these oxides, only a few materials have been developed as SOFC electrolytes due to numerous requirements of the electrolyte components. These requirements include fast ionic transport, negligible electronic conduction, and thermodynamic stability over a wide range of temperature and oxygen partial pressure. In addition, they must

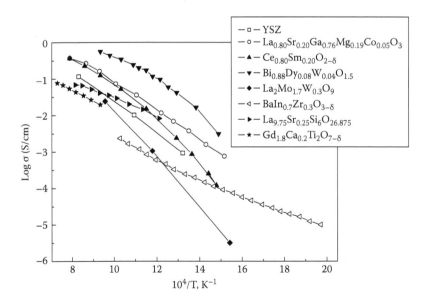

FIGURE 1.1 Oxygen ionic conductivity of various solid-state electrolytes. (Data from Kharton, V.V. et al., *Solid State Ionics*, 174, 135, 2004.)

have thermal expansion compatible with that of electrodes and other construction materials, negligible volatilization of components, suitable mechanical properties, and negligible interaction with electrode materials under operation and fabrication conditions. These requirements limit the choice of electrolyte materials to zirconia and ceria fluorites and $LaGaO_3$-based perovskites. The zirconia electrolyte has been the most favored electrolyte for SOFCs, especially those operated at temperatures above 800°C. Several books and numerous review articles have been written in recent years on the preparation and properties of zirconia electrolytes [1–4]. The ceria electrolyte is attractive for low-temperature (400–600°C) SOFCs and as electrolyte interlayers for intermediate-temperature (600–800°C) SOFCs with zirconia electrolytes. In this chapter, the characteristics of zirconia and ceria, as well as those of $LaGaO_3$-based electrolytes which are attractive for intermediate-temperature SOFCs, will be reviewed.

1.2 OXYGEN ION CONDUCTION

Unlike zirconia, pure stoichiometric ceria forms the fluorite structure over the whole temperature range from room temperature to the melting point, 2,400°C. This fluorite structure is built on the basis of a Ce^{4+} cation face centered cubic (FCC) packing with oxygen ions located in the tetrahedral sites of the structure. To obtain high oxygen ion conduction properties part of the Ce^{4+} must be substituted by another cation with a lower valence state, such as Gd^{3+}, Sm^{3+}, or Y^{3+}. To achieve the electrical neutrality, vacancies are simultaneously created in the anion network. Therefore, the ionic conductivity of substituted ceria, i.e., doped ceria, at a given temperature depends on the nature and the concentration of the aliovalent cation. For example, the dissolution of samaria into the fluorite phase of CeO_2 can be written by the following defect equation in Kröger–Vink notation

$$Sm_2O_3 \xrightarrow{\;CeO_2\;} 2Sm'_{Ce} + 3O_O^{\times} + V_O^{\cdot\cdot} \qquad (1.1)$$

Each additional samaria formula unit creates one oxygen vacancy. The concentration of the vacancies is given by the electrical neutrality condition, $2[Sm'_{Ce}] = [V_O^{\cdot\cdot}]$, inferring that the vacancy concentration is linearly dependent on the dopant level. The ionic conductivity, σ, can be calculated by

$$\sigma = en\mu \qquad (1.2)$$

where e is the charge, μ the mobility of oxygen vacancy, and n the number of mobile oxygen ion vacancies. Increasing the dopant concentration will lead to the introduction of more vacancies into the lattice and should result in higher conductivity. Unfortunately, this correlation only applies at low concentrations of dopant and it is found that at higher concentrations the oxygen ion conductivity is limited. In the case of oxygen ion conductors, such as doped ceria and stabilized zirconia, Equation (1.2) results in Equation (1.3) which indicates that the conductivity increases with increasing fraction of mobile oxide ion vacancies [5]. To move through the crystal, the oxygen ions must be able to move into an unoccupied equivalent site with a minimum of hindrance, thus

$$\sigma = \frac{A}{T}[V_{\ddot{O}}](1-[V_{\ddot{O}}])\exp\left(-\frac{E}{RT}\right) \tag{1.3}$$

where E is the activation energy for conduction, R the gas constant, T absolute temperature, and A the preexponential factor. Therefore, the conductivity of doped ceria and stabilized zirconia varies as a function of dopant concentration and shows a maximum at a specific concentration. A simple analysis would predict that conductivity achieves a maximum when half of the oxygen lattice sites are vacant, but this is not the case. The isothermal conductivity does increase as the level of substitution increases, but the maximum is observed at relatively low additions of dopant because of the interactions of the substitutional cations with the charge-compensating oxygen vacancy it introduces. Shown in Figure 1.2 is the conductivity of yttria-doped zirconium and cerium oxides as a function of dopant concentration [6–8]. Doped bismuth oxides also have similar behavior.

The maximum conductivity of fluorite-structured oxygen ion conductors is not only a function of dopant concentration, but also of dopant radius [9]. For stabilized zirconia, the conductivity increases as the radius becomes close to that of Zr^{4+} ($r_{VIII} = 0.084$ nm). The best value is reached for Sc^{3+} ($r_{VIII} = 0.087$ nm) with a conductivity of 0.1 Scm^{-1} at 800°C for $(Sc_2O_3)_{0.1}(ZrO_2)_{0.9}$. For the sake of comparison,

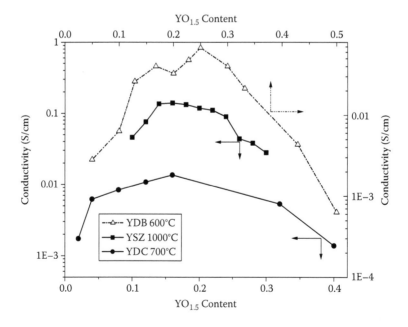

FIGURE 1.2 Composition dependence of conductivity for yttria-stabilized zirconia (YSZ) measured at 1000°C [7], yttria-doped bismuth oxide (YDB) at 600°C [6], and yttria-doped ceria (YDC) at 700°C [8].

the conductivity for $(Y_2O_3)_{0.1}(ZrO_2)_{0.9}$ at the same temperature is close to 0.03 Scm^{-1} (r_{VIII} = 0.102 nm). Similarly, the best conductivity of bismuth-oxide electrolytes is observed with erbia dopant. These show that the highest ionic conductivity would be obtained in a lattice with the lowest difference between the dopant and host radius. It is therefore inferred that the oxygen ion conductivity in fluorite could be enhanced by a decrease of the ionic radius mismatch. The lowest mismatch implies the lowest internal stress in the lattice and thus the lowest activation energy for oxygen ion migration. Accordingly, gadolinia should give the highest conductivity as the dopant in ceria electrolytes although some controversies have arisen concerning which dopant has given the best conductivity.

1.3 ZIRCONIA ELECTROLYTES

The zirconia-based oxides are the most widely used electrolytes for SOFCs operated at high (800–1000°C) and intermediate (600–800°C) temperatures. Most of the SOFC systems currently being developed employ yttria- and scandia-stabilized zirconia electrolytes. This is because of their high ionic conductivity, high mechanical and chemical stability, and compatibility toward other components used in the SOFC. The zirconia is also abundant and relatively low in cost. With respect to the other solid electrolytes, zirconia ceramics exhibit a minimum electronic contribution to total conductivity in the oxygen partial pressure range (>10^{-30} atm) most important for practical applications as fuel cells. This approximate oxygen partial pressure range is from 100 to 200 atm down to 10^{-25} to 10^{-20} atm [1]. It should be noted that in reducing environments the n-type electronic transport in stabilized zirconia is higher than that of LaGaO$_3$-based electrolytes, but zirconia shows lower p-type electronic conduction and, thus, higher performance under oxidizing conditions. Note also that the performance of lanthanum gallate at low-oxygen partial pressures is limited by reduction and gallium oxide volatilization rather than the n-type electronic conductivity.

1.3.1 YTTRIA-STABILIZED ZIRCONIA

Zirconia exists in three different structures of monoclinic, tetragonal, and cubic phases [10]. The monoclinic zirconia, m-ZrO$_2$, is the stable form at room temperature, the cubic phase, c-ZrO$_2$, is stable at the high temperature, while the tetragonal phase, t-ZrO$_2$, is the transition state between m-ZrO$_2$ and c-ZrO$_2$. Upon heating, the m→t-ZrO$_2$ and t→c-ZrO$_2$ phase transformation occur at 1170 and 2370°C, respectively. These phase transformations are martensitic in nature and reversible on cooling although the temperature at which the t→m-ZrO$_2$ transformation occurs is somewhat lower, 950–1000°C. The monoclinic and tetragonal phase transformations are accompanied by a large volume change leading to disintegration of the ceramic body during cooling from high sintering–temperatures to subsequent low usage–temperatures. Dopants, such as yttria, are added to stabilize the high-temperature cubic and tetragonal structures to low temperature by forming solid solutions with ZrO$_2$. For example, 8.0 mol% Y$_2$O$_3$ is believed to be the lowest concentration to stabilize the cubic phase to room temperature. The composition range over which tetragonal and cubic structures exist is narrow and temperature dependent, and is determined by the type of the dopant. For compositions below the minimum amount of the dopant required to stabilize the cubic structure, the phase assemblage is complex and may consist of two or more phases in addition to variants of the same phase with a different degree of dopant distribution. For example, Ioffe et al. [11] reported the existence of the cubic phase down to 1.4 mol% Y_2O_3. However, this seems to contradict the recent literature on phase diagrams, where the equilibrium phase assemblage at 1000°C for 4.5 to 8.0 mol% Y_2O_3 should be a mixture of cubic and tetragonal phases. The two-phase mixture can be beneficial as in partially stabilized zirconia, in which transformation toughening improves the mechanical properties.

When Y^{3+} cations are used to substitute Zr^{4+} at the corresponding lattice sites, they also create vacancies in the oxygen sublattice since Y^{3+} cations have a lower valence than Zr^{4+}. The vacancy production can be shown in Kröger–Vink notation similar to Equation 1.1.

$$Y_2O_3 \xrightarrow{ZrO_2} 2Y'_{Zr} + V_O^{\cdot\cdot} + 3O_O^\times \qquad (1.4)$$

The vacancy is considered to be transportable and thus the reason for oxygen ion conduction in the stabilized zirconia. The composition with 8.0 mol% Y_2O_3 (8YSZ, or yttria-stabilized zirconia) has traditionally dominated because the ionic conductivity exhibits a maximum at that yttria content [12–16] (Figure 1.3), which is supposed to be the minimum amount of yttria needed for the stabilization of the fluorite type cubic zirconia down to room temperature [10, 17, 18]. This optimal concentration and the corresponding conductivity are greatly dependent on the processing history and microstructural features, such as dopant segregation, impurities, kinetically limited phase transitions, and formation of ordered microdomains [1].

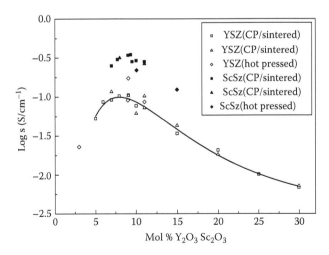

FIGURE 1.3 Conductivity of yttria- and scandia-stabilized zirconia in air at 1000°C. (Data from Fergus, J.W., *J. Power Sources*, 162, 30, 2006.)

As shown in Figure 1.3, the conductivity usually increases with dopant concentration when the composition is smaller than 8.0 mol% Y_2O_3. The increase is due to the increase of the amount of charge carriers, i.e., the oxygen vacancies. At a vacancy concentration of about 3%, which corresponds to an amount of 7.0 mol% Y_2O_3, electrostatic interactions take place among defects of Y'_{Zr} and $V^{..}_O$, which results in an increase of the activation energy. At 8.0 mol% Y_2O_3, there is an equilibrium between the increase of oxygen vacancies and the increase of electrostatic interactions between the created defects, i.e., the vacancies and the substituted dopant cations, Y'_{Zr}. On further increase of the dopant concentration beyond 8.0 mol% Y_2O_3, the interaction among the defects becomes the dominating factor, giving rise to a gradual decrease in conductivity because the mobility of the oxygen ions is hindered [12,19].

The activation energy for conductivity is believed to be caused by oxygen-vacancy migration and defect-pair disassociation. Badwal [10] has observed a change in the slope of the Arrhenius plots, toward decreasing activation energy with an increase in temperature in various YSZ systems. In compositions of interest as solid electrolytes, the curvature in the Arrhenius plots decreases with decreasing dopant content (Figure 1.4). Compositions with low dopant content show almost straight line behavior with increasing or decreasing temperatures. Table 1.1 gives the activation energy in the high- and low-temperature ranges for a number of compositions. The change in the conductivity with the dopant content at a constant temperature and the conductivity behavior as a function of temperature are considered to be caused by factors including degrees of interaction between defect pairs, ordering of vacancies, different defect configurations, and complex defect pair associations.

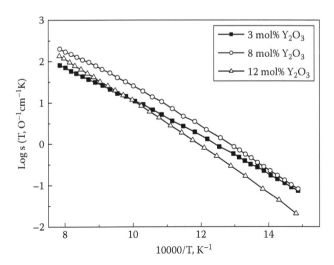

FIGURE 1.4 Arrhenius plots for three Y_2O_3-ZrO_2 compositions showing different curvatures with increasing temperatures. (Data from Badwal, S.P.S., *Solid State Ionics*, 52, 23, 1992.)

TABLE 1.1

Activation Energy (Ea) and Conductivity (σ) Values at High- and Low-Temperature Range for Some Zirconia-Based Electrolytes [10]

Composition	Ea, eV		σ (1000°C), S·cm^{-1}	
	400–500°C	850–1000°C	(a)	(b)
$3Y_2O_3$	0.95	0.80	0.056	0.049
$8Y_2O_3$	1.10	0.91	0.164	0.137
$10Y_2O_3$	1.09	0.83	0.13	0.13
$12Y_2O_3$	1.20	1.04	0.068	0.068
$7.8Sc_2O_3$	1.35	0.78	0.32	0.21

Note: (a) Immediately after temperature reaches 1000°C and (b) after 5000-min annealing.

1.3.2 SCANDIA-STABILIZED ZIRCONIA

Scandia-zirconia solutions are of considerable interest as candidate materials for intermediate-temperature SOFCs because of their high ionic conductivity compared with any of the known stabilized zirconia electrolytes [20]. However, several reports on the ionic conductivity and phase analysis of the scandia zirconia system indicate that it is complex with multicomponent phase assemblage for most of the compositions for SOFCs. There are many discrepancies for the phase diagram, which may arise from the difficulty in obtaining equilibrium in the Sc_2O_3-ZrO_2 system [10, 21]. The amount of Sc_2O_3 required to fully stabilize the cubic structure is about 8.0 to 9.0 mol%. In compositions below about 6.0 to 7.0 mol%, monoclinic zirconia coexists with the fluorite-related phase [22], while for the content of 8.0 mol% or above, a poorly conducting ordered rhombohedral phase, which is termed as the β phase, has been reported [23]. The β-phase is known to be a rhombohedral distortion of the fluorite-type structure. The composition has been variously given as 12.5 mol% [24] and as 12.73 mol% Sc_2O_3 [22]. The stability range for the cubic phase is also a function of the temperature. It is 9 to 15.0 mol% at 1200°C. At room temperature, zirconia with 10.0 mol% Sc_2O_3 shows a mixture of β-phase and cubic phase. High-temperature x-ray diffraction indicated that the β-phase transformed to the cubic phase at 1000°C [25].

The observed phase depends not only on the composition but also the preparation method. Yamamoto et al. [25] and Boulc'h and Djurado [26] have fabricated Sc_2O_3-ZrO_2 solid solutions with sol-gel and spray pyrolysis methods. For the composition of 5.0 mol% Sc_2O_3, the ceramic contains 60% cubic and 40% tetragonal phase when prepared with the spray pyrolysis technology and sintered at 1500°C for 2 h, but a similar composition of 4.9 mol% Sc_2O_3 contains only 11% cubic phase when prepared with the sol-gel process and sintered at 1700°C for 5 h. A single cubic phase of ZrO_2 with 8.0 mol% Sc_2O_3 (8ScSZ) is observed with the sol-gel method [25], whereas monoclinic- and fluorite-related phases are present when 8ScSZ is prepared via solid-state reaction at 1700°C [27]. Badwal [28] has also reported that

TABLE 1.2

m-ZrO_2 Content in Sc_2O_3-ZrO_2 Solid Solutions Prepared with Solid-State Reaction and Coprecipitation and Sintered in Different Conditions [28]

Composition mol% Sc_2O_3	Sintering Temperature (°C) and Time (h)	m-ZrO_2 (%)
Coprecipitation		
7.8	2000 (20)	0
7	1900 (7)	0
6	1900 (5.5)	0
5	1900 (5.5)	<4
4.5	1900 (7)	8–10
Solid-State Reaction		
7.8	1700 (15)	12
7.8	1700 (15); 1900 (3)	0
6.8	1700 (15)	24
6.8	1700 (15); 1900 (3)	6
5.9	1700 (15)	30
5.9	1700 (15); 1900 (3)	16
4.9	1700 (15)	44
4.9	1700 (15); 1900 (3)	25
4.4	1700 (15)	47

the preparation techniques can influence the phase equilibrium, which may play a major role in determining the electrolyte conductivity. Shown in Table 1.2 is the phase composition of the Sc_2O_3-ZrO_2 system with different Sc_2O_3 ratios prepared by coprecipitation and solid-state reaction.

The higher conductivity of scandia-stabilized zirconia (ScSZ) is attributed to the smaller mismatch in size between Zr^{4+} and Sc^{3+} ions, as compared with that between Zr^{4+} and Y^{3+} ions, leading to a smaller energy for oxygen migration, and thus higher conductivity, as shown in Figure 1.3 [29]. The activation energy for conduction in ScSZ tends to increase with decreasing temperatures, which is similar to that of YSZ. Shown in Figure 1.5 are the conductivity isotherms of the ScSZ electrolytes, at 750, 850, and 950°C as a function of the scandia content [30]. The conductivity maximum is reached with the 10.0 mol% Sc_2O_3, where it is 8.0 mol% Y_2O_3 for YSZ. The temperature dependence of the conductivity is shown in Figure 1.6. Scandia-stabilized zirconia (ScSZ) with 9.0 mol% Sc_2O_3 has the highest conductivity, 0.343 S·cm^{-1} at 1000°C [32], which is much higher than that of 8YSZ at the same temperature, 0.164 S·cm^{-1} [10]. However, the highest conductivity is not found at the lowest solubility limit for the cubic phase. Unlike YSZ, zirconia with higher Sc_3O_2 content (>8.0 mol%) also shows high conductivity. The difference between the YSZ and ScSZ systems may be the difference of ionic radii of Y^{3+} and Sc^{3+} ions. The ionic

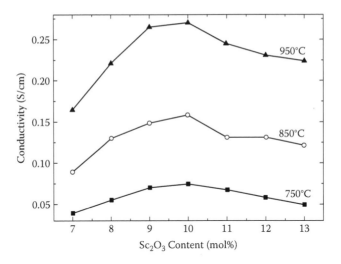

FIGURE 1.5 Conductivity isotherms in the zirconia–scandia system at 750, 850, and 950°C [30].

radius of Sc^{3+} is closer to that of Zr^{4+}, so the distortion of the lattice by the substitution of Sc^{3+} for Zr^{4+} does not severely affect the migration of oxygen ions. Therefore, a low ion migration enthalpy is expected for ScSZ. Consequently, in the high-temperature range, such as 850 to 1000°C, the activation energy for conduction of the ScSZ system is lower than that of the YSZ, as shown in Table 1.1. On the other hand, the association enthalpy between the oxide ion vacancies and dopant cation decreases with increasing dopant radius [34,35]. The highest association enthalpy

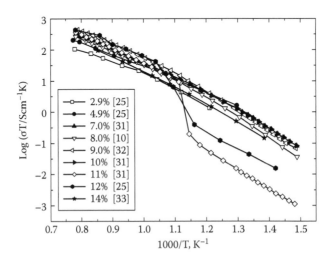

FIGURE 1.6 Conductivity of zirconia–scandia system with a different scandia ratio.

is therefore expected for the smallest dopant, Sc^{3+}, which leads to a high-activation energy for conduction in the low-temperature range, as also indicated in Table 1.1. At temperatures below 500°C, the conductivity of ScSZ is similar to, or even lower than, that of YSZ due to the tendency for the activation energy to increase with decreasing temperatures.

Another issue with ScSZ is that there is a decrease in conductivity at around 580°C as indicated by broken lines in two of the curves for higher scandia contents, 10.0 to 12.0 mol%. With high scandia contents, the cubic phase transforms to a lower conductivity rhombohedral phase, the β phase, at lower temperatures [25]. The phase change can be avoided by limiting the scandia content to 8 mole% [25] or by codoping with other oxides, such as those of bismuth [36] or ytterbium [37].

1.3.3   AGING OF ZIRCONIA ELECTROLYTES

Another issue of stabilized zirconia is the decrease in ionic conductivity due to aging during operation at high temperatures [25,31]. For example, the conductivity of the as-sintered 8ScSZ sample dropped from 0.13 Scm^{-1} to 0.012 Scm^{-1} after annealing at 1000°C for 1,000 h when measured at 800°C [25]. YSZ samples with low dopant concentration show a similar decrease (Table 1.1). For the Sc_2O_3-ZrO_2 system, some authors have attributed this degradation in conductivity to the development of the ordered rhombohedral β-phase.

Badwal et al. [38] found that 8.0 mol% Sc_2O_3-ZrO_2 composition was not cubic, but instead had tetragonal symmetry (t′-phase) as confirmed with convergent beam electron diffraction patterns and dark field images [32, 38]. They also pointed out that the dopant-rich t′-phase was metastable in nature, having a higher conductivity and did not transform under stress. The t′-phase segregates to a dopant-rich cubic phase and a low-dopant tetragonal phase on heating in the range $800 < T < 1200$°C [28]. This is in good agreement with the explanation offered by Gauckler and Sasaki [39] and Goodenough [40]. In addition, Goodenough proposes that when phase segregation occurs, the mobile oxygen vacancies become trapped in the nonpercolating phase, so the conductivity decreases with time. This is consistent with the observed increase in resistance of both 7.0 and 7.8 mol% Sc_2O_3-ZrO_2 at both 850 and 1000°C, but a very low degradation rate for 9.0 mol% or higher Sc_2O_3 at 1000°C (Figure 1.7). The t′-phase has also been observed in the Y_2O_3-ZrO_2 system, but for much lower Y_2O_3 content. Nomura et al. [29] used Raman spectroscopy to correlate the change in conductivity with the transformation of ZrO_2, combining the change of the ionic conductivity. Figure 1.8 shows that the ratio of peak intensity at 260 cm^{-1} (I_{260}) (tetragonal) and at 640 cm^{-1} (I_{640}) (cubic), which corresponds to the amount of the tetragonal phase, correlated with the decrease in conductivity for both 8YSZ and 8ScSZ. The magnitude of the decrease in conductivity during aging is much larger for ScSZ than YSZ. For example, after 5,000 h, the conductivity of ScSZ, which was initially about twice that of YSZ, was the same as that of YSZ. This implies that the amount of the t′-phase may contribute more to the conductivity of ScSZ, compared to that of YSZ [12], or ScSZ contains more t′-phase than YSZ. On the other hand, the aging effect of the electrical properties is also interpreted by the presence of different defect associations in

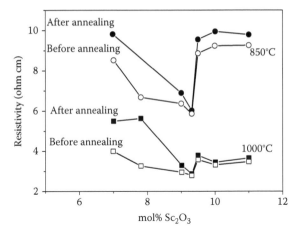

FIGURE 1.7 Resistivity plots for ScSZ as a function of the dopant content both before and after annealing at 850 and 1000°C [31].

the Y_2O_3-ZrO_2 system [30]. Increasing the dopant content can suppress the aging effect. For example, no degradation of the conductivity has been observed in the 11ScSZ, 12ScSZ, and 13ScSZ specimens, and a small degradation is observed in compositions containing Sc_2O_3 between 9.0 and 11.0 mol% [15,31]. Kondoh et al. [41] attributed the decreased conductivity to the short range ordering of oxygen ion vacancies around zirconium ions with aging, so resistance to degradation during aging can be improved by ensuring that the defect structure of stabilized

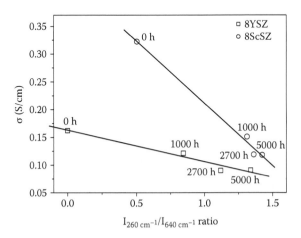

FIGURE 1.8 Change of electrical conductivity against Raman scattering intensity ratio I_{260}/I_{640} at 260 cm^{-1} and at 640 cm^{-1} [29].

zirconia does not relax with aging. Their results indicate that the appropriate amount of In_2O_3 doping, 5.0 mol%, into 8YSZ can effectively suppress the aging effect.

1.3.4 GRAIN-BOUNDARY EFFECT ON CONDUCTIVITY

Grain-boundary conduction is very important because grain boundaries in YSZ exhibit a blocking effect with regard to the ionic transport [42–44]. From an electrical point of view, a grain boundary consists of a grain-boundary core and two adjacent space-charge layers. The potential of the grain-boundary core of acceptor-doped ZrO_2 is positive; the enrichment of oxygen vacancies in the grain-boundary core is most probably responsible for the positive potential. The positively charged grain-boundary core expels oxygen vacancies while it attracts acceptor cations, thus causing the oxygen vacancy depletion and the acceptor accumulation in the space-charge layer [45–48].

A Schottky barrier model explains the grain-boundary electrical properties. In this model, a discontinuous grain-boundary impurity phase and direct grain-to-grain contacts are assumed. The ionic transport across the grain boundaries only takes place through the direct grain-to-grain contacts, rather than through the intermediate grain-boundary phase. The grain-boundary impurity phase blocks the ionic transport by decreasing the conduction path width and constricting current lines; this is the extrinsic cause of the grain-boundary blocking effect. The grain-to-grain contacts are electrically resistive in nature, which is due to the oxygen vacancy depletion in the space-charge layer, and is considered to be the intrinsic as well as the decisive cause of the grain-boundary blocking effect [43,45,49]. The intrinsic grain-boundary blocking effect becomes less pronounced at high temperatures and in the case of high-dopant concentrations. As a result, the grain-boundary contribution increases with decreasing temperatures, which is particularly important for intermediate-temperature SOFCs. For example, for YSZ materials produced by several different methods, the fraction of the total resistance due to grain boundary resistance is negligible at 900°C, but increases to ~0 to 40% at 700°C, and then further to ~10 to 65% at 500°C [42].

The Schottky barrier model shows that the space charge potential is typically 0.25 V for 8YSZ and decreases with decreasing grain size. Consequently, the concentration of oxygen vacancies in the space-charge layer increases, resulting in increasing conductivity with decreasing grain size. For example, processing YSZ to produce grain sizes less than 10 nm results in conductivities that were 50% higher than those of materials with larger grain sizes [50]. An increased conductivity with decreased grain size was also reported by Tien [51] for $Ca_{0.16}Zr_{0.84}O_{1.84}$ at temperatures below 800°C, but the grain-boundary conductivity is still about two orders of magnitude lower than the bulk conductivity, even when the grain size is as small as 41 nm. On the other hand, grain-boundary transport becomes especially important for nano-structured materials due to their high proportion of grain-boundary area. Thus, the benefits of small particle sizes in reducing processing temperatures must be balanced against increased grain-boundary resistance, particularly at lower operating temperatures.

The grain-boundary conductivity could be improved by the addition of small amounts of transition-metal oxides, especially at intermediate temperatures. For example,

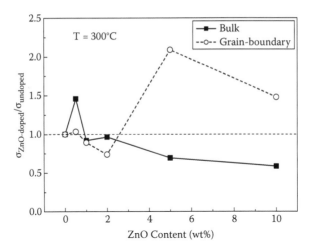

FIGURE 1.9 The grain boundary and bulk conductivities at 300°C for 8YSZ ceramics doped with ZnO with reference to the undoped YSZ sintered at the same temperature as a function of ZnO content [52].

the grain-boundary conductivity can be significantly improved by a small addition of ZnO (Figure 1.9). The grain-boundary conductivity reaches the maximum around 5 wt% ZnO addition. The ZnO addition has the added benefit of the densification process by reducing sintering temperature.

On the other hand, the grain boundary conductivity can be greatly decreased with impurities such as silica. With a high-purity material, the oxygen vacancy depletion in the space-charge layer accounts for the grain-boundary electrical properties. However, with a normal-purity material, the grain-boundary resistance is increased not only by the intrinsic blocking effect, but also by current constriction due to the grain-boundary impurity phase, such as silica. Badwal and Rajendran [53] found that the addition of only 0.2 wt% silica was sufficient to decrease the grain-boundary conductivity of YSZ by a factor greater than 15. Therefore, eliminating the effect of impurities on the grain boundary can benefit the ionic conductivity of the electrolyte. A number of strategies have been suggested to control the silica-rich grain boundary, including specialized heat treatments, the application of pressure, and various additives. Systematic variation of the boundary coverage of silicon by controlling grain size at low silicon concentrations for fully dense, highly pure specimens of nano- and microcrystalline calcia-doped or yttria-doped zirconia showed a significant increase of the conductivity at interfaces with decreasing crystallite size down to the nano-meter size [54,55]. Ralph et al. [56] reported that ionic conductivity could be enhanced by trapping impurities through the formation of low temperature compounds with a small amount of second phase additives. Some oxides can form eutectic compounds with impurities in the grain boundary, which cannot only enhance sintering, but also can be trapped to certain areas in the grain boundary upon cooling.

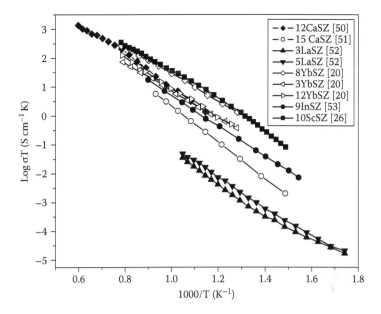

FIGURE 1.10 Conductivity of zirconia stabilized with various dopants.

1.3.5 OTHER DOPING AND CODOPING

In addition to Y_2O_3 and Sc_2O_3, zirconia has been doped with other oxides, such as CaO [57], La_2O_3 [58], and Yb_2O_3 [25]. Figure 1.10 shows that Sc and Yb doping leads to the highest conductivities [25,31,57–60]. The variation of conductivity of zirconia stabilized with different dopants is reasonable because the maximum conductivity of doped zirconia is observed when the concentration of acceptor-type dopant is close to the minimum necessary to completely stabilize the cubic fluorite-type phase [16–18], and further additions of dopant result in a decrease of conductivity due to the increasing association of the oxygen vacancies and dopant cations into complex defects of low mobility, especially when the difference between the radius of Zr^{4+} and the dopant cation is large [6,34,35]. Yamamoto et al. [18] have investigated the conductivity and migration enthalpy of Ln^{3+} stabilized zirconia as a function of Ln^{3+} radius. Because the radius of Zr^{4+} is smaller than that of Ln^{3+} dopants (lathanides, yttrium, and scandium) the maximum conductivity appeared when zirconia is stabilized with Sc^{3+}, which has the smallest radius among all the Ln^{3+}. The radius of Yb^{3+} is just slightly larger than Sc^{3+}, so the conductivity of YbSZ is similar to that of ScSZ. Bivalent metal oxides, such as MgO [61] and ZnO [62], are also used to stabilize zirconia, but they are less effective than trivalent rare earth metal dopants due to a higher tendency to form defect association and lower thermodynamic stability of bivalent metal doped zirconia [16,17].

Codoping has been used to improve the electrochemical properties, chemical stability, thermal stability, or decreasing the cost of Ln^{3+} as a stabilizer. Usually, codoping involves adding a third cation to modifying the yttria or scandia-stabilized zirconia, such as in the Mn_2O_3-Sc_2O_3-ZrO_2 [63], Bi_2O_3-Sc_2O_3-ZrO_2 [36,64], and Ca_2O_3-Y_2O_3-ZrO_2 [57,68–70] ternary systems. For example, the addition of Bi_2O_3 can inhibit the transformation of ScSZ from cubic to the rhombohedral phase [36,64], as well as significantly decrease the sintering temperature [64]. The conductivity of ScSZ doped with 2 mol% Bi_2O_3 is as high as 0.16 Scm^{-1} at 600°C [36]. The addition of ZnO to YSZ can also decrease the sintering temperature and promote the densification [52]. For example, the addition of 5 wt% ZnO in 8YSZ increased the density from 89 to 96%. Liu and Lao [52] have found an increase of total conductivity of ZnO-doped YSZ, which they attributed to enhanced grain boundary conductivity due to ZnO providing an oxygen ion conductance channel across the silica-rich grain boundary. A similar explanation is used to explain the enhancement of grain-boundary conductivity of Al_2O_3-doped zirconia [71]. Alumina has both beneficial [71,72] and detrimental [73] effects to the conductivity of zirconia-based electrolytes, depending on the doping level. Large volume fractions of insulating alumina decrease the conductivity; minor additions increase the grain-boundary conductivity by scavenging the silica impurity phase at the grain boundary into new phases. The sintering behavior and mechanical properties can also be improved by adding the alumina [72]. The yttrium content, beyond which the conductivity will rapidly decrease, is higher in CeO_2-YSZ than in YSZ, indicating that ceria additions reduce the amount of defect association or the defect migration energy [66].

Lei and Zhu [63] found that adding 2.0 mol% Mn_2O_3 to 11ScSZ can inhibit the cubic-rhombohedral phase transformation in both oxidation and reduction atmospheres, and the codoped zirconia can reach nearly full density when sintered at temperatures as low as 850°C. The conductivity of $2Mn_2O_3$-11ScSZ sintered at 900°C is ~0.1 Scm^{-1} at 800°C. Figure 1.11 illustrates the conductivity of some zirconia-based ternary systems [32,42,57,63–67].

1.3.6   FABRICATION OF ZIRCONIA ELECTROLYTE FILMS

Zirconia electrolyte films can be fabricated by a number of processes including chemical, physical, and ceramic processing methods. The chemical process includes mainly chemical vapor deposition, electrochemical vapor deposition (EVD), sol-gel, and spray pyrolysis. Among these, EVD is a well-established method that was developed by Westinghouse Electric Corporation in 1977 to fabricate gas-tight thin layers for tubular SOFCs. In this process, a porous ceramic substrate divides a reactor into two chambers, one of which is filled with a metal compound reactant and the other with an oxygen source reactant. The film deposition on the porous substrate is completed with two steps. The first step involves pore closure by a reaction between the reactant metal chloride vapors and water vapor (or oxygen). The next step is film growth, proceeding due to the presence of an electrochemical potential gradient across the deposited film. In this step oxygen ions formed on the water vapor side of the substrate diffuse through the thin zirconia layer to the chloride side. The oxygen ions react with the chloride vapors to form the electrolyte. At the deposition temperatures of 1000 to 1200°C, YSZ dense films can be produced at growth rates in the range of 2.8 to 52 μm h^{-1} [74, 75]. The EVD process has been successfully used

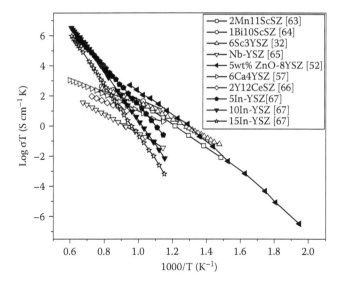

FIGURE 1.11 Conductivity of codoped zirconia.

by Westinghouse for the production of thousands of tubular cells for several multi-kilowatt power generation systems. In addition, it is also a widely used technique for depositing uniform and gas-tight layers of interconnect materials for SOFCs. Although EVD is well established, only simple stoichiometric compounds can be deposited, as each component has to be evaporated at different temperatures due to the different vapor pressures. The constituents have to be deposited from independently controlled sources, which adds complexity to the system. In addition, high reaction temperature, the presence of corrosive gases, and the relatively low deposition rates of this process increase the cost of SOFC systems.

Conventional ceramic processing techniques such as slurry coating, screen printing, tape casting, and sintering have also been used to prepare electrolyte films. These techniques have been extensively used for preparing electrolytes of a few tens of microns to >200 μm thickness. Slurry coating, which is also called "dip coating," can be applied to deposit electrolyte films for both tubular and planar SOFCs. A water suspension with a small amount YSZ fine powder is used to coat a porous anode or cathode substrate. The coated layer is then dried at room temperature, preheated at elevated temperatures, and finally sintered. This cycle is normally repeated 5 to 10 times and even more. Using an ethanol suspension of YSZ may significantly shorten the drying step. The uniformity of the coated layer can be modified with a spin coating method. With a suspension consisting of 10 wt% YSZ and ethanol, Xu et al. [76] prepared dense and crack-free films on a porous anode. The thickness of the film was from 1 to 10 μm, depending on the coating cycles. YSZ film was also successfully coated on large porous cathode tubes, which were 22 mm in diameter and 900 mm in length [77]. The thickness and gas tightness depended on the speed of tube withdrawal from the slurry, the number of dipping cycle, and the viscosity of the slurry.

FIGURE 1.12 Cross-sectional image of a five-layer cell with the anodic substrate Ni-YSZ, anodic functional layer Ni-ScSZ, electrolyte ScSZ, and interlayer GDC prepared with a tape-casting technique [79].

The slurry process has been enhanced with vacuum to fabricate planar SOFCs [78]. This method is of low cost and thus has been widely used to develop low-cost SOFCs. However, together with other liquid precursor methods such as sol-gel and spray pyrolysis, it is time, labor, and energy intensive because the coating–drying–sintering has to be repeated in order to avoid cracking formation.

Electrolytes for planar SOFCs are often prepared by the tape casting, which is also a ceramic processing technique and has already been developed to produce large-area thin zirconia films. The process starts with a highly viscous slurry consisting of a mixture of YSZ powder, organic binder, and plasticizer. The slurry is cast directly onto the joining substrate, followed by drying and sintering. This technique has been widely used to fabricate thin electrolyte films supported on electrode materials as well as monolithic electrolytes. In addition, it is used to fabricate the multilayer structure for SOFC. Shown in Figure 1.12 is a scanning electron microscope (SEM) micrograph of a cell consisting of five layers. The anode support of nickel-yttria stabilized zirconia (Ni-YSZ), anode functional layer of nickel-scandia stabilized zirconia (Ni-ScSZ), electrolyte thin film of ScSZ, and interlayer of gadolinia-doped ceria are prepared by multilayer tape casting and cofiring [79]. Tape casting and screen printing are considered very cost-effective and promising techniques. While these methods may be appropriate for small area cells, the large shrinkage associated with the removal of polymeric binders and plasticizers in subsequent sintering steps reduces the quality of large ($>10 \times 10$ cm^2) area cells.

Thermal spray, laser deposition, physical vapor deposition, and magnetron sputtering are physical processes that are used for fabrication of electrolyte thin films. Sputtering is a reliable technique for film deposition and is being used in industry for mass production. A sputtering process involves energetic noble gas ions (usually argon ions) bombarding the target material, and then atoms or ions disengage from the target surface and are deposited on a substrate separated a distance away from the target. Oxygen gas is usually introduced into the sputtering chamber to help the formation of fully oxidized films. Homogeneous and dense YSZ films with a few micrometers in thickness on porous and dense substrates have been successfully deposited by the sputtering technique. Conventional sputtering techniques include radio-frequency (RF) sputtering and reactive direct current (DC) magnetron sputtering. RF sputtering is frequently used to deposit YSZ films because of the ability to use either metallic or electrically insulating YSZ targets. However, the deposition rates of RF sputtering are slow, usually less than 1 μm/h. DC magnetron sputtering with metallic targets has a higher sputtering rate, typically of a few micrometers per hour. Murray et al. [80] deposited 8 μm-thick YSZ films on porous cathode substrates using DC magnetron sputtering. They also used this method to prepare a 0.5 μm-thick yttria-doped ceria interlayer, which performed well in a fuel cell at 650°C. However, the deposition rate of reactive DC magnetron sputtering may be slowed down by increasing the oxygen partial pressure, which is usually necessary to obtain fully oxidized YSZ films. A recently developed radical assisted sputtering (RAS) technique separates the sputtering and oxidation stages both spatially and electrically, and can obtain fully oxide films with improved deposition rates. In RAS, a substrate mounted on a rotating drum (100 rps) passes successively through a sputtering zone to deposit an insufficiently oxidized thin layer and a reaction zone to oxidize the thin layer by a radio frequency oxygen radical source. Figure 1.13 shows the cross-section of a 5 μm-thick YSZ film deposited on a NiO-YSZ substrate by RAS. Dense and uniform films were obtained after postannealing at 1200°C. Films deposited by sputtering usually have columnar grain structure. A postannealing process at about 1100 to 1300°C can densify the films. The lower-annealing temperatures are important for minimizing unwanted interfacial reactions between the YSZ film and

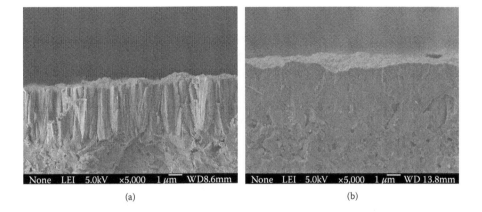

(a)   (b)

FIGURE 1.13 SEM images of a 5-μm YSZ film on porous Ni-YSZ substrate deposited by RAS-1100C (a) as-deposited and (b) YSZ film postannealed at 1200°C [81].

the cathode materials. Improved area specific resistance (ASR) of below 0.1 Ωcm^2 and open circuit voltages (OCV) of around 1 V have been successfully demonstrated [81].

1.4 CERIA ELECTROLYTES

Pure stoichiometric ceria is not a good oxygen ion conductor. The oxygen ion conductivity can be introduced dramatically by low valance doping. Thus the conductivity depends on the characteristics of the dopant elements and their concentrations.

1.4.1 EFFECT OF DOPANT RADIUS

Kevane et al. [82] reported in 1963 that doping calcia into ceria could introduce extrinsic ionic conduction. The ionic conductivity increases with CaO concentration up to 7.0 to 8.0 mol% after which it goes through a broad flat maximum and decreases again above 13 to 14 mol%. Similar results were also reported for dopants of other alkaline earth oxides. In the case of SrO and MgO, the solubility limit is reached before passing the maximum. Since long-term stability with CO_2 and H_2O for ceria doped with alkaline-earth oxides is much lower than that for ceria doped with rare-earth oxides and Y_2O_3, the latter have been widely investigated as electrolytes for SOFCs. In addition, the conductivities of ceria with trivalent dopants are much higher than those with bivalent dopants, as shown in Figure 1.14. Among those doped with rare-earth oxides and Y_2O_3, gadolinia- and samaria-doped ceria are widely considered to be the electrolytes for low and intermediate SOFCs because of their high conductivity and low activation energy.

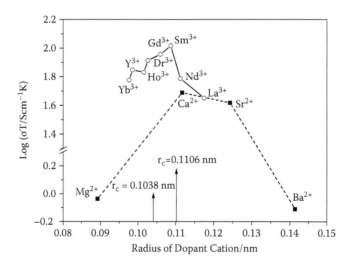

FIGURE 1.14 Ionic conductivity of doped ceria at 1073 K against the radius of dopant cation, r_c shown in the horizontal axis is for the critical radius of divalent or trivalent cation, respectively. (Data from Inaba, H. and Tagawa, H., *Solid State Ionics*, 83, 1, 1996.)

It has been noted that the conductivity and activation energy can be correlated with the ionic radius of the dopant ions, with a minimum in activation energy occurring for those dopants whose radius most closely matches that of Ce^{4+}. Kilner et al. [83] suggested that it would be more appropriate to evaluate the relative ion mismatch of dopant and host by comparing the cubic lattice parameter of the relevant rare-earth oxide. Kim [84] extended this approach by a systematic analysis of the effect of dopant ionic radius upon the relevant host lattice and gave the following empirical relation between the lattice constant of doped-ceria solid solutions and the ionic radius of the dopants.

$$a = 0.5413 + \sum_k (0.00220\Delta r_k + 0.00015\Delta z_k)m_k \tag{1.5}$$

where a (in nm) is the lattice constant of the ceria solid solution at room temperature, Δr_k (in nm) is the difference in ionic radius between the kth dopant and Ce^{4+}, which in eightfold coordination is 0.097 nm, Δz_k is the valence difference $(z_k - 4)$, which is -1 for rare-earth sesquioxides, and m_k is the mole percent of the kth dopant in the form of MO_x. If doped ceria is a simple solid solution, the lattice parameter should follow Vegard's rule, i.e., that a linear relationship exists between lattice parameter and the concentration of the dopant. When Vegard's slope approaches 0, the lowest mismatch is obtained between the ionic radius of dopant and cerium. The lowest mismatch corresponds to the smallest activation energy and highest conductivity. From Kim's formula, the following expressions for Vegard's slopes can be derived:

$$S_{V3,k} = [0.0022(r_k - 0.097 \text{ nm}) - 0.00015 \text{ nm}] \cdot m \tag{1.6}$$

for trivalent dopants such as rare-earth oxides and yttria, and

$$S_{V2,k} = [0.0022(r_k - 0.097 \text{ nm}) - 0.00030 \text{ nm}] \cdot m \tag{1.7}$$

for divalent dopants such as alkaline-earth oxides. These formulae give values of the critical radius, r_c, which is the radius giving a Vegard's slope of zero. Kim's expression implies that $r_c = 0.1106$ nm for divalent ions and 0.1038 nm for trivalent ions. Shown in Figure 1.14 are the electrical conductivities of ceria doped with 10 mol% rare-earth and alkaline-earth oxides against the radius of dopant ion [34]. As predicted by Kim's expression, calcia-doped ceria showed the highest conductivity among alkaline-earth oxide doped electrolytes since Ca^{2+} ion has the radius closest to 0.1106 nm. The electrical conductivities for electrolytes doped with MgO and BaO are exceptionally low, which may be ascribed to the insufficient solubility of these oxides in ceria. The highest conductivity as shown in Figure 1.14 is Sm-doped ceria among rare-earth oxide doped electrolytes. However, the ionic radius, which is closest to r_c, 0.1038 nm, is that of Gd^{3+}, not Sm^{3+}. In contradiction with the above finding, the highest conductivity is exhibited by Gd-doped ceria according to Steele's work [85]. These differences may be due to differences in dopant distribution caused by different cooling rates used during the fabrication process. The low mobility of cations in the fluorite lattice means that the high-temperature cation distributions, including segregation, can be "frozen-in" at elevated temperatures. The value taken for the critical ionic radius, therefore, should not be regarded as an absolute intrinsic property but can be used to correlate trends in the ionic conductivity behavior. It should be noted that different r_c values have been reported in the literature. Hong and Virkar [86] derived their r_c-value, 0.1024 nm, from plotting measured values of Vegard's slopes for a number of cations against the cation radius. Li et al. [87] measured the lattice constants of solutions of oxides of Sc, In, Y, Gd, and La in ceria. Their plots infer that $r_c = 0.101$ nm. From the work of Zhen et al. [88], it can be derived that 0.10130 nm $< r_c < 0.10145$ nm somewhere between the ionic radius of Y^{3+} (0.1019 nm) and Gd^{3+} (0.1053 nm) and closest to that of Gd^{3+}.

Different from the critical radius, Mori et al. [89] proposed a concept of effective crystallographic index to maximize the oxide ionic conductivity of doped ceria. The index, I, is defined as

$$I = (r_c/r_o) \times (r_d/r_h) \tag{1.8}$$

where r_c, r_o, r_d, and r_h are the average ionic radius of cation, the effective oxygen ionic radius, the average ionic radius of dopant and ionic radius of Ce^{4+}. They suggest that a high ionic conductivity is expected when the index approaches 1. When the association enthalpy between dopant and oxygen vacancy is minimized, the index goes toward 1 for doped ceria with a nondistorted fluorite structure. They also experimentally showed that the conductivity increases with increasing effective index when it is lower than 1. Figure 1.15 plots the relation between the index and conductivity of samaria- and lanthana-doped ceria. It should be noted that the oxygen vacancy levels of sample 2 ($Sm_{0.21}Ce_{0.79}O_{1.891}$), sample 8 (($Sm_{0.75}Sr_{0.2}Ba_{0.05})_{0.175}Ce_{0.825}O_{1.891}$),

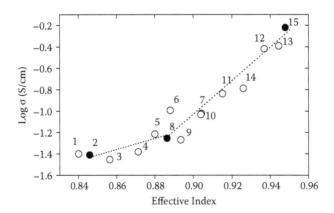

FIGURE 1.15 Relationship between effective index and conductivity [89]. (1) $Sm_{0.2}Ce_{0.8}O_{1.90}$, (2) $Sm_{0.21}Ce_{0.79}O_{1.891}$, (3) $Sm_{0.25}Ce_{0.75}O_{1.875}$, (4) $(Sm_{0.5}Ca_{0.5})_{0.175}Ce_{0.825}O_{1.87}$, (5) $(Sm_{0.5}Ca_{0.5})_{0.2}Ce_{0.8}$ $O_{1.85}$, (6) $(Sm_{0.5}Ca_{0.5})_{0.225}Ce_{0.775}O_{1.84}$, (7) $(Sm_{0.5}Ca_{0.5})_{0.25}Ce_{0.75}O_{1.81}$, (8) $(Sm_{0.75}Sr_{0.2}Ba_{0.05})_{0.175}Ce_{0.825}$ $O_{1.891}$, (9) $La_{0.125}Ce_{0.875}O_{1.94}$, (10) $La_{0.15}Ce_{0.85}O_{1.925}$, (11) $La_{0.175}Ce_{0.825}O_{1.91}$, (12) $(La_{0.8}Sr_{0.1})_{0.175}$ $Ce_{0.825}O_{1.90}$, (13) $(La_{0.8}Sr_{0.2})_{0.175}Ce_{0.825}O_{1.89}$, (14) $(La_{0.77}Sr_{0.2}Ba_{0.03})_{0.175}Ce_{0.825}O_{1.892}$, (15) $(La_{0.75}Sr_{0.2}$ $Ba_{0.05})_{0.175}Ce_{0.825}O_{1.891}$.

and sample 15 ($(La_{0.75}Sr_{0.2}Ba_{0.05})_{0.175}Ce_{0.825}O_{1.891}$) are the same, but the conductivity increases with increasing effective index. In addition, the activation energy of doped ceria with a large index value is lower than that with a small index value. Further, they have shown that the improvement of ionic conductivity was associated with microstructure change and the microdomain size of low-conductive ceria was bigger than that of high-conductive ceria. This concept was applicable to samaria- and lanthana-doped ceria as well as to alkaline earth oxide codoped ceria in limited dopant concentration. When considering doped ceria with a wider range of dopant concentration, however, a contradictory result is obtained. For example, Figure 1.2 shows the maximum conductivity at 15 mol% $YO_{1.5}$, whereas the effect index increases with yttria content. In addition, this concept is not applicable to ceria codoped with rare-earth oxides, since a maximum conductivity is usually found at a composition between the single dopants, whereas the effective index increases linearly with the concentration.

1.4.2 GD-DOPED CERIA

Since the radius of Gd^{3+} is the closest to r_c, gadolinia-doped ceria (GDC) solid solutions have been recognized to be leading electrolytes for use in intermediate- and low-temperature SOFCs. Gd doping will generate equations similar to Equation (1.1) and Equation (1.3), the conductivity therefore depends on the gadolinia content. However, because the concentration of free vacancies is a function of temperature, it is difficult to select the optimal dopant concentration for all situations. At higher-dopant concentrations oxygen vacancy–vacancy interactions increase the average migration enthalpy. Hohnke [90] reported conductivities and activation energies for different gadolinia dopant concentrations in the range of 1 to 20 mol% and at temperatures between 400 and 800°C where he observed the highest conductivity and lowest activation energy for 6 mol% gadolinia-doped ceria. Steele [85] observed the peak conductivity at 25 mol% gadolinia, while Zha et al. [91] and Zhang [92] observed the highest conductivity and lowest activation energy at 15 mol% gadolinia. Shown in Figure 1.16 is the comparison of conductivity as a function of gadolinia content. The controversy in the peak conductivity might be due to the microstructure difference caused by the different fabrication processes as well as to an amorphous glassy phase in the grain boundaries caused by impurities. The impurity of starting materials and the synthesis procedure will certainly have a great effect on the conductivity since impure ceramic electrolyte samples exhibiting significant grain-boundary resistances and variation in the sintering conditions can produce changes in the bulk conductivity by an order of magnitude as the extent of dopant segregation is modified.

The contribution of bulk (grain interior) and grain boundary to the total conductivity of GDC can be clearly identified with impedance spectra, especially those measured at low temperature. Shown in Figure 1.17 is the typical impedance spectra measured at 350°C in air [94] for clean GDC with a SiO_2 impurity of about 30 ppm. Figure 1.17 demonstrates that GDC with low-gadolinia contents (e.g., $Gd_{0.05}Ce_{0.95}O_{2-\delta}$) has a large grain-boundary arc, implying that the blocking effect of grain boundaries dominates the total conductivity. The grain-boundary effect in a clean GDC is usually attributed to intrinsic blocking behavior described as the space-charge effect.

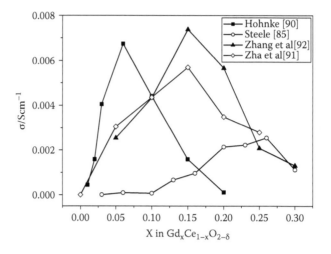

FIGURE 1.16 Conductivity at 500°C for $Gd_xCe_{1-x}O_{2-x/2}$ as a function of dopant concentration.

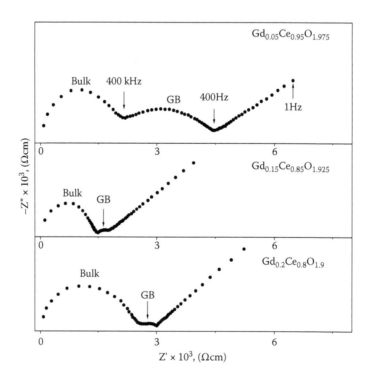

FIGURE 1.17 Impedance spectra at 350°C for clean GDC ceramics [94].

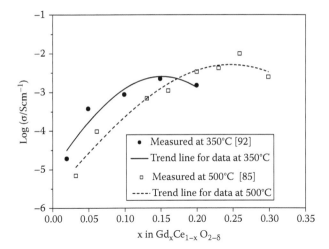

FIGURE 1.18 Grain-boundary conductivity of $Gd_xCe_{1-x}O_{2-\delta}$ as a function of Gd content [85, 92].

It is believed that at the grain boundaries, effectively negatively charged Gd'_{Ce} species segregate to form space-charged layers and capture positively charged $V_O^{..}$, and therefore resist the $V_O^{..}$ transport across the charged layers. Fortunately, the space-charged potential decreases as the dopant concentration increases [95–97], so the grain-boundary resistance decreases sharply with increasing Gd content. Zhang et al [94] reported that at 350°C, with the composition increasing from 5 to 20 mol% gadolinia, the contribution of the grain-boundary resistance to the total conduction decreases from 58.4 to 14.6%. Shown in Figure 1.18 is the grain-boundary conductivity as a function of dopant concentration. The conductivity increases with the concentration to the maximum at 25 mol% $GdO_{1.5}$ as reported by Steele [85] and 15 mol% $GdO_{1.5}$ by Zhang et al. [92] and decreases with high-dopant concentrations. As reported by Zhang et al., the grain-boundary resistance is too small to be clearly separated from the bulk resistance when the dopant concentration exceeds 20 mol%. Further, Kharton et al. [98] confirmed that the grain-boundary contribution to the total conductivity of GDC decreases with increasing average grain size for grain sizes larger than 2 to 3 μm, while the bulk conductivity of the materials prepared by a similar route is essentially independent of grain dimensions.

Equation (1.3) infers that the bulk conductivity of doped ceria should be a function of dopant content. However, the bulk conductivity of GDC shows little dependence on Gd content. Chen et al. [99] have deposited single-crystalline epitaxial $Gd_{0.2}Ce_{0.8}O_{1.9}$ thin films, which are highly epitaxial and contain only a small number of small-angle grain boundaries. Therefore, the measured single symmetric semicircle of the impedance spectrum for the film can be attributed to the bulk conductivity. In addition, the magnitude of the conductivity of the thin film is similar to that of the bulk, and they have the same activation energy. Further work by Huang et al. [100] found that upon decreasing the film thickness, impedance arcs associated with

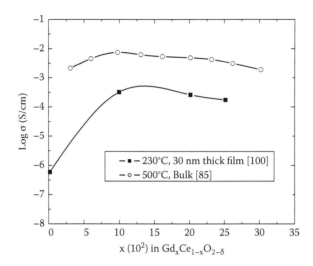

FIGURE 1.19 Bulk conductivity of GDC as a function of dopant concentration.

grain-boundary conduction gradually diminished. In the spectra of 50 nm (or less) films, only the arc corresponding to grain conduction remained, demonstrating that the grain-boundary impedance is largely eliminated. In the 20- and 50-nm films, whose thicknesses are comparable to grain size, bulk conductivity is significantly elevated and the corresponding activation energy decreases to 0.6 eV. A plot of film conductivity as a function of Gd concentration (Figure 1.19) infers that the bulk conductivity shows little dependence on Gd content. The bulk conductivity measured with GDC ceramics is also shown in Figure 1.19 for comparison.

As shown in Equation (1.3), the oxygen ion conductivity of doped ceria is a function of both dopant concentration and temperature. At a given concentration, the conductivity is usually represented by the equation:

$$\sigma T = \sigma_0 \exp\left(-\frac{E}{RT}\right) \tag{1.9}$$

This is the classical model for ionic conduction. For the trivalent dopants, the classical model suggests that at low temperatures the pre xponential term, σ_0, should be independent of temperature, while the activation energy would have the form $(\Delta H_m + \Delta H_a)$, where ΔH_m and ΔH_a represent the enthalpy of motion and association enthalpy of $(Re'_{ce} - V_O^{\cdot\cdot})$ defect complexes, respectively. At sufficiently high temperatures the defect complexes can be dissociated so that the activation energy of migration of free vacancies would simply be given by ΔH_m. Calculations by Kilner and Waters [101] indicate that defect associates would predominate up to about 1000 K for $\Delta H_a = 0.2$ eV. At intermediate temperatures, $\ln(\sigma T)$ versus $\frac{1}{T}$ plots should be curved as $[V_O^{\cdot\cdot}]$ is no longer independent of temperature. Therefore, the Arrhenius curve for GDC electrolyte usually cannot be fitted to a single straight line, but to at least two lines which exhibit different slopes. This infers a different activation energy in the low- and high-temperature range. At low temperatures, defect pairs form due to interaction between the oxide ion vacancies, $V_O^{..}$, and aliovalent cations, Gd'_{Ce}:

$$V_O^{..} + Gd'_{Ce} = (V_O^{..}Gd'_{Ce})$$

(1.10)

and

$$V_O^{..} + 2Gd'_{Ce} = (Gd'_{Ce}V_O^{..}Gd'_{Ce})$$

(1.11)

$(V_O^{..}Gd'_{Ce})$ is more likely to occur due to the expected random distribution of Gd'_{Ce} in the solid solution. The slope representing the low-temperature line corresponds to the sum of motion enthalpy and association enthalpy of the defect pair, $(V_O^{..}Gd'_{Ce})$, while the high-temperature slope only to the motion enthalpy. The curvature point is 400°C as obtained with GDC ceramics sintered at 1400°C from clean $Gd_{0.1}Ce_{0.9}O_{1.95}$ powders [85]. Below 400°C, σ_0 is 1.00×10^6 S·cm^{-1}·K and E is 0.77 eV; and above 400°C, they are 1.09×10^5 S·cm^{-1}·K and 0.64 eV, respectively. These results show that the $(Gd'_{Ce}V_O^{..})$ complexes are essentially dissociated around 400°C. Steele [85] believed that the low-dissociation temperature was consistent with the relatively small association enthalpy, 0.13 eV. With clean materials, Zhang et al. [92] observed a curvature temperature of 550°C for ceria doped with gadolinia from 5.0 to 30 mol%. A much high-curvature point, ~650°C, is reported by Zha et al. [91]. The activation energy in the high-temperature range and the conductivity at 600°C are listed in Table 1.3 for selected GDC. A similar association enthalpy is observed by Zhang et al., but it is much higher, ~0.30 eV, as estimated with Zha's work [91]. The contradiction arises possibly from the cleanness of the sample. Steele's and Zhang's results were obtained from a clean $Gd_{0.1}Ce_{0.9}O_{1.95}$ powder with SiO_2 lower than 30 ppm. The predominant constituent of the grain-boundary region in impure ceria electrolyte is SiO_2. At low temperatures the grain-boundary resistance increases due to the formation of a thin siliceous film that blocks the oxygen ion as well as the electron transfer, resulting in lower ionic conductivity and higher activation energy. At elevated temperatures, however, due to the higher activation, the boundary resistance can become

TABLE 1.3

Ionic Conductivity Data for Selected Gadolinia-Doped Ceria (GDC)

Composition	E(eV)	$\sigma\cdot600°C$ (10^{-3}Scm^{-1})	Reference
$Gd_{0.1}Ce_{0.9}O_{1.95}$	0.64	25.3	85
$Gd_{0.2}Ce_{0.8}O_{1.9}$	0.78	18.0	85
$Gd_{0.1}Ce_{0.9}O_{1.95}$	0.61	14.6	112
$Gd_{0.1}Ce_{0.9}O_{1.95}$	0.71	13.0	92
$Gd_{0.2}Ce_{0.8}O_{1.9}$	0.72	18.0	92
$Gd_{0.1}Ce_{0.9}O_{1.95}$	0.86	19.3	91
$Gd_{0.2}Ce_{0.8}O_{1.9}$	0.86	16.9	91

negligible. This temperature is suggested to be 500°C by Steele based on conductivity data obtained with pure samples. When less-pure samples are used, much lower grain-boundary conductivities and higher activation energy for grain-boundary conductivity will be expected at low temperatures. This will certainly enhance the difference between the slopes in high- and low-temperature ranges, indicating enlarged dissociation enthalpy.

1.4.3 EFFECT OF SiO₂ IMPURITY

The effects of SiO_2 impurity on the ionic conductivity and activation energy of GDC were further investigated by Zhang et al. [94] who found that the total conductivity decreased significantly with increasing SiO_2 content. Shown in Figure 1.20 are the impedance spectra on $Gd_{0.2}Ce_{0.8}O_{1.9}$ ceramics with different amounts of SiO_2. At temperatures below 450°C, the resistance increase is predominantly attributed to the grain-boundary resistance. The higher the SiO_2 content, the larger the resistance. An increase in the bulk resistance was also observed and was attributed to SiO_2 dissolution in $Gd_{0.2}Ce_{0.8}O_{1.9}$ crystallites. At high temperatures, which are necessary for GDC ceramic formation, SiO_2 might enter into GDC crystallites at either substitutional or interstitial sites:

$$SiO_2 \xrightarrow{\ CeO_2\ } Si_{Ce}^{\times} + 2O_O^{\times} \tag{1.12}$$

$$SiO_2 + 2V_O^{\cdot\cdot} \xrightarrow{\ CeO_2\ } Si_i^{\cdot\cdot} + 2O_O^{\times} \tag{1.13}$$

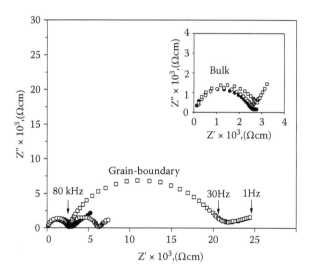

FIGURE 1.20 Impedance spectra measured at 350°C in air for $Gd_{0.2}Ce_{0.8}O_{1.9}$ with (●) 30 ppm, (○) 200 ppm, and (□) 3000 ppm SiO_2. Inset is the impedance at high frequencies, showing the effect of SiO_2 content on the bulk conductivity [94].

The substitutional reaction occurs when Si is introduced to GDC by directly replacing Ce in the parent fluorite structure. If this does happen, a large lattice distortion is expected since the radius of silicon ion (Si^{4+}) is much lower than r_c. The lattice distortion will increase the activation energy for oxygen ion migration and the association enthalpy of complex defects, resulting in high-lattice resistance. The interstitial reaction happens by introducing Si^{4+} to a site that is normally empty, while O^{2-} is also introduced to an empty site, the oxygen ion vacancy. If the interstitial route is favored, it is more likely since Si^{4+} is much smaller than Ce^{4+}; the dissolution will fill oxygen vacancies, and eventually reduce the conductivity.

As discussed earlier, at the grain boundaries, the space-charged layers formed by the negatively charged Gd'_{Ce} block the movement of $V_O^{\cdot\cdot}$ across the boundaries [95,96]. Guo et al. [97] suggest that the space-charge potential decreases with the dopant, i.e., gadolinia, concentration. Therefore, a low grain-boundary resistance is observed with a high amount of gadolinium dopant. This situation does not change when SiO_2 is present. As shown in Figure 1.21 the grain-boundary resistance decreases with Gd_2O_3 concentration when 3,000 ppm SiO_2 is added. However, the grain boundary dominates the total resistance for the sample with different amounts

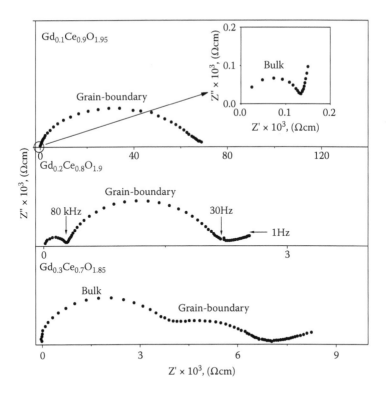

FIGURE 1.21 Impedance spectra measured at 350°C in air for GDC ceramics with 3000 ppm SiO_2. Inset is the impedance indicating the bulk conductivity of $Gd_{0.1}Ce_{0.9}O_{1.95}$ [94].

of dopant, which is not the case for a clean sample containing <30 ppm SiO_2 (see Figure 1.17). The thin siliceous film may not be continuous, so the boundary path is not fully blocked and the grain-boundary resistance decreases with dopant concentration and increases with the impurity level.

Steele [85] suggested that, for a clean sample, grain-boundary resistance dominates the total conductivity at low temperatures. At temperatures above 500°C, the intrinsic grain-boundary contribution could not be detected. However, for an impure GDC sample, the grain-boundary conductivity is much lower than the lattice conductivity and usually dominates up to around 1000°C resulting in a relatively high-activation energy (~1 eV) for the total conductivity. The impurity can arise from several routes, including from the original ores as SiO_2 is ubiquitous in minerals, from furnace refractories during high-temperature sintering procedures, from glassware used for precursor fabrication, and from the silicone grease used in the apparatus to establish input gas mixtures for SOFC test assemblies. Therefore, attempts have been made to mitigate the negative siliceous-impurity effect, i.e., increase the grain-boundary conductivity by using additives such as ferric oxides.

Zhang et al. [102–104] have investigated the effect of ferric oxide addition to GDC solid solutions. They used up to 5.0 mol% $FeO_{1.5}$ and found 0.5 mol% $FeO_{1.5}$ addition had the best effect on scavenging any SiO_2 impurity. Shown in Figure 1.22 is the comparison of impedance for three samples: a clean $Gd_{0.2}Ce_{0.8}O_{1.9}$ (<30 ppm SiO_2), an impure $Gd_{0.2}Ce_{0.8}O_{1.9}$ with 0.2 mol% SiO_2, and the impure $Gd_{0.2}Ce_{0.8}O_{1.9}$ with 0.5 mol% $FeO_{1.5}$. The result suggests that SiO_2, as mentioned above, is extremely detrimental to the grain-boundary conduction. The arc corresponding to the grain-boundary resistance is successfully suppressed by adding 0.5 mol% $FeO_{1.5}$ to the

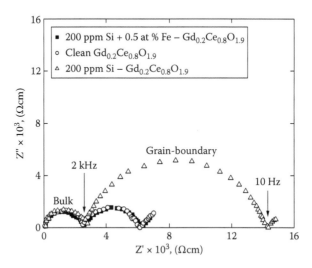

FIGURE 1.22 Impedance plots at 350°C for clean $Gd_{0.2}Ce_{0.8}O_{1.9}$, $Gd_{0.2}Ce_{0.8}O_{1.9}$ with 200 ppm SiO_2, and $Gd_{0.2}Ce_{0.8}O_{1.9}$ with 200 ppm SiO_2 and 500 ppm $FeO_{1.5}$. The impedance spectrum of the one with SiO_2 and $FeO_{1.5}$ almost overlaps with that of the clean GDC, whereas the spectrum of GDC with 200 ppm SiO_2 has an enlarged grain-boundary arc [102].

impure GDC. The suppressed arc is almost equal to that of the clean sample. This suggests that the negative impurity effect is fully removed by using ferric-oxide additive. It should be noted that the ferric-oxide addition does not increase the bulk resistance. In addition to the effect on conductivity, ferric oxide can reduce the sintering temperature of GDC. The grain-boundary resistance experienced a rapid decrease with sintering temperature from 1100 to 1250°C and reached a relative broad minimum over 1250 to 1500°C. Cho et al. [105] examined the effect of CaO addition on grain-boundary conductivity in $Gd_{0.2}Ce_{0.8}O_{1.9}$ containing 500 ppm SiO_2 as an impurity. The addition of 2% CaO increased the grain-boundary conductivity about 50 times without affecting the bulk conductivity significantly. The CaO addition can decrease the grain-boundary resistivity to the level of clean GDC. They attributed the conductivity enhancement to boundary-structure change from a random structure into a faceted one with a zig-zag shape.

Jurado [106] found that adding 1 mol% TiO_2 to GDC with 500 ppm SiO_2 increased the grain-boundary conductivity two times and slightly increased the bulk conductivity. Different from Fe_2O_3, CaO, and TiO_2, adding Al_2O_3 exhibits a detrimental effect on the conductivities, especially on grain-boundary conductivity, which is possibly related to the formation of $GdAlO_3$ [107]. Deterioration in grain-boundary conductivity was also observed when 0.5 mol% MnO_2 was used [103]. However, a contradictory effect is reported for cobalt oxide. For GDC with SiO_2 impurity, Zhang et al. [103] reported that the cobalt oxide addition reduced the grain-boundary conductivity to a level of less than 1/10 of that without cobalt oxide. Meanwhile, for clean GDC, cobalt oxide did not significantly affect the conductive behavior. Lewis et al. [108] also reported low grain-boundary conductivity with cobalt for GDC with a high level of SiO_2 impurity. However, others have reported that addition of a small amount of cobalt to GDC enhanced the grain-boundary conductivity, similar to iron addition, by forming a thin layer of amorphous material at the grain boundaries, which acts as a cleaner of impurities segregated at the grain boundary. For example, Perez-Coll et al. [109] reported that, with commercial ceria-gadolinia powders as the precursors, addition of cobalt reduces the densification temperatures and significantly affects both microstructural contributions of impedance spectra by enhancing the bulk and grain-boundary conductivities and lowering their activation energies. As shown in Figure 1.23 the grain-boundary conductivities are significantly enhanced with 2 mol% Co addition, but the beneficial effect decreases with increasing sintering temperature. This could be due to depletion of the Co content in grain boundaries of samples fired at high temperatures because either the solubility of Co in GDC increases with temperature (although this needs verification) or the volatilization of cobalt oxides occurs above 1400°C.

Although the metal oxide addition seems to be an effective approach for reducing the effect of harmful siliceous phases at the grain boundary, the resulting increased conductivity does not exceed that of a clean GDC. In addition, the enhancement in grain-boundary conductivity might be accompanied by an increase of electronic conductivity, which can lead to current leakage during SOFC operation. For example, Perez-Coll et al. [109] have reported that the addition of Co can decrease the sintering temperature and increase the grain-boundary conductivity significantly, but the p-type conductivity is also elevated more than 20 times (Figure 1.24). Recently,

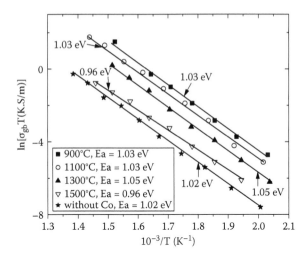

FIGURE 1.23 Grain-boundary conductivity of $Gd_{0.2}Ce_{0.8}O_{1.9}$ with 2% Co and sintered at 900, 1100, 1300, and 1500°C. Data for a sample sintered without addition of Co is also shown for comparison [109].

Kim et al. [110] proposed a new approach to improve the grain-boundary conduction without any additives. The conductivity of $Gd_{0.1}Ce_{0.9}O_{1.95}$ containing 500 ppm SiO_2 is increased about four times by postsintering heat treatment at 1350°C for 20 h, and the conductivity increases with heating time. Similar work has also been reported by Zhang et al. [111] who found that aging at 1000°C significantly affected

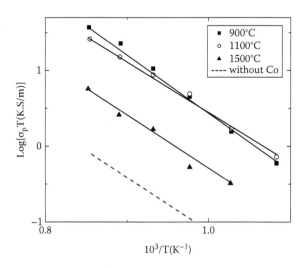

FIGURE 1.24 p-type conductivity of $Gd_{0.2}Ce_{0.8}O_{1.9}$ with 2% Co sintered at 900, 1100, and 1500°C. The dashed line is for a $Gd_{0.2}Ce_{0.8}O_{1.9}$ sample sintered without addition of Co [109].

electrical conductivity of $Gd_xCe_{1-x}O_{2-\delta}$ ceramics. For compositions with $x < 0.2$, high-temperature aging led to an increase in the total conductivity resulting mainly from an improved grain-boundary behavior. However, for the compositions with $x > 0.2$, high-temperature aging had a detrimental effect on both the grain and grain-boundary conductivities.

1.4.4 SM-DOPED CERIA

Although Steele [85] found $Gd_{0.1}Ce_{0.9}O_{1.95}$ to have the highest conductivity among the doped ceria electrolytes, as shown in Figure 1.14, other work [34] found that $Sm_{0.2}Ce_{0.8}O_{1.9}$ has the highest conductivity among the series $Ln_{0.2}Ce_{0.8}O_{1.9}$, where $Ln = La, Nd, Sm, Eu, Gd, Y, Ho, Tm,$ and Yb, so there has been interest in samaria-doped ceria (SDC) as an electrolyte for SOFCs. The compositional dependence of conductivity at 600°C for SDC is shown in Figure 1.25. The conductivity increases with x, reaches the maximum at $x = 0.15$–0.2 [91,113–116], and decreases with x at high-dopant concentration. Yahiro et al. [113] found the peak conductivity at $x = 0.2$, whereas Zha et al. [91] observed the optimal dopant concentration of $x = 0.15$. This controversy is also reported for GDC [85,91,92]. All the studies indicate that the conductivity increases with increasing samaria concentration to a maximum, but the explanation for this maximum differs among researchers.

Some researchers believe the maximum results from two opposite effects related to oxygen vacancy [113,116]. Firstly, samarium substitution causes a significant increase in oxygen vacancies within its solubility limit to maintain charge neutrality, which increases the ionic conductivity, but at higher dopant levels the interaction between dopant cation (Sm'_{ce}) and oxygen vacancy $(V_O^{\cdot\cdot})$ leads to a decrease in the vacancy mobility, and a subsequent decrease in the ionic conductivity. Jung et al. [114] attribute the increase of conductivity to the decrease in activation energy for

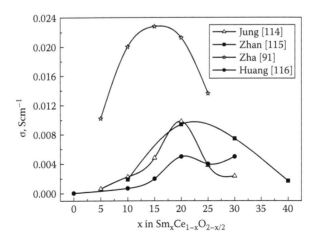

FIGURE 1.25 Composition dependence of conductivity at 600°C for $Sm_xCe_{1-x}O_{2-x/2}$ [91,114–116].

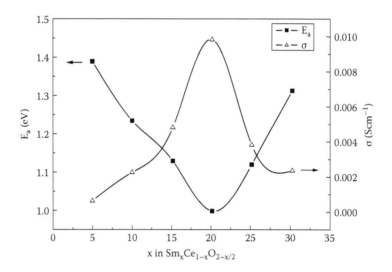

FIGURE 1.26 Composition dependence of the activation energy and conductivity at 600°C for $Sm_xCe_{1-x}O_{2-x/2}$ [114].

conduction due to the defect interaction. As shown in Figure 1.26, the activation energy shows the minimum value at the concentration where the conductivity shows the highest level. Similar conductive behavior is also observed with yttria- and gadolinia-doped ceria, which has the lowest energy and highest conductivity at about 6.0 mol% $YO_{1.5}$ and $GdO_{1.5}$ [90,117]. According to Faber et al. [117], the decrease of activation energy results from attractive interactions between immobile dopant cations and mobile oxygen vacancies caused by their effective charges in the lattice. When the amount of dopant increases, the oxygen vacancies interact with more dopant cations. The energy of an oxygen vacancy in a saddle point between sites with interactions with dopant will be lower than that in a saddle point between sites without any interactions. As observed, the minimum activation energy occurs with a dopant concentration of 20 mol% $SmO_{1.5}$. At higher dopant concentrations, deeper traps are formed owing to the dopant cations closely neighboring each other, and thus the activation energy increases.

Zhan et al. [115] attribute the maximum conductivity at $x = 0.2$ to a grain-boundary effect. For $Sm_xCe_{1-x}O_{2-x/2}$, the bulk conductivity decreases gradually with samarium content due to the increase in the association enthalpy and the oxygen ion migration enthalpy, while the grain boundary conductivity increases substantially because of the decreasing grain boundary effect. The two opposite effects result in the highest total conductivity for $Sm_{0.2}Ce_{0.8}O_{1.9}$. As seen in Figure 1.27, at T > 400°C, the grain conductivity decreases gradually with an increase in samarium content while the grain-boundary conductivity increases sharply. Therefore, it is the two opposite effects that lead to the maximum of the total conductivity at $x = 0.2$.

The Arrhenius plots of the conductivities in air (Figure 1.28) show that the conductivity does not obey a single Arrhenius relationship over the entire temperature range.

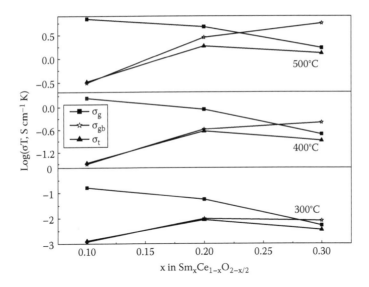

FIGURE 1.27 Composition dependence of the apparent lattice conductivity, σ_g, grain-boundary conductivity, σ_{gb}, and the total conductivity, σ_t, for $Sm_xCe_{1-x}O_{2-x/2}$ measured at 300, 400, and 500°C [115].

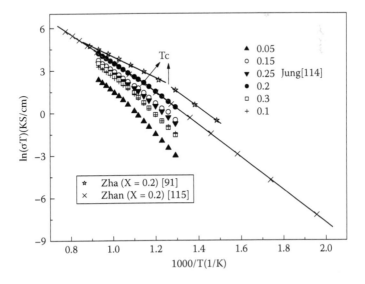

FIGURE 1.28 Arrhenius plots of conductivities of $Sm_xCe_{1-x}O_{2-x/2}$ for $x = 0.05, 0.1, 0.15, 0.2, 0.25$, and 0.3 [91, 114, 115].

A marked curvature has been found at the critical (curvature) temperature, T_C. Similar to that with GDC, when the temperature is below T_c, the defects in SDC associate and form defect pairs. Whereas at $T > T_c$, the pairs dissociate. Therefore, the curvature as indicated with the slope difference corresponds to the association energy. Zhan et al. [115] believe that oxygen ion conduction in $Sm_{0.2}Ce_{0.8}O_{1.9}$ is controlled by the grain boundary at low temperatures and by both the bulk (grain interior) and the grain boundary at high temperatures, leading to a curvature as shown in Figure 1.28. The addition of samarium can depress the grain-boundary effect, so the curvature becomes less pronounced with the increase in samarium content.

Jung et al. [114] proposed a model to explain the curvature in the Arrhenius plots. Below T_c, the oxygen vacancies condense into clusters of ordered vacancies; above T_c, all vacancies appear to be mobile. According to Jung et al. [114], the dopant cations may not only act just as traps for isolated vacancies, but also as nuclei for the formation of ordered-vacancy clusters. The simplest model for such a system of nucleation would consist of a dopant cation as a nucleation center with negative charge having a critical temperature T_c: below T_c, the oxygen vacancies with a positive charge are progressively trapped in the clusters with decreasing temperature; above T_c, the vacancies are dissolved into the matrix of the normal sites. From this model for the nucleation center of the vacancy cluster, one can obtain the trapping energy or dissociated energy. The composition dependence of the trapping energy is shown in Figure 1.29. The plot of trapping energy versus dopant concentration shows a minimum at $x = 0.2$. This tendency is clearly consistent with that of the activation energy as shown in Figure 1.26. Thus, the extra energy, i.e., the trapping energy, needed for an oxygen vacancy to move from a regular site to the saddle point decreases with the increase of the dopant cations and reaches a minimum at $x = 0.2$. However, at higher-dopant

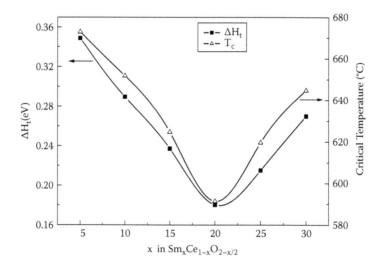

FIGURE 1.29 Composition dependence of the trapping energy, ΔH_t, and the critical temperature, T_c, for $Sm_xCe_{1-x}O_{2-x/2}$ [114].

concentrations the dopant ions become so close to each other as to form deeper traps, which increases the energy needed to bring the oxygen vacancy from a regular site to the saddle point. Figure 1.29 shows that a minimum in the critical temperature also occurs at $x = 0.2$. The composition dependencies of the critical temperature and the trapping energy are both consistent with that of the activation energy and show a minimum with a dopant concentration of $x = 0.2$, which supports the model.

1.4.5 YTTRIA-DOPED CERIA

The electrical conductivity and activation energy of $Y_xCe_{1-x}O_{2-x/2}$ with various x are reported by different authors [118–122, 131]. Ou et al. [118] synthesized $Y_xCe_{1-x}O_{2-x/2}$ ($x = 0.1$–0.25) by a carbonate coprecipitation method. The conductivity reaches the highest value at $x = 0.15$. Balazs and Glass [8] synthesized the $Y_xCe_{1-x}O_{2-x/2}$ ($x = 0$–0.4) by a solid-state reaction method. Increasing dopant concentration leads to an increase in ionic conductivity until it reaches the maximum value at $x = 0.16$. This result is consistent with that obtained by Ou [118], but other reports [121, 122] indicate that $Y_xCe_{1-x}O_{2-x/2}$ achieves the highest conductivity at $x = 0.2$ or 0.1. Shown in Figure 1.30 are selected data on the conductivity of yttria-doped ceria (YDC) at 500°C reported by different groups.

The doping concentration also affects the activation energy as shown in Figure 1.31 [118–120]. Balazs et al. [8] have shown that the activation energy decreases with the doping concentration to a minimum value at $x \approx 0.04$. This is consistent with the early report by Wang et al. [122] who observed an increase in activation energy with increasing dopant concentration for dopant levels higher than 0.05. The activation energy obtained by Herle and Ou is remarkably lower than that obtained by Balazs [8]. Herle et al. [119] measured the conductivity of $Y_xCe_{1-x}O_{2-x/2}$ at 300 to 1000°C in

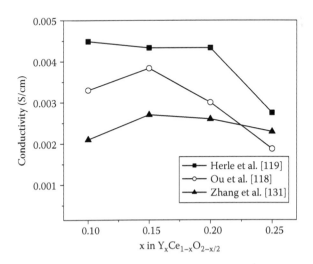

FIGURE 1.30 Conductivity at 500°C for $Y_xCe_{1-x}O_{2-x/2}$ with various yttria concentrations [118,119,131].

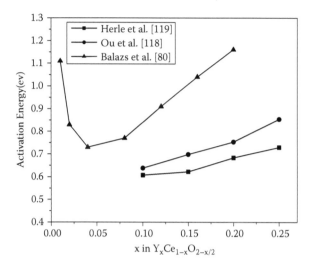

FIGURE 1.31 Variation of activation energy at different yttria concentrations.

air and found that the Arrhenius plots break into two straight-line portions, one at low temperatures (<500°C) and the other at high temperatures (>600°C). The curvature temperature, which corresponds to the point of intersection of the two straight lines, is similar with that of gadolinia- and samaria-doped ceria. However, Bellino et al. [123] reported a very low curvature temperature, 190°C, for $Y_{0.2}Ce_{0.8}O_{1.9}$ with small grain sizes.

YDC has been prepared by various methods, including solid-state reaction, coprecipitation, glycine-nitrate process and metal organic chemical vapor deposition (MOCVD). Table 1.4 shows that the properties depend on the preparation method [118,119,127,129,130]. Zha et al. [129] have studied the influence of sintering

TABLE 1.4

Selected Data on the Conductivity at 600°C for $Y_{0.2}Ce_{0.8}O_{1.9}$ Prepared by Various Methods

Method	Sintering Temperature (°C)	Relative Density (%)	Conductivity ($10^{-3}Scm^{-1}$)	Ref.
Glycine-nitrate	1500		9.0	127
Distillation	1500	92	5.0	129
Carbonate coprecipitation	1400	>95	7.0	118
Oxalate coprecipitation	1500	99.4	14	119
MOCVD	Deposited at 600		6.0	130

processes on the electrical conductivity and encountered difficulty in densifying nano-sized YDC particles at temperatures below 1600°C. Bellino et al. [123] used a fast-firing process to fire the YDC pellets at 700 to 1000°C, and the resulting samples showed remarkably high conductivities compared with samples prepared by a conventional sintering process.

Eguchi et al. [124] fabricated a cell based on $Y_{0.2}Ce_{0.8}O_{1.9}$ electrolyte with a thickness of 1.72 mm and obtained an open-circuit voltage (OCV) of 0.72 V at 800°C. Mehta et al. [125] prepared YDC film by RF sputtering, and raised the OCV to 0.84 V at the same temperature when the electrolyte was 1 mm in thickness. Kim et al. [126] used a sol-gel method to prepare the YDC film with the thickness of 1.6 mm for a cell that generated a maximum power density of 144 mWcm^{-2} at 800°C. Peng et al. [127] have reported a YDC-based cell with relative high maximum power density. Using Ni-YDC as the anode and $Sm_{0.5}Sr_{0.5}CoO_3$ as the cathode, the cell generated maximum powder density of 360 mWcm^{-2} at 650°C. In addition, YDC-salt composite electrolytes, such as YDC carbonates, have been shown to be superior to ceramic YDC electrolytes at intermediate temperatures [128]. The addition of salts has been shown to significantly increase the conductivity, which is in the range of 10^{-2} to 0.5 Scm^{-1} between 400 and 700°C, compared to 10^{-4} to 10^{-2} for pure YDC in the same temperature range. This increase in conductivity also led to increases in power density.

1.4.6 CODOPED CERIA ELECTROLYTES

It is believed that the ionic conductivity of doped ceria electrolytes is affected by not only the concentration and the distribution of oxygen vacancies, but also the lattice strain. Codoping might suppress the ordering of oxygen vacancy and therefore lower the activation energy of conduction and improve the ionic conductivity. For example, Wang et al. [132,133] investigated a set of Sm and Gd codoped ceria, $Sm_xGd_{0.15-x}Ce_{0.85}O_{1.925}$ (x = 0, 0.05, 0.1, 0.15) and found that, as shown in

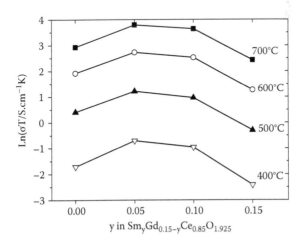

FIGURE 1.32 Effect of Sm-doping content on the conductivity in air for codoped ceria with nominal composition of $Sm_yGd_{0.15-y}Ce_{0.85}O_{1.925}$ [132,133].

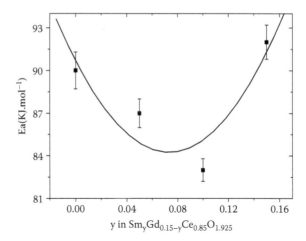

FIGURE 1.33 Effect of Sm-doping concentration (y) on the activation energy (Ea) of conduction for codoped ceria of $Sm_yGd_{0.15-y}Ce_{0.85}O_{1.925}$ [132,133].

Figure 1.32, in the temperature range of 400 to 700°C, the ionic conductivities for codoped ceria are higher than those for ceria with a single dopant, $Gd_{0.15}Ce_{0.85}O_{1.925}$ and $Sm_{0.15}Ce_{0.85}O_{1.925}$. However, for $Sm_xGd_{0.2-x}Ce_{0.8}O_{1.9}$ ($0 \leq x \leq 0.2$) at temperatures below 500°C, the conductivity of the codoped ceria is higher than those of the singly doped ceria, but at temperatures above 500°C, the conductivities of the codoped compositions are between those of $Gd_{0.15}Ce_{0.85}O_{1.925}$ and $Sm_{0.15}Ce_{0.85}O_{1.925}$. The activation energy for conductivity was always lower for the codoped ceria than those of single doped ceria (Figure 1.33). The authors suggest that, for $Sm_xGd_{0.15-x}Ce_{0.85}O_{1.925}$ samples, the ordering of oxygen vacancy might be suppressed as a result of codoping, which will lead to the decrease in activation energy of conduction and the increase in oxygen ion conductivity. However, for $Sm_xGd_{0.2-x}Ce_{0.8}O_{1.9}$, the deviation of the lattice parameter from pure CeO_2 is increased as a result of high-dopant concentration, which will lead to an increase in the activation energy of conduction and a decrease in conductivity. In early research conducted by Kim et al. [134], the conductivity of $Sm_xGd_{0.2-x}Ce_{0.8}O_{1.9}$ ($x = 0.01, 0.03, 0.05$) was about 30% higher than the single-doped ceria, $Gd_{0.2}Ce_{0.8}O_{1.9}$. In addition, they found that, the open-circuit voltage for SOFC with codoped ceria as the electrolyte was much higher than that with single-doped ceria.

Doped ceria electrolytes easily develop n-type electronic conduction at high temperatures and low oxygen partial pressures, which is a constraint for their use as an electrolyte material for SOFCs. Codoping might suppress the electronic conductivity and extend the electrolyte domain. For example, Maricle et al. [135] reported that the addition of 3.0 mol% Pr to GDC increases the electrolyte domain regime by two orders of oxygen partial pressure. Suppression of electronic conductivity has also been reported in 5.0 mol% Mg codoped GDC [136], 2.0 mol% Pr codoped $Sm_{0.1}Ce_{0.9}O_{1.95}$ [137] and a series of codoped GDC [134]. In these works, open-circuit voltage (V_{OC}, Figure 1.34) was reported to increase by using codopants.

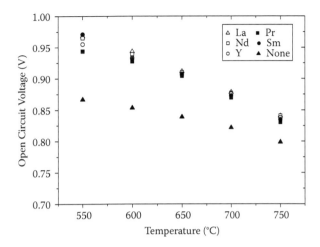

FIGURE 1.34 Comparison in open-circuit voltage of a unit cell with gadolinia-doped ceria electrolyte containing 3 mol% of Ln(Y, Sm, Nd, Pr, and La) codopant [134].

However, controversial results on the codoping effect have been reported. Sha et al. [138] reported that, for codoped $Y_xLa_{0.2-x}Ce_{0.8}O_{1.9}$, in the temperature range of 700 to 850°C, when $x = 0.06$, 0.10 and 0.14, much higher ionic conductivity is observed than those of the singly doped ceria with the same dopant concentration and when $x = 0.06$, maximal conductivity is obtained, but the activation energies of codoped compositions are slightly higher than that of corresponding singly doped ones. In the lower-temperature region, the conductivity and activation energy of codoped ceria are between that of singly doped ones. In contrast to the enhanced conductivity, Yoshida et al. [139,140] found that ceria doubly doped with La^{3+} and Y^{3+}, in which the average ionic radius of the dopants are adjusted closely to that of Sm^{3+}, do not show any synergistic effect on ionic conductivity, and only give the average value of the conductivities of the respective single-doped ceria. In a recent report on Y^{3+} and Mg^{2+} codoped ceria, Wang et al. [141] found that $Y_{0.065}Mg_{0.12}Ce_{0.815}O_{2-\delta}$ shows oxygen ion conductivity close to and even a little higher than that of $Gd_{0.1}Ce_{0.9}O_{1.95}$ prepared with a similar procedure. Y and Mg are much cheaper than Gd in raw materials and $Y_{0.065}Mg_{0.12}Ce_{0.815}O_{2-\delta}$ has better stability against reduction than that of GDC. Shown in Table 1.5 is a summary of codoping effects.

1.4.7 PREPARATION

1.4.7.1 Solid-State Reaction

The fabrication procedure affects the product's microstructure including grain size, grain-boundary width, and porosity. In addition, different procedures introduce various amounts of impurities to the product. Therefore, the electrical conductivity and activation energy are affected by the fabrication procedure since, as mentioned above,

TABLE 1.5
Codoping Effects

Codoped	Single Doped	Effect on Conduction Behaviors
$(Gd_{0.2}Sm_{0.2}Y_{0.2}Nd_{0.2}Dy_{0.2})_{0.1}Ce_{0.9}O_{1.95}$	$Gd_{0.1}Ce_{0.9}O_{1.95}$	σ (500–800°C) 20%~45% higher [142]
$Ln_xGd_{0.2-x}Ce_{0.8}O_{1.9}$, x = 0.01, 0.03, 0.05	$Gd_{0.2}Ce_{0.8}O_{1.9}$	V_{OC} (700°C) increased 0.04–0.06 V
Ln = Sm		
Ln = Y		σ (500–700°C) 30%~50% higher
Ln = Nd and La		σ increased linearly with x
Ln = Pr		σ fluctuated
		σ decreased with x [134]
$Sm_xGd_{0.2-x}Ce_{0.8}O_{1.9}$, x = 0.05, 0.1	$Gd_{0.2}Ce_{0.8}O_{1.9}$	σ increased at 200–500°C, decreased
	$Sm_{0.2}Ce_{0.8}O_{1.9}$	at 500–700°C, Ea decreased
$Sm_xGd_{0.15-x}Ce_{0.8}O_{1.925}$, x = 0.05, 0.1	$Gd_{0.15}Ce_{0.85}O_{1.9}$	σ increased, Ea decreased [132, 133]
	$Sm_{0.15}Ce_{0.85}O_{1.9}$	
$Mg_{0.05}Gd_{0.1}Ce_{0.8}O_{1.9}$	$Gd_{0.15}Ce_{0.85}O_{1.9}$	σ increased >60%
		V_{OC} increased [136]
$Y_xSm_{0.2-x}Ce_{0.8}O_{1.9}$, x = 0.05, 0.1, 0.15	$Sm_{0.2}Ce_{0.8}O_{1.9}$	σ (550–700°C) increased up to 40%,
	$Y_{0.2}Ce_{0.8}O_{1.9}$	Ea decreased [143]
$Pr_{0.02}Sm_{0.08}Ce_{0.9}O_{1.95}$	$Sm_{0.1}Ce_{0.9}O_{1.95}$	σ (550–750°C) increased up to 100%,
		V_{OC} decreased [44]
$Y_xLa_{0.2-x}Ce_{0.8}O_{1.9}$, x = 0.02, 0.06, 0.10, 0.14	$La_{0.2}Ce_{0.8}O_{1.9}$	σ increased at 700–850°C and was
	$Y_{0.2}Ce_{0.8}O_{1.9}$	between these two single- doped
		samples at 550–650°C [138]
$Y_{0.1}La_{0.1}Ce_{0.8}O_{1.9}$	$La_{0.2}Ce_{0.8}O_{1.9}$	σ was about the average conductivities
	$Y_{0.2}Ce_{0.8}O_{1.9}$	of the single-doped samples [139]
$Y_{0.065}Mg_{0.12}Ce_{0.815}O_2$	$Gd_{0.1}Ce_{0.9}O_{1.95}$	σ increased and comparable with that
		of $Gd_{0.1}Ce_{0.9}O_{1.95}$, increased stability
		in reduced atmosphere [141]

the conduction behavior depends on the characteristics of grain, grain-boundary, and porosity, as well as the impurity level. The first method used to prepare doped ceria is the solid-state reaction, which is widely used in synthesizing ceramics due to its high selectivity, high yields, absence of solvents, and simplicity.

It is catastrophic, however, as one needs to achieve powders with high-ionic conductivity since solid-state synthesis of doped ceria requires long-time calcining at high temperatures (>1000°C) and repeated milling, which can lead to contamination by the reagent or other impurities and result in the impurity forming at the grain boundary and blocking the ionic conduction. In addition, solid-state reaction is a multiphase reaction that proceeds in the interface and the rate-limiting step—diffusion between ionic phases—is determined by a considerable number of uncertain factors.

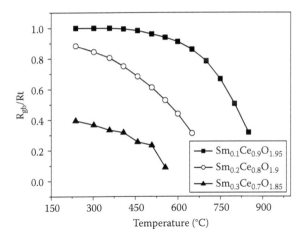

FIGURE 1.35 Contribution of the grain-boundary resistance (R_{gb}) to the total resistance (R_t) for samaria-doped ceria [115].

Therefore, the composition and structure of the product often exhibits nonstoichiometry and inhomogeneity, which can lead to decrease in grain conductivity.

Prepared by the solid-state reaction, the highest conductivity of 12×10^{-3} Scm^{-1} at 600°C, and the lowest activation energy in the high-temperature range, 0.85 eV, was observed for $Sm_{0.2}Ce_{0.8}O_{1.9}$ by Balazs and Glass [8]. The conductivity at 600°C can be as low as 5×10^{-3} Scm^{-1} and as high as 17×10^{-3} Scm^{-1} [124]. Compared with data obtained with other methods, the conductivity is relatively low and the activation energy is high. Zhan et al. [115] reported much worse performance for $Sm_{0.2}Ce_{0.8}O_{1.9}$ showing a conductivity of 9.4×10^{-3} Scm^{-1} at 600°C and an activation energy of 1.01 eV. Zhan also studied the effect of the grain-boundary resistance on the total resistance and found that the contribution of the grain-boundary resistance to the total resistance is overwhelming (>90%), leading to low total conductivity (Figure 1.35). This might infer that a high level of impurity is introduced to the ceramic during processing and blocks oxygen ion migration through the grain boundary.

1.4.7.2   Sol-Gel Process

The sol-gel process is one of the chemical synthesis methods, and it was adopted for synthesizing a large number of glass materials and novel ceramic oxide compositions in the 1960s. Through the past few decades, the sol-gel process has proven to be an important wet chemical route for laboratory synthesis and in some cases for eventual production of ceramics—generally oxide—and other powders. The most popular method of sol-gel powder preparation involves the following steps [145]: (a) sol preparation; (b) progress of network formation culminating in gelation, often by hydrolysis–condensation reactions involving alkoxides, but also by other means like olation (hydroxy bridge) and oxolation (oxo bridge) condensation mechanisms and kinetics—the basic factors controlling gel formation include pH control, temperature,

speed of mixing, and the condensation kinetics; (c) drying of gel monoliths sometimes at controlled humidity and temperature to obtain a desired miscrostructure; and
(d) heat treatment. It is generally very easy to obtain stoichiometric ceramic powders
with high homogeneity when using the sol-gel process since the precursors are usually mixed at an atomic level in solutions. Therefore, with doped ceria derived from
the sol-gel process it is possible to achieve a higher grain interior conductivity than
that derived from the solid-state reaction since the distribution of the dopant in the
solid solution as derived with the sol-gel process is more uniform.

Although the properties of materials produced by the sol-gel process are generally superior to those produced by solid-state processing, this is not always the
case. For example, using the sol-gel process, Pinol et al. [146] obtained a conductivity of only 11.4×10^{-3} Scm^{-1} at 600°C with an activation energy as high as
1.0 eV for $Gd_{0.2}Ce_{0.8}O_{1.9}$, which is similar to that reported by Huang et al. [147], who
prepared $Gd_{0.1}Ce_{0.9}O_{1.95}$ powder using the sol-gel process and obtained a conductivity of 3.6×10^{-3} Scm^{-1} and an activation energy of 1.08 eV. Huang et al. [116] also
reported a conductivity of 5.0×10^{-3} Scm^{-1} and an activation energy of 0.97 eV for
$Sm_{0.2}Ce_{0.8}O_{1.9}$. Such low conductivity and high activation energy are not expected for
the sol-gel process and could be due to impurities from solvents and reagents.

1.4.7.3   Coprecipitation Process

Many of the earliest syntheses of nano-particles were achieved by the precipitation of
sparingly soluble products from aqueous solutions followed by thermal decomposition of those products to oxides. Coprecipitation is the simultaneous precipitation of
multiple species and exhibits the following characteristics [148]: (a) the products of
precipitation reactions are generally sparingly soluble species formed under conditions of high supersaturation; (b) such conditions dictate that nucleation will be a
key step of the precipitation process and that a large number of small particles will
be formed; (c) secondary processes, such as Ostwald ripening and aggregation, will
dramatically affect the size, morphology, and properties of the products; and (d) the
supersaturation conditions necessary to induce precipitation are usually the result of
a chemical reaction. The coprecipitation of metal cations as carbonates, bicarbonates,
or oxalates, followed by their subsequent calcination and decomposition, is a common method for producing crystalline nano-particulate oxides. The calcination will,
however, almost invariably lead to agglomeration or, at high temperatures, aggregation and sintering. Fortunately, nano-particulate hydroxide, carbonate, and oxalate
precursors tend to decompose at relatively low temperatures (<400°C) due to their
high surface areas, thereby minimizing agglomeration and aggregation. The advantage of the coprecipitation method is that it can be used to obtain stoichiometric fine
or ultrafine powder via direct reaction in the solution. The oxalate and carbonate
coprecipitation are two methods often used to prepare doped ceria.

In the oxalate coprecipitation method, oxalate acid reacts with metal cations to
form the precipitate, which is subsequently calcined to obtain the products. The
advantage of oxalate acid as a precipitant is that, unlike hydroxides, oxalates are less
sensitive to the treatment conditions, such as washing and drying. In addition, the

yield is approximately 100%, which is higher than that of carbonate coprecipitation. Because of the high homogeneity of the product, the powder derived from oxalate coprecipitation has high bulk conductivity. An early report of this method comes from Herle et al. [149], who obtained high conductivity and low activation energy. The process has been modified by Zha et al. [91] and resulted in much higher performance than those prepared with solid-state reaction and sol-gel techniques. For example, the conductivity at 600°C is 21.2×10^{-3} Scm^{-1} for $Sm_{0.2}Ce_{0.8}O_{1.9}$, which is much higher than that obtained with the other two methods. In addition, the activation energy is as low as 0.67 eV, comparing to about 1.0 eV for sol-gel and solid-state reaction methods. Improved conductivity performance has also been observed by Peng et al. [150].

Nevertheless, the oxalate coprecipitation method has some problems. For example, this method usually results in rodlike doped ceria particles, which are agglomerations of smaller particles with irregular shapes. Hence, the green density of the compact body is relatively low, so it is difficult to fabricate a dense electrolyte film or membrane. In addition, the poor flow of the rodlike powder makes forming difficult.

Similarly, in the carbonate coprecipitation method ammonium carbonate is usually used as precipitant and a carbonate precursor can be obtained. Major advantages of carbonate coprecipitation over other wet-chemical synthetic routes, typically sol-gel, hydrothermal treatment, and oxalate coprecipitation, lie in the following aspects [151]: (a) rare-earth carbonates readily form solid solutions, which ensures high cation homogeneity (atomic level) in the precursors and in the solid solution oxides; (b) unlike hydroxyl precipitates, carbonate precursors are nongelatinous, which allows low agglomeration of the resultant powders; and (c) though rare-earth oxalates also easily form solid solutions, their high solubility in the solvent frequently leads to irregularly shaped large (several microns) particles.

Most of the literature focuses on the aspects of sinterability and microstructure, but limited data on the electrical properties is available. Tok [152] reported a conductivity of 18.3×10^{-3} Scm^{-1} at 600°C for $Gd_{0.1}Ce_{0.9}O_{1.95}$, and we measured a high conductivity of 22×10^{-3} scm^{-1} for $Sm_{0.2}Ce_{0.8}O_{1.9}$ at the same temperature. Their activation energies are relatively low—less than 0.7 eV. Although conductivity data reported for doped ceria prepared with carbonate precipitation is varied from different authors [153–155], the conductivity is generally high and the activation energy is usually low for ceria electrolytes fabricated with this method.

Because powders from the carbonate coprecipitation method are nano-sized, spherical particles with high specific area, they have high sinterability, i.e., high density can be obtained at relatively low sintering temperatures. The lower-sintering temperature offers the possibility for obtaining ceramics with submicron grain size, since conventional solid-state reaction usually results in coarsening of the grains. Christie et al. [156] observed very high grain-boundary conductivity for samples with submicron grains. More interesting, the impedance data show that low grain resistance for small grain sized material is associated with a decrease in electrode polarization resistance. This phenomenon and result would suggest a new approach to the production of doped ceria for use as the solid electrolyte in the low-temperature SOFC.

1.4.7.4 Glycine Nitrate Process

In recent years, novel combustion synthesis processes capable of producing ultra-fine powders of oxide ceramics in a surprisingly short time at a lower calcination temperature with improved powder characteristics have been developed. The powder characteristics, such as crystallite size, surface area, and nature (hard or soft) of agglomeration, are primarily governed by enthalpy or flame temperature generated during combustion, which are dependent on the nature of the fuel and the fuel-to-oxidant ratio. Rapid evolution of a large volume of the gaseous products during the combustion dissipates the heat of combustion and limits the rise of temperature, thus reducing the possibility of premature local partial sintering among the primary particles. The gas evolution also helps in limiting the inter-particle contact, and thus results in a more easily fireable product.

One of the lowest cost amino acids, glycine (NH_2CH_2COOH), is known to act as a complexing agent for a number of metal ions as it has a carboxylic acid group at one end and an amino group at the other end. The zwitterionic character of a glycine molecule can effectively complex metal ions of varying ionic sizes, which helps in preventing their selective precipitation and maintains compositional homogeneity among the constituents. On the other hand, glycine can also serve as a fuel in the combustion reaction, being oxidized by nitrate ions.

With the glycine nitrate process, single- or multicomponent oxide powders can be prepared rapidly and simply. The method immediately produces high-surface-area, compositionally homogeneous powders, usually with low levels of residual carbon. When the stoichiometry is properly adjusted, gaseous combustion products are environmentally attractive, being composed of H_2O, CO_2 and N_2. These advantages are mainly due to the nature of the glycine nitrate combustion reaction, which is rapid and self-sustaining, and occurs at high temperatures. In addition, the powder exhibits good flow and low apparent density, which is good for dry pressing of thin films. This process is very successful in fabricating ceria-based SOFCs for laboratory research [144,159].

As shown in Figures 1.36(a) and 1.36(b), the glycine nitrate process–derived powder is highly porous. The pore size ranges from tens of nanometers to several micrometers. The powder is thus named as "foam" powder, which shows extremely low fill densities of less than 1/100 of the theoretical density value. The low fill density makes it possible to prepare thin films of doped ceria by dry pressing, in which the films are processed by means of punches in a hardened metal die. Figures 1.36(c) and 1.36(d) show a cross-sectional view of an 8 μm-thick GDC film fabricated by dry pressing.

The conductivity of doped ceria prepared by the glycince nitrate process is generally lower than that by carbonate precipitation and higher than that by sol-gel and solid-state reaction. Xia and Liu [112] reported a conductivity of 16.7×10^{-3} Scm^{-1} at 600°C for $Gd_{0.1}Ce_{0.9}O_{1.95}$ with a very low activation energy of 0.61 eV. Although the conductivity value varies among different reports, the coprecipitation and glycine nitrate methods generally result in doped ceria with higher conductivity and lower activation energy than that derived from solid-state reaction and sol-gel techniques.

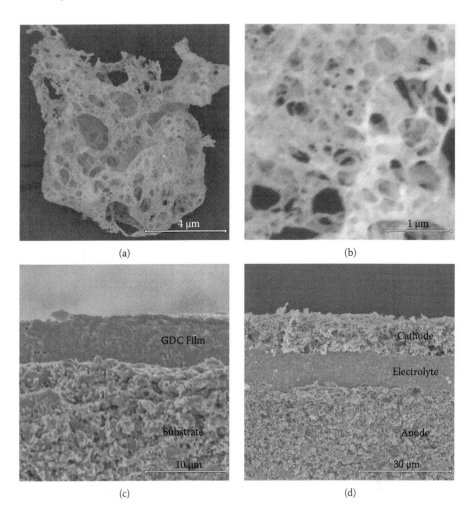

FIGURE 1.36 SEM micrographs of (a) a highly porous "foam" GDC particle, (b) a portion of the particle shown in (a), (c) a cross-section of an 8-μm thick GDC film supported on a NiO-GDC substrate, and (d) a cross-section of a fuel cell consisting of a 15-μm thick GDC electrolyte, a Ni-GDC anode, and an $Sm_{0.5}Sr_{0.5}CoO_3$ cathode [159].

Figure 1.37 and Figure 1.38 show comparisons of conductivity and activation energy for $Sm_{0.2}Ce_{0.8}O_{1.9}$ and $Gd_{0.1}Ce_{0.9}O_{1.95}$ prepared with different processes.

1.4.8 ELECTRONIC CONDUCTIVITY AND CELL VOLTAGE

Pure and doped ceria are mixed ionic and electronic conductors under reducing atmospheres, due to considerable p-type electronic conductivity under reducing conditions. The electronic conduction is due to the reduction of Ce^{4+} to Ce^{3+} under

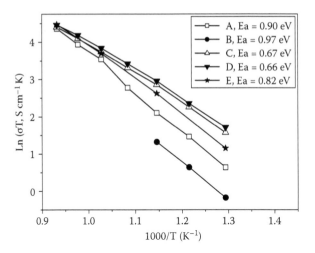

FIGURE 1.37 The Arrhenius plot of $Ce_{0.8}Sm_{0.2}O_{1.9}$ from different methods: (A) solid-state reaction [95]; (B) sol-gel process [116]; (C) oxalate coprecipitation [91]; (D) carbonate coprecipitation, our work; and (E) glycine-nitrate process [157].

low-oxygen pressure conditions, which can be expressed as Equation (1.14):

$$O_O^\times + 2Ce_{Ce}^\times \rightleftharpoons 2Ce_{ce}' + \frac{1}{2}O_2 + V_O^{\cdot\cdot} \tag{1.14}$$

$$k = [Ce_{Ce}']^2[V_O^{\cdot\cdot}]p_{O_2}^{1/2} \tag{1.15}$$

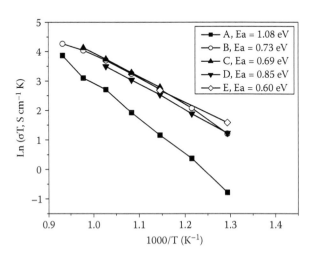

FIGURE 1.38 Arrhenius plot of $Ce_{0.9}Gd_{0.1}O_{1.95}$ from different methods: (A) sol-gel process [147]; (B) oxalate coprecipitation [91]; (C) carbonate coprecipitation [152]; (D) carbonate coprecipitation [155]; and (E) glycine-nitrate process [112].

where k is the equilibrium constant of the reaction in Equation (1.14), so the electronic conductivity can be written as:

$$\sigma_e = \mu_e e[Ce'_{Ce}] \tag{1.16}$$

In pure ceria, taking the electroneutrality condition into consideration and using the reasonable approximation:

$$[Ce'_{Ce}] = 2[V_O^{..}] \tag{1.17}$$

Combining Equation (1.15) to (1.17):

$$\sigma_e = \mu_e e[Ce'_{Ce}]\alpha p_{O_2}^{-1/6} \tag{1.18}$$

In doped ceria, the concentration of oxygen vacancies is almost constant because the concentration of oxygen vacancies generated by Equation (1.14) is negligible compared with that generated by Equation (1.19); see also Equation (1.1):

$$M_2O_3 \xrightarrow{CeO_2} 2M'_{Ce} + 3O_O^x + V_O^{..} \tag{1.19}$$

So the electronic conductivity is proportional to the $-1/4$ power of p_{O2}:

$$\sigma_e = \mu_e e[Ce'_{Ce}]\alpha p_{O_2}^{-1/4} \tag{1.20}$$

Matsui et al. [160] investigated the dependence of electronic conductivity on the partial pressure of oxygen at 400 to 800°C of samaria-doped ceria, as shown in Figure 1.39. The behavior of electronic conductivity shows good agreement with Equation (1.20), except at 800°C, and Matsui et al. suggested this deviation is due to the association of vacancies. As shown in Figure 1.39(a), the electronic conductivity is lower than the ionic conductivity at relatively high oxygen partial pressure, so the total conductivity is predominantly ionic. Also in Figure 1.39(a), the points at which $\sigma_e = 0.01\sigma_i$ and $\sigma_e = \sigma_i$ shift to lower oxygen partial pressure with a decrease in temperature, which indicates that reducing the operating temperature can suppress the electronic conduction. This might be the most important reason that doped ceria–based SOFCs are usually considered to operate at temperatures below 600°C. Steele [85,161] thought 500°C is the most appropriate temperature. According to Matsui's study, the electronic conductivity can also be suppressed by humidifying the fuel gas when doped ceria was used as electrolyte.

The electronic conductivity of doped ceria has been studied both theoretically and experimentally [162–173]. The electronic conductivity of $Gd_{0.2}Ce_{0.8}O_{1.9}$ has been reported to be 5.45×10^{-6} Scm^{-1} at 700°C ($p_{O2} = 120$ ppm) [162], 6.87×10^{-5} Scm^{-1} at 800°C ($p_{O2} = 100$ ppm) [162] and 8.18×10^{-6} Scm^{-1} at 800°C in air [163]. Xiong et al. [164] measured the conductivity of $Y_{0.2}Ce_{0.8}O_{1.9}$ to be $\sim 10^{-5}$ Scm^{-1} in air at 700°C with and found that the activation energy is 2.25 eV at low oxygen partial pressure ($\log p_{O2}$ [MPa] $= 10^{-7}$), but only 1.24 eV at relatively high oxygen partial pressure ($\log p_{O2}$ [MPa] $= 10^{-1}$). The variation in activation energy corresponds to the

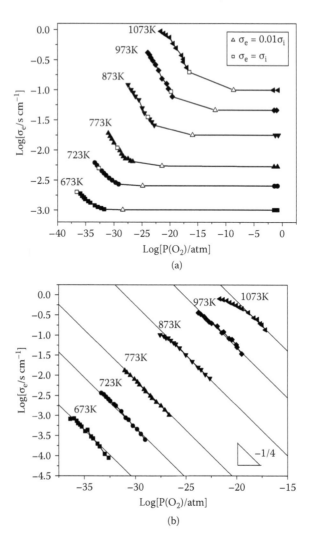

FIGURE 1.39 Oxygen partial pressure dependency of (a) total conductivity and (b) electronic conductivity of $Sm_{0.2}Ce_{0.8}O_{1.9}$ [160].

n-type electronic conduction in the low oxygen partial pressure range and the p-type conduction at high oxygen partial pressure range, respectively. In addition, dopants influence the electronic properties. The oxygen partial pressure for $\sigma_e = \sigma_i$ at 800°C is 10^{-15} Pa for the dopant of La [173], 10^{-10} Pa for Sm [174]10^{-15} Pa for Sm [173], 10^{-8} Pa for Gd [174], and 10^{-12} Pa for Gd [175]. The reason for the difference between different dopants is not clearly understood at this time.

The open-circuit voltages (V_{OC}) of SOFCs with doped-ceria electrolytes are relatively low. SOFC can be considered an oxygen concentration cell. The open-circuit voltage of the cell can be theoretically expressed by the Wagner equation [176]:

$$E = \frac{1}{4F} \int_{\mu_{O_2,II}}^{\mu_{O_2,I}} t_{ion} d\mu_{O_2} \tag{1.21}$$

where F is the Faraday constant, t_{ion} is the transference number of oxygen ions, and

$$t_{ion} = \frac{\sigma_{O_2-}}{\sigma_{O_2-} + \sigma_e} = \frac{\sigma_{O_2-}}{\sigma_{O_2-} + \sigma_e^0 pO_2^{-1/4}} \tag{1.22}$$

$\mu_{O_2,I}$ and $\mu_{O_2,II}$ are the chemical potentials of oxygen on the anode and cathode side surface of the electrolyte, respectively. By applying:

$$\mu_{O_2} = \mu_{O_2}^0 + RT \ln\left(\frac{p_{O_2}}{p_{O_2}^0}\right) \tag{1.23}$$

where O_2 is considered to be an ideal gas. In a pure ionic conducting electrolyte, $t_{ion} = 1$, and Equation (1.24) becomes the Nernst equation.

$$E = \frac{RT}{4F} \int_{\ln pO_2'}^{\ln pO_2''} \frac{\sigma_{O_2-}}{\sigma_{O_2-} + \sigma_e^0 pO_2^{-1/4}} d\ln pO_2 \tag{1.24}$$

Actually, V_{OC} of practical cells always deviate from the theoretical calculation [177,178]. Miyashita [179] proposed two empirical equations for V_{OC}, and the equations match well with the experimental results.

$$V_{OC} = V_{th} - \frac{Ea}{2e} \tag{1.25}$$

$$V_{OC} = V_{th} - (1 - t_{ion}) \tag{1.26}$$

where V_{th} is the Nernst voltage, and Ea is the activation energy of the oxygen ions.

The deviation of V_{OC} for the practical SOFC from the theoretical voltage, V_{th}, originates from the internal shorting of the cell under open-circuit conditions. The internal shorting current imposes overpotentials at the electrodes, and consequently, evaluating the V_{OC} should take into consideration the electrodes process. Electrodes with low-catalytic activity will cause a lower V_{OC} than that with high activity. For example, Matsui et al. [177] fabricated two cells that differed only in the cathode material—one cell used $La_{0.6}Sr_{0.4}CoO_3$ as the cathode and the other used Pt as the cathode. The cell using $La_{0.6}Sr_{0.4}CoO_3$ had a higher V_{OC} than the one with Pt, because of the lower overpotential or higher catalytic activity of the $La_{0.6}Sr_{0.4}CoO_3$ cathode.

A simple expression of V_{OC} of a cell under internal shorting was proposed by Zhang et al. [178].

$$V_{OC} = E_N - i_L(R_a + R_c + R_{int}) \tag{1.27}$$

where E_N is the Nernst potential of the cell reaction, i_L is the cell internal shorting current density, R_{int} represents the internal ohmic resistance of the cell, and R_a and R_c stand for anodic and cathodic polarization resistance, respectively. The value of the internal shorting current density under open-circuit conditions depends on many factors, such as the operating temperature, the oxygen partial pressure distribution in the electrolyte, the ionic conductivity and electronic conductivity of electrolyte, and the electrode polarization resistance [180]. Zhang et al. [178] calculated the leakage current density by measuring the amount of water produced by a cell under open circuit condition. As shown in Table 1.6, the cell with a $Sm_{0.5}Sr_{0.5}CoO_3$ cathode shows significant current leakage compared with that using a Pt mesh only. The leakage current density dropped by about 30% when the cell was operated at 0.5 Acm^{-2} at 600°C instead of 650°C indicating that the ceria-based electrolyte becomes more electrolytic under lower temperature operating conditions, which agrees with the model simulation [181]. Dalslet et al. [181] evaluated the influence of anode and cathode polarization on the open-circuit voltage. Both anode and cathode polarization could affect the V_{OC} by way of shielding the electrolyte from low- or high-oxygen partial pressure. Similarly, humidifying the fuel gas could elevate the V_{OC} [160]. Zhang et al. [178] also studied the V_{OC} of cells with different electrolyte thickness. V_{OC} increases with increasing thickness, which agrees with the results reported by Matsui et al. [177]. Increasing the electrolyte thickness will increase the resistance of both oxide ion and electron transport, so the internal shorting current decreases with the electrolyte thickness. Consequently, the electrode overpotential decreases with the thickness of the electrolyte.

1.4.9 FUEL CELL PERFORMANCE

Due to its high oxygen ion conductivity, doped ceria has been considered as the electrolyte for SOFCs operated at low and intermediate temperatures since the 1990s.

TABLE 1.6
Leakage Current Density Under Open-Circuit Condition at Different Temperatures and 0.5 Acm^{-2} at 600°C [178]

Temperature (°C)	i_L (Acm^{-2})	i_L^* (Acm^{-2})
650	1.401	0.095
600	0.854	0.079
550	0.405	0.067
500	0.189	0.064
600*	0.595	0.079

Note: i_L = with cathode; i_L^* = Pt mesh only; 600* = operating at 0.5 Acm^{-2} at 600°C.

However, early reports show a very low performance. For example, Hatchwell et al. [182] obtained power densities of only 12 mWcm^{-2} at 600°C with $Gd_{0.2}Ce_{0.8}O_{1.9}$ as the electrolyte and $La_{0.6}Sr_{0.4}CoO_3$ as the cathode. The lower power density is possibly caused by high resistance of the 0.4 mm thick electrolyte. Steele et al. [183,184] have theoretically shown that the use of thinner electrolyte films would allow the operating temperature to be lowered to 500°C, where a 400 mWcm^2 power density is expected. This prediction is based on the assumption that the electrolyte component contributes no more than 0.15 Ωcm^2 to the total cell resistance. This target value can be obtained with a 15 μm-thick $Gd_{0.1}Ce_{0.9}O_{1.95}$ electrolyte at 500°C. However, with a 5 to 10 μm $Gd_{0.1}Ce_{0.9}O_{1.95}$ electrolyte prepared by tape casting, when using a Ni-YSZ anode and a $La_{0.6}Sr_{0.4}Co_{0.2}Fe_{0.8}O_3$ cathode, a power output of only 120 mWcm^{-2} was obtained at 650°C [185]. The gap in performance between the prediction and experimental result is possibly due to the poor catalytic activity of the Ni-YSZ anode at low temperature. With Ni-GDC as the anode, Doshi et al. [186] reported a power output of over 140 mWcm^{-2} at 500°C. The significant increase is due to the use of Ni-GDC cermet, which provides adequate performance at 500°C for fuels such as hydrogen, syngas, and methane.

Another difficulty that has restricted exploitation of the attractive properties at 500°C has been the need to develop alternative cathode compositions that function effectively at lower temperatures. With $Sm_{0.5}Sr_{0.5}CoO_3$ as the cathode, Xia et al. [144,187] obtained power densities of around 400 mWcm^{-2} at 600°C, using samaria- and gadolinia-doped ceria as the electrolytes. The target performance of 400 mWcm^{-2} at 500°C has recently been achieved by using $Ba_{0.5}Sr_{0.5}Co_{0.8}Fe_{0.2}O_3$ as the cathode and samaria-doped ceria as the electrolyte [188]. As shown in Figure 1.40,

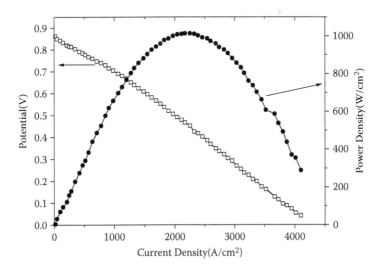

FIGURE 1.40 Cell voltage and power density as functions of current density at 600°C for fuel cell with $Sm_{0.15}Ce_{0.85}O_{2-\delta}$ electrolyte, $Ba_{0.5}Sr_{0.5}Co_{0.8}Fe_{0.2}O_{3-\delta}$ cathode, and Ni+$Sm_{0.15}Ce_{0.85}O_{2-\delta}$ anode. The cell is operated with 3% H_2O-humidified H_2 as the fuel [188].

TABLE 1.7

Maximum Power Output (W_M) and Open-Circuit Voltage (V_{OC}) for SOFCs with Doped-Ceria Electrolytes

Cell Components	Fuel and Cell Temperature	V_{OC} and W_M	Reference
$Gd_{0.2}Ce_{0.8}O_{1.9}$, 400 μm, electrode-supported tube cell Ni-YSZ and $La_{0.6}Sr_{0.4}CoO_3$ electrode	H_2 600°C	0.911 V 12 mWcm^{-2}	182
$Gd_{0.1}Ce_{0.9}O_{1.95}$, 5–10 μm, tape casting Ni-YSZ and $La_{0.6}Sr_{0.4}Co_{0.2}Fe_{0.8}O_3$ electrode	CO_2–H_2 650°C	0.68 V 120 mWcm^{-2}	185
$Ce_{0.8}Cd_{0.2}O_{1.9}$, 30 μm, tape casting Ni-$Ce_{0.8}Cd_{0.2}O_{1.9}$ and ANLC-1 electrode	H_2 500°C	0.96 V 140 mWcm^{-2}	186
$Sm_{0.2}Ce_{0.8}O_{1.9}$, 30 μm, screen printing Ni-$Sm_{0.2}Ce_{0.8}O_{1.9}$ and $Sm_{0.5}Sr_{0.5}CoO_3$ electrode	H_2 600°C	0.86 V 397 mWcm^{-2}	187
	CH_4 600°C	0.71 V 304 mWcm^{-2}	187
$Sm_{0.15}Ce_{0.85}O_{2-\delta}$, 20 μm, dry pressing Ni-$Sm_{0.15}Ce_{0.85}O_{2-\delta}$ and $Ba_{0.5}Sr_{0.5}Co_{0.8}Fe_{0.2}O_{3-\delta}$ electrode	H_2 600°C	0.87 V 1010 mWcm^{-2}	188
$Sm_{0.2}Ce_{0.8}O_{1.9}$, 30 μm, dry pressing Ni-$Sm_{0.2}Ce_{0.8}O_{1.9}$ and $Sm_{0.5}Sr_{0.5}CoO_3$ electrode	Biomass-produced gas, 650°C	0.83 V 700 mWcm^{-2}	189
$Sm_{0.2}Ce_{0.8}O_{1.9}$, 30 μm, dry pressing $Sm_{0.2}Ce_{0.8}O_{1.9}$-coated Ni and $Sm_{0.5}Sr_{0.5}CoO_3$ electrode	CH_4 600°C	0.90 V 350 mWcm^{-2}	190

a power density of over 1000 mWcm^{-2} has been recorded at 600°C using humidified H_2 as the fuel. Shown in Table 1.7 are the performances of some fuel cells with doped ceria as the electrolyte.

Doped ceria is also used as the electrolyte for single-chamber SOFCs operated at low and intermediate temperatures. Hibino et al. [191] obtained a power density of 416 mWcm^{-2} at 500°C using 30 vol% methane, 18 vol% ethane, 14 vol% propane and nitrogen as the fuel. The electrolyte was 0.15 mm-thick $Ce_{0.8}Sm_{0.2}O_{1.9}$ and the electrodes were Ni-$Ce_{0.8}Sm_{0.2}O_{1.9}$ and $Sm_{0.5}Sr_{0.5}CoO_3$. When PdO was introduced into the anode, the power density was increased to 644 mWcm^{-2} using CH_4 as fuel at the same temperature [192]. Suzuki et al. [193] used $La_{0.8}Sr_{0.2}Co_{0.2}Fe_{0.8}O_3$ as the cathode and C_3H_8 as the fuel, but the power density was only 210 mWcm^{-2} at 650°C. The low performance might be due to the thicker electrolyte (0.5 mm). In 2004, Shao and Haile [188] reported excellent performance of the single chamber SOFC with $Ba_{0.5}Sr_{0.5}Co_{0.8}Fe_{0.2}O_{3-\delta}$ as the cathode. The interfacial polarization resistance between the cathode and $Ce_{0.85}Sm_{0.15}O_{1.925}$ electrolyte was remarkably lower. When using a gas mixture which consisted of propane, oxygen, and helium as the fuel, the peak power density reached 440 mWcm^{-2} at a furnace temperature of 500°C. In a later report, they were able to use the heat released by the partial oxidation of hydrocarbons at the anode to sustain the cell temperature without external heating [194]. A layer of 7 wt% Ru-containing CeO_2 was coated onto the anode surface to enhance the catalytic activity for the propane oxidation. The cell maintained a temperature of 580°C in the absence of external heating. A V_{OC} of about 0.7 V was obtained with a maximum power density of 247 mWcm^{-2} and a short-circuit current density of 1.1 Acm^{-2}. Buergler et al. [195] improved the anode material for single chamber SOFCs. Pd was used as a promoter for partial oxidation of methane. The addition of Pd to the anode increased the OCV. Recently, Yano et al. [196] fabricated a $Ce_{0.9}Gd_{0.1}O_{1.9}$ electrolyte of 15-μm thickness which can operate at as low as 300°C. When using butane, ethanol, and dimethyl ether as the fuel, power density reached 58.7, 44.2, and 36.6 mWcm^{-2}, respectively. Shown in Table 1.8 is the performance of some single-chamber SOFCs with doped ceria as the electrolytes. It should be noted that the cell temperature is usually much higher (up to 200°C) than the temperature of the furnace which is used to heat the single chamber SOFCs.

TABLE 1.8

Performance of Single-Chamber SOFCs with Doped-Ceria Electrolytes

Cell Components	Fuel and Cell Temperature	V_{OC} and W_M	Ref.
$Ce_{0.8}Sm_{0.2}O_{1.9}$, 0.15 mm, Ni-$Ce_{0.8}Sm_{0.2}O_{1.9}$ and $Sm_{0.5}Sr_{0.5}CoO_3$ electrode	30% CH_4+18% C_2H_6 +14%C_3H_8(vol) 500°C	0.92 V 416 mWcm^{-2}	191
$Ce_{0.8}Sm_{0.2}O_{1.9}$, 0.15 mm, 7%PdO+30%$Ce_{0.8}Sm_{0.2}O_{1.9}$(wt)-Ni and $Sm_{0.5}Sr_{0.5}CoO_3$ electrode	CH_4 500°C	0.90 V 644 mWcm^{-2}	192
$Ce_{0.8}Sm_{0.2}O_{1.9}$, 0.5 mm, Ni-$Ce_{0.8}Sm_{0.2}O_{1.9}$ and $La_{0.8}Sr_{0.2}Co_{0.2}Fe_{0.8}O_3$ electrode	C_3H_8 650°C	0.78V 210 mWcm^{-2}	193
$Sm_{0.15}Ce_{0.85}O_{2-\delta}$, 20 μm, Ni-$Sm_{0.15}Ce_{0.85}O_{2-\delta}$ and $Ba_{0.5}Sr_{0.5}Co_{0.8}Fe_{0.2}O_{3-\delta}$ electrode	C_3H_8 500°C	0.70 V 440 mWcm^{-2}	184
$Sm_{0.15}Ce_{0.85}O_{1.925}$, 20 μm, 7%wtRu-CeO_2+Ni-$Sm_{0.15}Ce_{0.85}O_{1.92}$ and $Ba_{0.5}Sr_{0.5}Co_{0.8}Fe_{0.2}O_{3-\delta}$ electrode	C_3H_8 580°C	0.70 V 247 mWcm^{-2}	194
$Ce_{0.9}Gd_{0.1}O_{1.95}$, 0.29 mm, 0.01%wtPd-Ni-$Ce_{0.9}Gd_{0.1}O_{1.95}$ and $Sm_{0.5}Sr_{0.5}CoO_3$ electrode	CH_4 600°C	0.68 V 468 mWcm^{-2}	195
$Ce_{0.9}Gd_{0.1}O_{1.9}$, spin coating, 15 μm Ni-$Ce_{0.8}Sm_{0.2}O_{1.8}$ and $Sm_{0.5}Sr_{0.5}CoO_3$ electrode	C_4H_{10} 300°C	0.90 V 58.7 mWcm^{-2}	196
	Ethanol 300°C	0.76 V 44.2 mWcm^{-2}	196
	Dimethyl ether 300°C	0.74 V 36.6 mWcm^{-2}	196

1.4.10 INTERLAYER FOR ZIRCONIA-BASED SOFCs

Reducing cost is one of the main challenges for implementation of SOFCs. Reducing the operating temperature to around or less than 700°C can lower cost and improve long-term durability because (a) degradation of materials is retarded, (b) some relatively inexpensive metallic materials can be employed, and (c) much more flexible stack designs can be adopted. In addition to reducing electrolyte thickness, minimizing the polarization resistance at electrode–electrolyte interfaces is of great important to reduce the operating temperatures, so new cathode materials with a higher electrocatalytic activity than those of the conventional $La_{1-x}Sr_xMnO_3$ perovskites are needed. Strontium-doped rare earth cobaltite oxides, such as $La_{1-x}Sr_xCo_{1-y}Fe_yO_3$ (LSCF), possess excellent mixed electronic-ionic conduction characteristics, relatively high ionic conductivity, and high catalytic activity for oxygen reduction. The cobaltite-based perovskites are therefore widely considered as cathode materials for reduced temperature SOFCs with doped ceria, (La,Sr)(Ga,Mg)O$_3$ and stabilized zirconia electrolytes. However, cobaltite-based perovskites react with zirconia during fuel cell production and operation. The reaction occurs even at around 800°C to form $La_2Zr_2O_7$ and $SrZrO_3$ [197,198], which exhibit high ohmic resistance. Cobaltite perovskites are compatible with doped ceria up to 1200°C [199], so ceria has been proposed as a protective layer to prevent reaction between the cathode and the stabilized zirconia electrolyte, and has been shown to be effective in improving performance [200–205].

The use of an interlayer to prevent the formation of $La_2Zr_2O_7$ and $SrZrO_3$ and the associated increase in bulk resistance has been evaluated with microstructural and electrical characterization. For example, Nguyen et al. [200] reported that the use of a GDC layer with a $La_{0.6}Sr_{0.4}CoO_{3-\sigma}$ as the cathode and YSZ as the electrolyte decreased the bulk resistance of a single cell at 800°C from 0.58 Ωcm^2 to 0.14 Ωcm^2.

Shown in Figure 1.41 are the impedance spectra at 700°C for $Sm_{0.5}Sr_{0.5}CoO_3$ (SSC) cathodes screen-printed on electrolyte substrates of $Sc_{0.1}Ce_{0.01}Zr_{0.89}O_{1.95}$ (SSZ) with and without a $Gd_{0.1}Ce_{0.9}O_{1.95}$ interlayer. The complex impedance plots show a single depressed arc, which corresponds to interfacial polarization resistance. The arc of SSC-SSZ is extremely large compared to that of SSC-GDC-SSZ, so SSC prepared directly on SSZ electrolyte would not provide satisfactory performance, but with GDC as the interlayer, the arc is almost as small as that of SSC-GDC [201]. The interfacial polarization resistance (area-specific resistance, ASR) at 700°C is estimated to be 0.26 Ωcm^2 for SSC-GDC, 20 Ωcm^2 for SSC-SSZ, and 0.50 Ωcm^2 for SSC-GDC/SSZ, so the ASR for the sample with an interlayer is only 2.5% of that without the GDC interlayer. With a ceria interlayer, SSC prepared on SSZ exhibits a very low overpotential: 50 mV at 700°C under a cathodic polarization current of 300 mA/cm^{-2}. Without the interlayer, the overpotential is higher than 400 mV even at a current density as low as 30 mA/cm^{-2}, which shows that the interlayer is effective not only in preventing SSC reaction with SSZ electrolyte, but also in improving the cathodic reaction of SSC. Shiono et al. [202] reported that the ASR of $La_{0.6}Sr_{0.4}CoO_{3-\delta}$ on SSZ electrolyte with a GDC interlayer is about 40% lower than that on GDC electrolyte, let alone directly on SSZ electrolyte. In another report, Rossignol et al. [203] found that $Pr_{0.5}Sr_{0.5}CoO_3$ and $Gd_{0.5}Sr_{0.5}CoO_3$ cathodes on YSZ electrolyte protected by a GDC layer give ASR values between 0.1 and 0.2 Ωcm^2 at 750°C. The performance is higher than that of

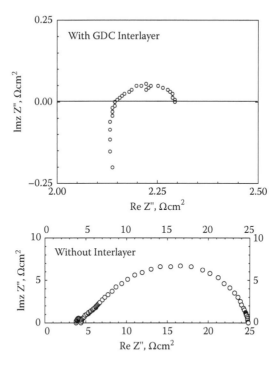

FIGURE 1.41 Complex impedance plots of SSC cathodes on SSZ electrolytes with and without GDC interlayer [201].

the corresponding cathode on GDC, which achieved the same resistance at 650°C. Long-term testing of $Pr_{0.5}Sr_{0.5}CoO_3$ and $Gd_{0.5}Sr_{0.5}CoO_3$ cathodes on YSZ electrolytes with GDC interlayers showed stable ASR values for 500 h at 800°C and minimal degradation after thermal cycling between room temperature and 800°C.

By combining an SSZ electrolyte film with strontium-doped $LaCoO_3$ as the cathode and a GDC interlayer, Nguyen et al. [200] increased the maximum power density at 800°C from 48 mWcm^{-2} to 940 mVcm^{-2}. Duan et al. [204] reported improved fuel cell performance for a single cell with $Ba_{0.5}Sr_{0.5}Co_{0.8}Fe_{0.2}O_3$ (BSCF) as the cathode, YSZ as the electrolyte, and GDC as the interlayer between YSZ and BSCF. The maximum power densities at 800°C are 0.44 and 1.56 mWcm^{-2} for cells without and with interlayer, respectively, and the corresponding interfacial polarization resistances at open-circuit condition are 1.20 and 0.308 Ωcm^2. Matsuzaki and Yasuda [205] reported that the voltage characteristics of an anode-supported single cell with LSCF as the cathode and SDC as the interlayer showed good performance up to the fuel utilization of 80% at which the direct current (DC) energy conversion efficiency is 46.2% (LHV). They suggest the DC energy conversion efficiency would be as high as 55.7% (LHV) for a fuel cell with CH_4 as the fuel at the same performance when H_2 is used as the fuel.

A shortcoming of the interlayer is that doped ceria reacts easily with stabilized zirconia, forming $(Zr,Ce)O_2$-based solid solutions [206]. The product is formed at the interface and exhibits an oxygen conductivity that is almost two orders of magnitude lower than either YSZ or GDC. Additionally, defects formed in the Ce-rich region, resulting from the differences in diffusivity of the counterdiffusion cations (Kirkendahl effect), negatively influence the performance of the electrolyte. To avoid extensive inter-diffusion at the interface, various interlayer fabrication techniques and procedures are used to minimize the solid-state reaction. Nguyen et al. [200] reported a conventional wet process, slurry dip-coating technique, to prepare the GDC interlayer with com-mercially available GDC powder (Rohdia). Their crystalline analysis and impedance spectroscopy show that the characteristics of SSZ/GDC interfaces are governed by the firing temperature of the GDC interlayers. At a firing temperature of 1300°C, a large amount of the $(Zr,Ce)O_2$-based solid solution is formed at the interface and the ohmic resistance is greatly increased. Between 1100 and 1200°C, the formation of the solid solutions and their effects on conductivity are nearly the same as each other, but less than that at 1300°C. Considering the effects of the firing temperature of the GDC film upon the performance of the cobaltite-based cathode, 1200°C is believed by Nguyen to be the optimum firing temperature of the GDC interlayer on SSZ electrolyte. Different behavior is observed by Shiono et al. [202] for the GDC interlayer prepared with the same technique and also with commercial GDC powder (Anan Kasei). When the SSZ electrolyte is presintered, no reaction is observed with electron probe microanalysis (EPMA) analysis up to 1320°C for 2 h. Duan et al. [204] reported that, with the slurry coating technique, severe solid-state reaction was observed between YSZ and GDC at 1300°C, and suggested 1250°C to be the best sintering temperature. Mai et al. [207] have shown that the structural properties of the doped ceria interlayer strongly influ-ence the electrochemical performance of SOFCs with LSCF cathodes. They found the best performance with fine powders and an optimal sintering temperature of 1250°C to maximize the density of the interlayer, while minimizing the formation of a solid solution between GDC and YSZ (Figure 1.42). To increase the density, cobalt oxide is

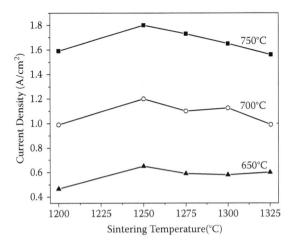

FIGURE 1.42 Current densities (A/cm^2) at 700 mV of SOFCs with GDC interlayers sintered at various temperatures [207].

used as a sintering aid for GDC, but cracks can evolve due to the high shrinkage of the layer relative to the underlying dense electrolyte. To remove the cracks, reactive sputtering is used to deposit a dense GDC interlayer on the YSZ electrolyte, resulting in an improved performance of the SOFCs. They suggested that the improved performance is not only a result of an increase in conductivity of the interlayer due to a lower porosity, but mainly a result of the improved buffering properties of a dense GDC interlayer, which inhibits $SrZrO_3$ formation, and avoids the formation of a solid solution between GDC and YSZ due to the low deposition temperature. Nguyen et al. [201] also reported that controlling the microstructure of interlayers is important for the performance of cobaltite-based cathodes.

1.5 LaGaO$_3$-BASED ELECTROLYTES

Solid solutions based on the perovskite-type oxide $LaGaO_3$—in particular, the oxide doped with Sr and Mg (LSGM)—exhibit a superior ionic conductivity higher than the conventional stabilized zirconia electrolytes [208], and have been investigated as potential candidates as alternative electrolyte materials for SOFCs operated at intermediate temperature [209, 210]. Higher conductivity is found only if LSGM is composed of cubic perovskites, and secondary phases lead to a decrease in conductivity [211]. An ideal cubic structure of ABO_3 perovskite consists of 12-fold coordinated large cations of A and 6-fold coordinated small cations of B [212–214]. The smaller cation of the octahedral B sites may be an alkaline, alkaline earth, and transition metal element; the larger A site may be occupied by a lanthanide, an alkaline earth, or an alkali atom. Perovskites are made up of a cubic close packing of A-cations and oxide ions accommodating B-cations in oxygen octahedral interstitial sites. Thus, the $LaGaO_3$ is the base compound in which the La and Ga sites can be doped with divalent ions with proper sizes. Sr^{2+} and Mg^{2+} are favorable dopants due to their low solution energies. Oxygen vacancies are formed according to the following reactions.

La site doped with Sr^{2+},

$$2SrO \xrightarrow{LaGaO_3} 2Sr'_{La} + V_O^{\cdot\cdot} \tag{1.28}$$

Ga site doped with Mg^{2+},

$$2MgO \xrightarrow{LaGaO_3} 2Mg'_{Ga} + V_O^{\cdot\cdot} \tag{1.29}$$

As shown in Equation (1.28), increasing the amount of Sr dopant will increase the amount of oxygen vacancies, so oxygen ion conductivity increases with increasing Sr content. Defect association is often observed in solid electrolytes in which oxygen vacancies are created with aliovalent dopants. Fortunately, the associations with divalent dopants at the La sites have very low binding energies in $LaGaO_3$. Further, a zero binding energy is predicted for the $Sr'_{La} V_O^{\cdot\cdot}$ defect pair, suggesting a high "free" vacancy concentration. The extreme low energy would be a major factor in promoting the observed high oxygen ion conductivity in the gallate, but the solid solubility of Sr into La sites of $LaGaO_3$ is poor and secondary phases of $SrGaO_3$ or La_4SrO_7 are formed when Sr content becomes higher than 10 mol%. Therefore, the amount

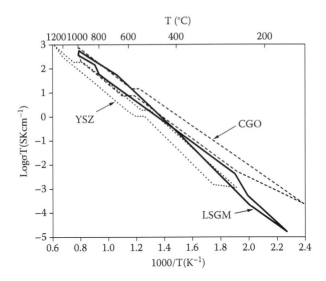

FIGURE 1.43 Conductivity range for LSGM, YSZ, and GDC. The range is plotted after Fergus with data estimated from Figures 2, 5, and 7 in reference [12].

of oxygen vacancies introduced by La site doping is limited to a low level [212,215]. However, additional oxide ion vacancies can be formed by B-site doping as indicated in Equation (1.29), so the conductivity is further increased by increasing the amount of doped Mg, which can attain a maximum of 20 mol% Mg doped on Ga sites. The lattice parameter also increases by doping Mg for Ga sites, since the ionic radius of Mg is larger than that of Ga. The solid solubility of Sr into $LaGaO_3$ lattice seems to reach a limit around 10 mol% without Mg; however, it increases up to 20 mol% by doping Mg for Ga site [212]. This seems to be due to the enlarged crystal lattice. In any case, the conductivity of LSGM is higher than that of YSZ and similar to that of GDC as shown in Figure 1.43. Although the conductivity of LSGM is somewhat lower than that of GDC, LSGM does not have an easily reducible ion such as Ce^{4+}, and thus is superior to GDC for use in low oxygen partial pressures. The highest oxide ion conductivity in $LaGaO_3$-based oxides is obtained at the compositions of $La_{0.8}Sr_{0.2}Ga_{0.85}Mg_{0.15}O_3$ and $La_{0.8}Sr_{0.2}Ga_{0.8}Mg_{0.2}O_3$ [12]. It should be noted that the oxygen pressure dependence of total conductivity of LSGM at reduced oxygen partial pressure is quite complex, suggesting possible decomposition or a decrease in the ionic conductivity due to association of oxygen vacancies [216].

Doping with transition metals such as cobalt and iron [217–224] has been proven to be effective in increasing the conductivity of LSGM. Figure 1.44 shows the conductivity of LSGM with 5.0 and 10 mol% Co or Fe doping at B sites [220]. Both dopants are effective in increasing the conductivity but Co is more effective than Fe, especially at low temperatures. Arrhenius plots with dopants have a lower slope at low temperature, inferring that electronic conduction is dominant in the low-temperature region since electronic transport is accompanied by lower activation energy. At higher temperatures, the slope becomes steeper, indicating that ionic conduction,

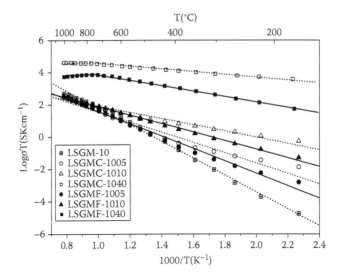

FIGURE 1.44 Conductivity of $La_{1-x}Sr_x(Ga_{0.8}Mg_{0.2})_{1-y}(Co\ or\ Fe)_yO_3$ in air [220].

which increases rapidly with increasing temperature due to higher activation energy, becomes dominant. When the dopant content is increased above 20%, however, conductivity increases at a lower rate or even drops in the high-temperature region which would be expected if the concentration of carriers remains constant. Shown in Figure 1.45 is the effect of Co doping on the electronic and oxygen ion conductivity.

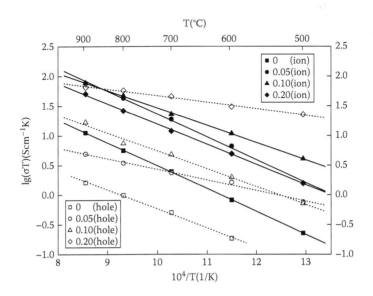

FIGURE 1.45 Conductivity of $La_{0.8}Sr_{0.2}Ga_{0.85-y}Mg_{0.15}Co_yO_3$ in air [221].

Both the oxygen ion and hole conductivities increase due to the cobalt doping, but the increase of the hole conductivity due to the cobalt substitution is more significant than that of the oxygen ionic conductivity, especially at high temperatures [221]. Too much doping increases hole conductivity, which can lead to current leakage and be detrimental to fuel cell performance. Thus, design of an optimal electrolyte depends on a balance between reducing ionic resistance by increasing dopant level and elevating electronic resistance by decreasing dopant level. For example, although total conductivity increased with increasing Fe amount [225], the transport number of the oxide ion decreased with increasing Fe amount. This infers that the increase rate for electronic conductivity is much higher than that for oxide ion as a result of iron addition. In particular, the decrease in ionic transport number is significant at 30 mol% Fe [226]. The optimal Fe-doping amount for Ga sites is around 20%, which reaches the highest fuel cell power density of about 1300 mWcm^{-2} at 1000°C. Other parameters, such as the electrolyte thickness, should be adjusted when doping the LSGM electrolyte. For example, the electrolyte thickness for optimal efficiency increases with increasing cobalt dopant level [227]. Nickel-doped LSGM has also been studied and it shows similar behavior to cobalt [228,229].

In general, LSGM powders are synthesized using solid-state reaction [230,231], self-propagating high-temperature synthesis (SHS) [232], and the sol-gel method [220,233–236]. Compared with the solid-state reaction, SHS is completed in a much shorter time, and a pure, homogeneous, very fine powder is obtained cost effectively and quickly. The sol-gel process can also produce fine powders with high homogeneity. Fabrication of thin films of LSGM is more difficult than for zirconia and ceria electrolytes since LSGM phase is usually formed at temperatures much higher than the other two electrolytes. However, dense membranes have been fabricated through electrostatic assisted vapor deposition on porous $La_{0.82}Sr_{0.18}MnO_3$ [237]. In one typical process, a sol precursor was atomized to form a charged aerosol and then deposited around 500 to 600°C, followed by sintering at 1000°C for 2 h. However, a second phase, such as La_4SrO_7, was observed suggesting the stoichiometry may have changed. Thin LSGM films can be deposited on different substrates by the pulsed-laser ablation (PLA) technique, for which mass spectrometry was combined in the PLA device to optimize the process parameters [238]. Tape casting has been applied to fabricate LSGM films as thin as 130 μm. The as-synthesized LSGM films exhibit the pure cubic perovskite structure [239]. To minimize ohmic resistance loss, the thickness of the LSGM film must be reduced. By applying a wet process, 10 μm-LSGM films have been achieved on porous YSZ [240], but did not lead to improvement in cell performance. The specific ohmic resistance of the film is much higher than that of theoretical bulk LSGM, which could be due to the presence of pinholes and much smaller grain sizes. Another possible reason is the reaction of LSGM with the electrode (YSZ-NiO) during the high-temperature firing process, which is confirmed by EDX analysis showing that nearly all of the strontium and most of the magnesium migrated into the YSZ substrate, leading to a higher resistance loss [240]. One way to solve this problem is to add a barrier layer between the electrode and LSGM, such as lanthanum-doped ceria [241–242] or strontium-doped ceria materials [243]. The interlayers can reduce the reaction between LSGM electrolyte and the NiO anode, while the LSGM layer could block the electronic conductivity of the doped ceria.

Anode-supported SOFCs with LSGM electrolytes (5-μm thick) and SDC interlayers (2-μm thick) were fabricated by pulsed laser deposition [243].

Early work by Ishihara et al. [244] showed that a single SOFC cell made from $La_{0.8}Sr_{0.2}Ga_{0.8}Mg_{0.2}O_3$ electrolyte with a Ni anode and a $Sm_{0.6}Sr_{0.4}CoO_3$ cathode can produce a power output of 440 mWcm^{-2} at 800°C. Maric et al. [245] later observed 425 mWcm^{-2} at 800°C in a cell using a $La_{0.6}Sr_{0.4}CoO_{3-\delta}$ cathode and a Ni-doped ceria anode on a 500-μm thick $La_{0.9}Sr_{0.1}Ga_{0.8}Mg_{0.2}O_3$ electrolyte. Further work by Huang et al. [246] showed that 900 mWcm^{-2} at 800°C was attainable from a fuel cell fabricated using a 600-μm thick $La_{0.8}Sr_{0.2}Ga_{0.83}Mg_{0.17}O_{2.815}$ electrolyte with a lanthanum oxide doped ceria interlayer. Most recently, Fukui et al. [239] reported that a thin film (130 μm) $La_{0.9}Sr_{0.1}Ga_{0.8}Mg_{0.2}O_3$ electrolyte cell exhibited a power density of 700 mWcm^{-2} at 800°C. Pena-Martinez et al. [247] reported that the lowest electrode overpotential, highest power–density, and lowest polarisation resistance values were obtained for the $Ba_{0.5}Sr_{0.5}Co_{0.8}Fe_{0.2}O_3/La_{0.9}Sr_{0.1}Ga_{0.8}Mg_{0.2}O_{2.85}/La_{0.75}Sr_{0.25}Cr_{0.5}Mn_{0.5}O_{3-\delta}$(LSCrM) system, i.e., 122 mWcm^{-2} at 800°C with a 1.5-mm thickness electrolyte. A theoretical extrapolation indicates a possible maximum power density around 1,600 mWcm^{-2} at 3.6 Acm^{-2} for a 120-μm thick LSGM electrolyte layer. Wan et al. [248] reported that a single cell with a 200-μm thick $La_{0.8}Sr_{0.2}Ga_{0.83}Mg_{0.17}O_{2.815}$ electrolyte and 20-μm thick interlayer produced 1400 mWcm^{-2} peak power density at 800°C with no obvious degradation of the output power in 30 days. A tubular electrolyte design reported by Du and Sammes [249] was based on extruded lanthanum gallate electrolytes with an interlayer based on doped ceria and achieved a repeatable and constant power of over 2.5 W per cell at 800°C and 0.7 V with a maximum power density of 482 mWcm^{-2}. With thin film electrolytes, excellent performance is reported for the maximum power densities of 3,270, 1,951, 612, and 80 mWcm^{-2} at 700, 600, 500, and 400°C, respectively [243]. Recently, SOFC stacks (1-kW class) with cobalt-doped LSGM electrolyte have been built and tested [250]. They provided a stable alternating current (AC) power output of 1 kW with an electrical efficiency of 45% LHV based on ac output and 48% LHV on DC output, which is considered to be excellent for such a small power generation system.

1.6 SUMMARY AND CONCLUSIONS

Yttria-stabilized zirconia (YSZ) is the state-of-the-art electrolyte for SOFCs due to its desirable chemical stability in both oxidizing and reducing atmospheres, low electronic conductivity, and high mechanical strength. The oxygen ion conductivity of YSZ is high enough for SOFCs operated at temperatures above 800°C. For SOFCs operated at intermediate temperatures, YSZ electrolyte should be fabricated into thin films to reduce the electrolyte resistance to an acceptable level. For example, at a temperature of 700°C YSZ should be fabricated at a thickness down to ~15 μm to achieve a typical area specific resistance value of 0.15 Ωcm^2. In principle, YSZ-based SOFCs can be operated at temperatures as low as 600°C when an electrolyte membrane around 2-μm thick is used. However, such a thin film will possibly reduce the durability of the electrolyte component under fuel cell operating conditions. In addition, at present it seems that the minimum thickness for dense impermeable films that can be reliably mass produced using relatively cheap ceramic fabrication routes is around 10 to 15 μm. Therefore, scandia-stabilized zirconia (ScSZ) and doped LaGaO$_3$ (LSGM), which exhibit higher oxygen ion conductivity than the traditional YSZ material, are attracting much attention as the electrolytes for intermediate-temperature SOFCs, especially those operated at temperatures around 600°C. For SOFCs operated at low temperatures, doped ceria electrolytes should be used. The use of gadolinia-doped ceria (GDC) should allow the cell-operating temperature to be lowered to around 500°C. In addition, ceria electrolytes are often used as an interlayer for zirconia and LaGaO$_3$-based SOFCs. It should be noted that continuous research is critical to develop fuel cells based on ScSZ, LSMG, and DCO electrolytes. Long-term evaluation concerning the physical, chemical, mechanical, and electrochemical properties of these materials should be addressed.

REFERENCES

1. V.V. Kharton, F.M.B. Marques, and A. Atkinson, *Solid State Ionics* 174 (2004) 135.
2. S.C. Singhal and K. Kendall, *High Temperature Solid Oxide Fuel Cells, Fundamentals, Design, and Application*, 2003 Elsevier, ISBN 1856173879.
3. R.M. Ormerod, *Chem. Soc. Rev.* 32 (2003) 17.
4. N.Q. Minh and T. Takahashi, *Science and Technology of Ceramic Fuel Cells*, 1995, ISBN 0 444 89568 X.
5. J.A. Kilner, *Solid State Ionics* 129 (2000) 13.
6. N.M. Sammes, G.A. Tompsett, H. Nafe, and F. Aldinger, *J. Eur. Ceram. Soc.* 19 (1999) 1801.
7. Y. Arachi, H. Sakai, O. Yamamoto, Y. Takeda, and N. Imanishai, *Solid State Ionics* 121 (1999) 131.
8. G.B. Balazs and R.S. Glass, *Solid State Ionics* 76 (1995) 155.
9. J.C. Boivin and G. Mairesse, *Chem. Mater.* 10 (1998) 2870.
10. S.P.S. Badwal, *Solid State Ionics* 52 (1992) 23.
11. A.I. Ioffe, D.S. Rutman, and S.V. Karpachov, *Electrochim. Acta* 23 (1978) 141.
12. J.W. Fergus, *J. Power Sources* 162 (2006) 30.
13. D.W. Strickler and W.G. Carlson, *J. Am. Ceram. Soc.* 47(3) (1964) 122.
14. J.M. Dixon, L.D. LaGrange, U. Mergen, C.F. Miller, and J.T. Porter, *J. Electrochem. Soc.* 110(4) (1963) 276.
15. C. Haering, A. Roosen, and H. Schichl, *Solid State Ionics* 176 (2005) 253.
16. T.H. Etsell and S.N. Flengas, *Chem. Rev.* 70 (1970) 339.
17. V.V. Kharton, E.N. Naumovich, and A.A. Vecher, *J. Solid State Electrochem.* 3 (1999) 61.
18. O. Yamamoto, Y. Arachi, H. Sakai, Y. Takeda, N. Imanishi, Y. Mizutani, M. Kawai, and Y. Nakamura, *Ionics* 4 (1998) 403.
19. A. Nakamura and J.B. Wagner Jr., *J. Electrochem. Soc.* 127(11) (1980) 2325.
20. S.P.S. Badwal and K. Foger, *Mater. Forum* 21 (1997) 187.
21. M.R. Thornber, D.J.M. Bevan, and E. Sumerville, *J. Solid State Chem.* 1 (1970) 545.
22. P. Duwez, F.H. Brown Jr., and F. Odell, *J. Electrochem. Soc.* 98 (1951) 365.
23. M.J. Bannister and P.F. Skilton, *J. Mater. Sci. Lett.* 2 (1983) 561.
24. F.M. Spiridonov, L.N. Popova, and R.Ya.Popil'Skll, *J. Solid State Chem.* 2 (1970) 430.
25. O. Yamamoto, Y. Arati, Y. Takeda, N. Imanishi, Y. Mizutani, M. Kawai, and Y. Nakamura, *Solid State Ionics* 79 (1995) 137.
26. F. Boulc'h and E. Djurado, *Solid State Ionics* 157 (2003) 335.

27. S.P.S. Badwal, *J. Mater. Sci.* 18 (1983) 3117.

28. S.P.S. Badwal, *J. Mater. Sci.* 22 (1987) 4125.

29. K. Nomura, Y. Mizutania, M. Kawaia, Y. Nakamuraa, and O. Yamamoto, *Solid State Ionics* 132 (2000) 235.

30. C. Haering, A. Roosen, H. Schichl, and M. Schnfller, *Solid State Ionics* 176 (2005) 261.

31. S.P.S. Badwal, F.T. Ciacchi, and D. Milosevic, *Solid State Ionics* 136–137 (2000) 91.

32. S.P.S. Badwal, F.T. Ciacchi, S. Rajendran, and J. Drennan, *Solid State Ionics* 109 (1998) 167.

33. I. Kosacki, H.U. Anderson, Y. Mizutani, and K. Ukai, *Solid State Ionics* 152–153 (2002) 431.

34. H. Inaba and H. Tagawa, *Solid State Ionics* 83 (1996) 1.

35. M. Mogensen, N.M. Sammes, and G.A. Tompsett, *Solid State Ionics* 129 (2000) 63.

36. S. Sarat, N. Sammes, and A. Smirnova, *J. Power Sources,* 160 (2006) 892.

37. R. Chiba, T. Ishii, and F. Yoshimura, *Solid State Ionics* 91 (1996) 249.

38. S.P.S. Badwal and J. Drennan, *Solid State Ionics* 53–6 (1992) 769.

39. L.J. Gauckler and K. Sasaki, *Solid State Ionics* 75 (1995) 203.

40. J.B. Goodenough, *Annual Review of Materials Research* 33 (2003) 91.

41. J. Kondoh, H. Shiota, S. Kikuchi, Y. Tomii, Y. Ito, and K. Kawachi, *J. Electrochem. Soc.* 149 (2002) J59–J72.

42. K. Yamahara, C.P. Jacobson, S.J. Visco, and L.C. De Jonghe, *Electrochem. Soc. Proc.* 2003–07, SOFC VII (2003) 187.

43. X. Guo and J. Maier, *J. Electrochem. Soc.* 148 (3) (2001) E121.

44. X. Guo and R. Waser, *Prog. Mater. Sci.* 51 (2006) 151.

45. K.L. Kliewer and J.S. Koehler, *Phys. Rev.* A140 (1965) 1226.

46. M.F. Yan, R.M. Cannon, and H.K. Bowen, *J. Appl. Phys.* 54 (1983) 764.

47. J. Maier, *J. Phys. Chem. Solids* 46 (1985) 309.

48. S.M. Mukhopadbyay and J.M. Blakely, *J. Am. Ceram. Soc.* 74 (1991) 25.

49. D. Bingham, P.W. Tasker, and A.N. Cormack, *Philos. Mag.* A 60 (1989) 1.

50. X. Xin, Z. Lu, Z. Ding, X. Huang, Z. Liu, X. Sha, Y. Zhang, and W. Su, *J. Alloys Compd.* 425 (2006) 69.

51. T.Y. Tien, *J. Appl. Phys.* 35 (1964) 122.

52. Y. Liu and L.E. Lao, *Solid State Ionics* 177 (2006) 159.

53. S.P.S. Badwal and S. Rajendran, *Solid State Ionics* 70–71 (1994) 83.

54. P. Mondal, A. Klein, W. Jaegermann, and H. Hahn, *Solid State Ionics* 118 (1999) 331.

55. X. Guo, *Computational Materials Science* 20 (2001) 168.

56. J.M. Ralph, J.A. Kilner, and B.C.H. Steele, *Mat. Res. Soc. Symp. Proc.* 575 (2000) 309.

57. J.H. Gong, Y. Li, Z.L. Tang, and Z.T. Zhang, *Mater. Lett.* 46 (2000) 115.

58. M.H. Zhou and A. Ahmad, *Int. J. Appl. Ceram. Technol.* 3 (2006) 218.

59. P. Thangadurai, V. Sabarinathan, A. Chandra Bose, and S. Ramasamy, *J. Phys. Chem. Solids* 65 (2004) 1905.

60. D.K. Hohnke, *J. Phys. Chem. Solids* 41 (1980) 777.

61. E.N.S. Muccillo and M. Kleitzb, *J. Eur. Ceram. Soc.* 15 (1995) 51.

62. Y. Nakada and T. Kimura, *J. Am. Ceram. Soc.* 80 [2] (1997) 401.

63. Z. Lei, and Q. Zhu, *Mater. Lett.* 61 (2007) 1311.

64. M. Hirano, T. Oda, K. Ukai, and Y. Mizutani, *Solid State Ionics* 158 (2003) 215.

65. J.-H. Lee and M. Yoshimura, *Solid State Ionics* 124 (1999) 185.

66. J.-H. Lee, S.M. Yoon, B.-K. Kim, J. Kim, H.-W. Lee, and H.-S. Song, *Solid State Ionics* 144 (2001) 175.

67. H. Naito, H. Yugami, and H. Arashi, *Solid State Ionics* 90 (1996) 173.

68. Y. Li, Z.L. Tang, and Z.T. Zhang, *J. Mater. Sci. Lett.* 18 (1999) 443.

69. J.H. Gong, Y. Li, Z.L. Tang, and Z.T. Zhang, *J. Mater. Sci.* 35 (2000) 3547.

70. M.M. Bu'cko, *Mater. Sci. Poland* 24 (2006) 39.

71. D. Lybye, Y.-L. Liu, M. Mogensen, and S. Linderoth, *Electrochemical Society Proceedings*, 2005–07, SOFC IX, 2005, pp. 954–96.

72. Y. Ji, J. Liu, Z. Lu, X. Zhao, T.M. He, and W.H. Su, *Solid State Ionics* 126 (1999) 277.

73. X. Guo, *Phys. Stat. Sol.* 183 (2001) 261.

74. U.B. Pal and S.C. Singhal, *J. Electrochem. Soc.* 137 (1980) 2937.

75. J.P. Dekker, V.E.J. van Dieten, and J. Schoonman, *Solid State Ionics* 51 (1992) 143.

76. X.Y. Xu, C.R. Xia, S.G. Huang, and D.K. Peng, *Ceramic International* 31 (2005) 1061.

77. M. Kuroishi, S. Furuya, K. Hiwatashi, K. Omoshiki, A. Ueno, and M. Aizawa, in *Solid Oxide Fuel Cells VII*, eds. H. Yokokawa and S.C. Singhal, The Electrochemical Society, Pennington, N.J., PV2001–16, 2001, p.1073.

78. F. Tietz, H.P. Buchkremer, and D. Stover, *Solid State Ionics*, 152–153 (2002) 373.

79. Z.R. Wang, S.R. Wang, J.Q. Qian, J.D. Cao, and T.L. Wen, 16th International Conference on Solid State Ionics, July 1–6, 2007, Shanghai, Book of Program and Abstract, p. 176.

80. E.P. Murray, T. Tsai, and S.A. Barnett, *Nature*, 400 (1999) 649.

81. W.J. Ji, Y.H. Gong, B. Xie, and H.Q. Wang, 16th International Conference on Solid State Ionics, July 1–6, 2007, Shanghai, Book of Program and Abstract, p. 479.

82. C.J. Kevane, E.L. Holverson, and R.D. Watson, *J. Appl. Phys.* 34 (1963) 2083.

83. J.A. Kilner, in: R. Metselaar, H.J.M. Heijligers, J. Schoonman (eds.), *Solid State Chemistry*, Elsevier Science, 1983, p.189.

84. D.J. Kim, *J. Am. Ceram. Soc.* 72 (1989) 1415.

85. B.C.H. Steele, *Solid State Ionics* 129 (2000) 95.

86. S.J. Hong and A.V. Virkar, *J. Am. Ceram. Soc.* 78 (1995) 433.

87. P. Li, I.W. Chen, J.E. Penner-Hahn, and T.Y. Tien, *J. Am. Ceram. Soc.* 74 (1991) 958.

88. Y.S. Zhen, S.J. Milne, and J.R. Brook, *Sci. Ceram.* 14 (1998) 1025.

89. T. Mori, J. Drennan, J.H. Lee, J.G. Li, and T. Ikegami, *Solid State Ionics* 154–155 (2002) 461.

90. D.K. Hohnke, *Solid State Ionics* 5 (1981) 531.

91. S.W. Zha, C.R. Xia, and G.Y. Meng, *J. Power Sources* 115 (2003) 44.

92. T.S. Zhang, J. Ma, H. Cheng, and S.H. Chan, *Mater. Res. Bull.* 41 (2006) 563.

93. H.B. Li, C.R. Xia, Z.X. Zhou, M.H. Zhu, and G.Y. Meng. *J. Materials Sci.* 41 (10) (2006) 3185.

94. T.S. Zhang, J. Ma, S.H. Chan, P. Hing, and J.A. Kilner, *Solid State Sci.* 6 (2004) 565.

95. J. Maier and B. Bunsenges, *Phys. Chem.* 90 (1986) 26.

96. X. Guo, *Solid State Ionics* 99 (1997) 247.

97. X. Guo, W. Sigle, and J. Maier, *J. Am. Ceram. Soc.* 86 (2003) 77.

98. V.V. Kharton, F.M. Figueiredo, L. Navarro, E.N. Naumovich, A.V. Kovalevsky, A.A. Yaremchenko, A.P. Viskup, A. Carneiro, F.M.B. Marques, and J.R. Frade, *J. Mater. Sci.* 36 (2001) 1105.

99. L. Chen, C.L. Chen, D.X. Huang, Y. Lin, X. Chen, and A.J. Jacobson, *Solid State Ionics* 175 (2004) 103.

100. H. Huang, T.M. Gur, Y.J. Saito, and F. Prinz, *Appl. Phys. Lett.* 89 (2006) 143107.

101. J.A. Kilner and C.D. Waters, *Solid State Ionics* 6 (1982) 253.

102. T.S. Zhang, J. Ma, S.H. Chan, and J.A. Kilner, *J. Electrochem. Soc.* 151(10) (2004) J84.
103. T.S. Zhang, J. Ma, Y.J. Leng, S.H. Chan, P. Hing, and J.A. Kilner, *Solid State Ionics* 168 (2004) 187.
104. T.S. Zhang, J. Ma, S.H. Chan, and J.A. Kilner, *Solid State Ionics* 176 (2005) 377.
105. P.S. Cho, S.B. Lee, D.S. Kim, J.H. Lee, D.Y. Kim, and H.M. Park, *Electrochem. Solid-State Lett.* 9(9) (2006) A399.
106. J.R. Jurado, *J. Materials Sci.* 36 (2001) 1133.
107. T.S. Zhang, Z.Q. Zeng, H.T. Huang, P. Hing, and J.A. Kilner, *Mater. Let.* 57 (2002) 124.
108. G.S. Lewis, A. Atkinson, B.C.H. Steele, and J. Drennan, *Solid State Ionics* 152–153(2002) 567.
109. D. Perez-Coll, P. Nunez, J.C.C. Abrantes, D.P. Fagg, V.V. Kharton, and J.R. Frade, *Solid State Ionics* 176 (2005) 2799.
110. D.K. Kim, P.S. Cho, J.H. Lee, D.Y. Kim, and S.B. Lee, *Solid State Ionics*, 177 (2006) 2125.
111. T.S. Zhang, J. Ma, L.B. Kong, P. Hing, S.H. Chan, and J.A. Kilner, *Electrochem. Solid-State Lett.* 7(6) (2004) J13.
112. C.R. Xia and M.L. Liu, *Solid State Ionics* 152–153 (2002) 423.
113. H. Yahiro, Y. Egushi, and H. Arai, *J. Appl. Electrochem.* 18 (1988) 527.
114. G.B. Jung, T.J. Huang, and C.L. Chang, *J. Solid State Electrochem.* 6 (2002) 225.
115. Z.L. Zhan, T.L. Wen, H.Y. Tu, and Z.Y. Lu, *J. Electrochem. Soc.* 148(5) (2001) A427.
116. W. Huang, P. Shuk, and M. Greenblatt, *Solid State Ionics* 100 (1997) 23.
117. J. Faber, A. Geoffroy, A. Roux, A. Sylvestre, and P. Abelard, *Appl. Phys.* A 49 (1989) 225.
118. D.R. Ou, T. Mori, F. Ye, M. Takahashi, J. Zou, and J. Drennan, *Acta Mater.* 54 (2006) 3737.
119. J.V. Herle, T. Horita, T. Kawada, N. Sakai, H. Yokokawa, and M. Dokiya, *J. Eur. Ceram. Soc.* 16 (1996) 961.
120. M. Dudek and J. Molenda, *Mat. Sci. Poland* 24 (2006) 45.
121. T.S. Zhang, J. Ma, L.B. Kong, S.H. Chan, and J.A. Kilner, *Solid State Ionics* 170 (2004) 209.
122. D.Y. Wang, D.S. Park, J. Griffith, and A.S. Nowick, *Solid State Ionics* 2 (1981) 95.
123. M.G. Bellino, D.G. Lamas, N.E.W. de Reca, *Adv. Funct. Mater.* 16 (2006) 107.
124. K. Eguchi, T. Setoguchi, T. Inoue, and H. Arai, *Solid State Ionics* 52 (1992) 165.
125. K. Mehta, S.J. Hong, J.F. Jue, A.V. Virkar, in: S.C. Singhal, H. Iwahara (eds.), *Proceedings of the 3rd International Symposium on Solid Oxide Fuel Cells*, Electrochemical Society, Pennington, NJ, 1997, pp. 92–103.
126. S.G. Kim, S.P. Yoon, S.W. Nam, S.H. Hyun, and S.A. Hong, *J. Power Sources* 110 (2002) 222.
127. R.R. Peng, C.R. Xia, X.Q. Liu, D.K. Peng, and G.Y. Meng, *Solid State Ionics* 152 (2002) 561.
128. B. Zhu, X.R. Liu, and P. Zhou, *J. Mater. Sci. Lett.* 20 (2001) 591.
129. S.W. Zha, Q.X. Fu, Y. Lang, C.R. Xia, and G.Y. Meng, *Mater. Lett.* 47 (2001) 351.
130. H.B. Wang, H.Z. Song, C.R. Xia, D.K. Peng, and G.Y. Meng, *Mater. Res. Bull.* 35 (2000) 2363.
131. T.S. Zhang, J. Ma, H.T. Huang, P. Hing, Z.T. Xia, S.H. Chan, and J.A. Kilner, *Solid State Ionics* 5 (2003) 1505.
132. F.Y. Wang, S.Y. Chen, and S. Cheng, *Electrochem. Commun.* 6 (2004) 743.
133. F.Y. Wang, B.Z. Wan, and S. Cheng, *J. Solid State Electrochem.* 9 (2005) 168.
134. N.J. Kim, B.H. Kim, and D. Lee, *J. Power Sources* 90 (2000) 139.

135. D.L. Maricle, T.E. Swarr, and S. Karavolis, *Solid State Ionics* 52 (1992) 173.
136. F.Y. Wang, S.Y. Chen, W. Qin, S.X. Yu, and S. Cheng, *Catal. Today* 97 (2004) 189.
137. Y. Ji, J. Liu, T.M. He, J.X. Wang, and W.H. Su, *J. Alloy. Compd.* 389 (2005) 322.
138. X.Q. Sha, Z. Lu, X.Q. Huang, J.P. Miao, Z.H. Ding, X.S. Xin, and W.H. Su, *J. Alloy. Compd.* 428 (2007) 59.
139. H. Yoshida, H. Deguchi, K. Miura, M. Horiuchi, and T. Inagaki, *Solid State Ionics* 140 (2001) 191.
140. H. Yoshida, T. Inagaki, K. Miura, M. Inaba, and Z. Ogumi, *Solid State Ionics* 160 (2003) 109.
141. F.Y. Wang, S. Cheng, C.H. Chung, and B.Z. Wan, *J. Solid State Electrochem.* 10 (2006) 879.
142. J.V. Herle, D. Seneviratne, and A.J. McEvoy, *J. Eur. Ceram. Soc.* 19 (1999) 837.
143. X.Q. Sha, Z. Lu, X.Q. Huang, J.P. Miao, L. Jia, X.S. Xin, and W.H. Su, *J. Alloy. Compd.* 424 (2006) 315.
144. C.R. Xia and M.L. Liu, *Solid State Ionics*, 144 (2001) 249.
145. A.C. Pierre, *Introduction to Sol-gel Processing*, Kluwer Academic, 1998, chap.1–2, p. 1–85.
146. S. Pinol, M. Najib, D.M. Bastidas, A. Calleja, X.G. Capdevila, M. Segarra, F. Espiell, J.C. Ruiz-Morales, D. Marrero-Lopez, and P. Nunez, *J. Solid State Electrochem.* 8 (2004) 650.
147. K.Q. Huang, M. Feng, J.B. Goodenough, *J. Am. Ceram. Soc.* 81 (1998) 357.
148. B.L. Cushing, V.L. Kolesnichenko, and C.J. O'Connor, *Chem. Rev.* 104 (2004) 3893.
149. J.V. Herle, T. Horita, T. Kawada, N. Sakai, H. Yokokawa, and M. Dokiya, *Solid State Ionics* 86–88 (1996) 1255.
150. R.R. Peng, C.R. Xia, D.K. Peng, and G.Y. Meng, *Mate. Lett.* 58 (2004) 604.
151. J.G. Li, T. Ikegami, and T. Mori, *Acta Mater.* 52 (2004) 2221.
152. A.I.Y. Tok, L.H. Luo, F.Y.C. Boey, and J.L. Woodhead, *J. Mater. Res.* 21 (2006) 119.
153. Y.R. Wang, T. Mori, J.G. Li, and Y. Yajima, *Sci. Technol. Adv. Mater.* 4 (2003) 229.
154. H.B. Li, C.R. Xia, M.H. Zhu, Z.X. Zhou, and G.Y. Meng, *Acta. Mater.* 54 (2006) 721.
155. T.S. Zhang, J. Ma, L.H. Luo, and S.H. Chan, *J. Alloy. Compd.* 422 (2006) 46.
156. G.M. Christie and F.P.F. van Berkel, *Solid State Ionics* 83 (1996) 17.
157. R.R. Peng, C.R. Xia, Q.X. Fu, G.Y. Meng, and D.K. Peng, *Mater. Lett.* 56 (2002) 1043.
158. T.S. Zhang, P. Hing, H.T. Huang, and J. Kilner, *Solid State Ionics* 148 (2002) 567.
159. C.R. Xia and M.L. Liu, *J. Am. Ceram. Soc.*, 84(8) (2001) 1903.
160. T. Matsui, M. Inaba, A. Mineshige, and Z. Ogumi, *Solid State Ionics* 176 (2005) 647.
161. B.C.H. Steele, *J. Power Sources* 49 (1994) 1.
162. J. Liu and W. Weppner, *Ionics* 5 (1999) 115.
163. A. Tsoga, A. Naoumidis, and D. Stover, *Solid State Ionics* 135 (2000) 403.
164. Y.P. Xiong, K. Yamaji, T. Horita, N. Sakai, and H. Yokokawa, *J. Electrochem. Soc.* 149 (2002) E450.
165. S. Lubke and H.D. Wiemhofer, *Solid State Ionics* 117 (1999) 229.
166. E. Ruiz-Trejo and J. Maier, *J. Electrochem. Soc.* 154 (2007) B583.
167. E. Ruiz-Trejo, G. Tavizón, and H. García-Ortegab, *J. Electrochem. Soc.* 154 (2007) A70.
168. T. Shimonosono, Y. Hirata, and S. Sameshima, *J. Am. Ceram. Soc.* 88 (2005) 2114.
169. P. Jasinski, V. Petrovsky, T. Suzuki, and H.U. Anderson, *J. Electrochem. Soc.* 152 (2005) J27.

170. X.W. Qi, Y.S. Lin, C.T. Holt, and S.L. Swertz, *J. Mater. Sci.* 38 (2003) 1073.
171. W. Zhu, C.R. Xia, D. Ding, X.Y. Shi, and G.Y. Meng, *Mater. Res. Bull.* 41 (2006) 2057.
172. J.C.C. Abrantes, D. Perez-Coll, P. Nunez, and J.R. Frade, *Electrochim. Acta* 48 (2003) 2761.
173. T. Shimonosono, Y. Hirata, Y. Ehira, S. Sameshima, T. Horita, and H. Yokokawab, *Solid State Ionics* 174 (2004) 27.
174. H. Yahiro, K. Eguchi, and H. Arai, *Solid State Ionics* 36 (1989) 71.
175. B.C.H. Steele, K. Zheg, R.A. Rudkin, N. Kiratzis, M. Chrisitie, in: M. Dokiya, O. Yamonoto, H. Tagawa, S.C. Singhal (eds.), *Proceedings of the Fourth International Symposium on Solid Oxide Fuel Cells (SOFC-IV)*, June 1995, The Electrochemical Society, New Jersey, 1995, pp. 1028–1038.
176. C. Wagner, *Z. Phys. Chem.* B41 (1933) 42.
177. T. Matsuia, T. Kosakab, M. Inabab, A. Mineshigec, and Z. Ogumi, *Solid State Ionics* 176 (2005) 663.
178. X.G. Zhang, M. Robertson, C. Deces-Petit, W. Qu, O. Kesler, R. Maric, and D. Ghosh, *J. Power Sources* 164 (2007) 668.
179. T. Miyashita, *J. Mater. Sci.* 40 (2005) 6027.
180. R.T. Leah, N.P. Brandon, and P. Aguiar, *J. Power Sources* 145 (2005) 336.
181. B. Dalslet, P. Blennow, P.V. Hendriksen, N. Bonanos, D. Lybye, and D. Mogensen, *J. Solid State Electrochem.* 10 (2006) 547.
182. C. Hatchwell, N.M. Sammes, and I.W.M. Brown, *Solid State Ionics* 126 (1999) 201.
183. B.C.H. Steele and A. Heinzel, *Nature* 414 (2001) 345.
184. B.C.H. Steele, *Solid State Ionics* 134 (2000) 3.
185. M. Sahibzada, B.C.H. Steele, K. Zheng, R.A. Rudkin, and I.S. Metcalfe, *Catal. Today* 38 (1997) 459.
186. R. Doshi, V.L. Richards, J.D. Carter, X.P. Wang, and M. Krumpelta, *J. Electrochem. Soc.* 146(4) (1999) 1273.
187. C.R. Xia, F.L. Chen, and M.L. Liu, *Electrochemical and Solid-State Letters* 4(5) (2001) A52.
188. Z.P. Shao, and S.M. Haile, *Nature* 431 (2004) 170.
189. Y.H. Yin, W. Zhu, C.R. Xia, C. Gao, and G.Y. Meng, *J. Applied Electrochem.* 34(12) (2004) 1287.
190. W. Zhu, C.R. Xia, J. Fan, R.R. Peng, and G.Y. Meng, *J. Power Sources* 160(2) (2006) 897.
191. T. Hibino, A. Hashimoto, T. Inoue, J. Tokuno, S. Yoshida, and M. Sanob, *Science* 288 (2000) 2031.
192. T. Hibino, A. Hashimoto, M. Yano, M. Suzuki, S. Yoshida, and M. Sanob, *J. Electrochem. Soc.* 149(2) (2002) A133.
193. T. Suzuki, P. Jasinski, H.U. Andersona, and F. Dogan, *J. Electrochem. Soc.* 151(10) (2004) A1678.
194. Z.P. Shao, S.M. Haile, J. Ahn, P.D. Ronney, Z.L. Zhan, and S.A. Barnett, *Nature* 435 (2005) 795.
195. B.E. Buergler, M.E. Siegrist, and L.J. Gauckler, *Solid State Ionics* 176 (2005) 1717.
196. M. Yano, T. Kawai, K. Okamoto, M. Nagao, M. Sano, A. Tomita, and T. Hibinoa, *J. Electrochem. Soc.* 154 (8) (2007) B865.
197. T. Kawada, N. Sakai, H. Yokokawa, M. Dokiya, and I. Ansai, *Solid State Ionics* 50 (1989) 189.
198. K. Eguchi, T. Setoguchi, T. Inoue, and H. Arai, *Solid State Ionics* 52 (1992) 165.
199. M. Godickermeier and L.J. Gauckler, *J. Electrochem. Soc.* 145 (1998) 414.

200. T.L. Nguyen, K. Kobayashi, T. Honda, Y. Iimura, K. Kato, A. Neghisi, K. Nozaki, F. Tappero, K. Sasaki, H. Shirahama, K. Ota, M. Dokiya, and T. Kato, *Solid State Ionics* 174 (2004) 163.

201. T.L. Nguyen, T. Kato, K. Nozaki, T. Honda, A. Negishi, K. Kato, and Y. Iimura, *J. Electrochem. Soc.*, 153(7) (2006) A1310.

202. M. Shiono, K. Kobayashi, T.L. Nguyen, K. Hosoda, T. Kato, K. Ota, and M. Dokiya, *Solid State Ionics* 170 (2004) 1.

203. C. Rossignol, J.M. Ralph, J.M. Bae, and J.T. Vaughey, *Solid State Ionics* 175 (2004) 59.

204. Z.S. Duan, M. Yang, A.Y. Yan, Z.F. Houa, Y.L. Dong, Y. Chong, M.J.Cheng, and W.S. Yang, *J. Power Sources* 160 (2006) 57.

205. Y. Matsuzaki and I. Yasuda, *Solid State Ionics* 152–153 (2002) 463.

206. A. Tsoga, A. Naoumidis, and D. Stover, *Solid State Ionics* 135 (2000) 403.

207. A. Mai, V.A.C. Haanappel, F. Tietz, and D. Stöver, *Solid State Ionics* 177 (2006) 210.

208. T. Ishihara, H. Matsuda, and Y. Takita, *J. Am. Chem. Soc.* 116(9) (1994) 3801.

209. T. Ishihara, H. Matsuda, M. Azmi, and Y. Takita, *Solid State Ionics* 79 (1995) 147.

210. K.Q. Huang, R.S. Tichy, and J.B. Goodenough, *J. Am. Ceram. Soc.* 81(10) (1998) 2581.

211. K.Q. Huang, R.S. Tichy, and J.B. Goodenough, *J. Am. Ceram. Soc.* 81(10) (1998) 2565.

212. T. Ishihara, *Bull. Chem. Soc. Jpn.* 79 (2006) 1155.

213. K.Q. Huang and J.B. Goodenough, *J. Alloys. Compd.* 303–304 (2000) 454.

214. D. Lybye and K. Nielsen, *Solid State Ionics* 167 (2004) 55.

215. M.S. Islam, P.R. Slater, J.R. Tolchard, and T. Dinges, *Dalton Trans.* (2004) 3061.

216. J.W. Stevenson, T.R. Armstrong, L.R. Pederson, J. Li, C.A. Lewinsohn, and S. Baskaran, *Solid State Ionics* 115 (1998) 571.

217. T. Ishihara, M. Enoki, J.W. Yan, and H. Matsumoto, *Electrochemical Society Proceedings*, 2005–07, SOFC IX, 2005, pp. 1117–1126.

218. J. Bradley, P.R. Slater, T. Ishihara, and J.T.S. Irvine, *Electrochemical Society Proceedings*, 2003–07, SOFC VII, 2003, pp. 315–323.

219. J.W. Stevenson, K. Hasinska, N.L. Canfield, and T.R. Armstrong, *J. Electrochem. Soc.* 147 (2000) 321.

220. J.W. Stevenson, T.R. Armstrong, D.E. McCready, L.R. Pederson, and W.J. Weber, *J. Electrochem. Soc.* 144 (10) (1997) 3613.

221. B.A. Khorkounov, H. Nafe, and F. Aldinger, *J. Solid State Electrochem.* 10 (2006) 479.

222. T. Ishihara, S. Ishikawa, M. Ando, H. Nishguchi, and Y. Takiga, *Solid State Ionics* 173 (2004) 9.

223. T. Ishihara, J. Tabuchi, S. Ishikawa, J. Yan, M. Enoki, and H. Matsumoto, *Solid State Ionics*, 177 (2006): 1949.

224. M. Enoki, J. Yan, H. Matsumoto, and T. Ishihara, *Solid State Ionics*, in press.

225. T. Ishihara, Y. Tsuruta, Y. Chunying, T. Todaka, H. Nishiguchi, and Y. Takita, *J. Electrochem. Soc.* 150(1) (2003) E17.

226. T. Ishihara, M. Ando, M. Enoki, and Y. Takita, *J. Alloys Compd.* 408–412 (2006) 507.

227. T. Ishihara, S. Ishikawa, C.Y. Yu, T. Akbay, K. Hosoi, H. Nishiguchi, and Y. Takita, *Phys. Chem. Chem. Phys.* 5 (2003) 2257.

228. A.A. Yaremchenko, V.V. Kharton, E.N. Naumovich, D.I. Shestakov, V.F. Chukharev, A.V. Kovalevsky, A.L. Shaula, M.V. Patrakeev, J.R. Frade, and F.M.B. Marques, *Solid State Ionics* 177 (2006) 549.

229. T. Ishihara, S. Ishikawa, K. Hosoi, H. Nishiguchi, and T. Takitsa, *Solid State Ionics* 175 (2004) 319.

230. T. Ishihara, J.A. Kilner, M. Honda, N. Sakai, H. Yokokawa, and Y. Takita, *Solid State Ionics* 113–115 (1998) 593.

231. Z.C. Li, H. Zhang, B. Bergman, and X.Z. Zou, *J. European Ceramic Society* 26 (2006) 2357.

232. H. Ishikawa, M. Enoki, T. Ishihara, and T. Akiyama, *J. Alloys. Compd.* 430 (2007) 246.

233. H. Jena and B. Rambabu, *Materials Chemistry and Physics* 101 (2007) 20.

234. K.Q. Huang and J.B. Goodenough, *J. Solid State Chem.* 136(2) (1998) 274.

235. O. Schulz and M. Martin, *Solid State Ionics* 135 (2000) 549.

236. L.G. Cong, T.M. He, Y. Ji, P.F. Guan, Y.L. Huang, and W.H. Su, *J. Alloys Comp.* 348(1–2) (2003) 325–331.

237. K. Choy, W. Bai, S. Charojrochkul, and B.C.H. Steele, *J. Power Sources* 71 (1998) 361.

238. T. Mathews, P. Manoravi, M.P. Antony, J.R. Sellar, and B.C. Muddle, *Solid State Ionics* 135 (2000) 397.

239. T. Fukui, S. Ohara, K. Murata, H. Yoshida, K. Miura, and T. Inagaki, *J. Power Sources* 106 (2002) 142.

240. J.W. Yan, Z.G. Lu, Y. Jiang, Y.L. Dong, C.Y. Yu, and W.Z. Li, *J. Electrochem Soc.*, 149 (2002) A1132.

241. Y.B. Lin and S.A. Barnett, *Electrochemical and Solid-State Letters*, 9 (2006) A285.

242. Z.M. Bi, B.L. Yi, Z.W. Wang, Y.L. Dong, H.J. Wu, Y.C. She, and M.J. Cheng, *Electrochemical and Solid-State Letters*, 7(2004): A105.

243. J. W. Yan, H. Matsumoto, M. Enoki, and T. Ishihara, *Electrochemical and Solid-State Letters*, 8 (2005) A389–A39.

244. T. Ishihara, M. Honda, T. Shibayama, T.H. Minami, H. Nishiguchi, Y. Takita, *J. Electrochem Soc.* 145 (1998) 3177.

245. R. Maric, S. Ohara, T. Fukui, H. Yoshida, M. Nishimura, and T. Inagaki, K. Miura, *J. Electrochem. Soc.* 146 (1999) 2006.

246. K.Q. Huang, J.H. Wan, J. Goodenough, *J. Electrochem. Soc.* 148 (2001) A788.

247. J. Pena-Martinez, D. Marrero-Lopez, D. Perez-Coll, J.C. Ruiz-Morales, and P. Nunez, *Electrochimica Acta* 52 (2007) 2950.

248. J.H. Wan, J.Q. Yan, and J.B. Goodenough, *J. Electrochem Soc.*, 152 (2005) A1511.

249. Y.H. Du and N. Sammes, *Ionics* 9 (2003) 7–14.

250. T. Inagaki, F. Nishiwaki, J. Kanou, S. Yamasaki, K. Hosoi, T. Miyazawa, M. Yamada, and N. Komada, *J. Alloys. Compd.* 408–412 (2006) 512.

2 Anodes

Zhe Cheng, Jeng-Han Wang, and Meilin Liu

CONTENTS

2.1 INTRODUCTION

The primary function of the anode for a solid oxide fuel cell (SOFC) is to promote the electrochemical oxidation of fuels. When a hydrocarbon fuel such as methane is used as the fuel, additional functions of the anode may include internal reforming or partial oxidation of the fuel. The chemical and electrochemical processes often take place preferentially at certain surface and interfacial sites or triple-phase boundaries (TPBs). The resistance to these electrode processes, or the anode polarization, is determined not only by the intrinsic surface catalytic activities toward fuel oxidation and reforming, but also by the microstructure, morphology, and transport properties of the electrode materials. The critical challenge in the design of an efficient anode is to optimize the charge and mass transfer along surfaces, across interfaces, and through the bulk electrode. For an anode-supported SOFC, the anode also acts as a structural support for the entire cell. In this case, the mechanical integrity is important as well.

There are a number of informative reviews on anodes for SOFCs [1–5], providing details on processing, fabrication, characterization, and electrochemical behavior of anode materials, especially the nickel-yttria stabilized zirconia (Ni-YSZ) cermet anodes. There are also several reviews dedicated to specific topics such as oxide anode materials [6], carbon-tolerant anode materials [7–9], sulfur-tolerant anode materials [10], and the redox cycling behavior of Ni-YSZ cermet anodes [11]. In this chapter, we do not attempt to offer a comprehensive survey of the literature on SOFC anode research; instead, we focus primarily on some critical issues in the preparation and testing of SOFC anodes, including the processing-property relationships that are well accepted in the SOFC community as well as some apparently contradictory observations reported in the literature. We will also briefly review some recent advancement in the development of alternative anode materials for improved tolerance to sulfur poisoning and carbon deposition.

2.1.1 ANODE REQUIREMENTS

The general requirements for an SOFC anode material include [1–3] good chemical and thermal stability during fuel cell fabrication and operation, high electronic conductivity under fuel cell operating conditions, excellent catalytic activity toward the oxidation of fuels, manageable mismatch in coefficient of thermal expansion (CTE) with adjacent cell components, sufficient mechanical strength and flexibility, ease of fabrication into desired microstructures (e.g., sufficient porosity and surface area), and low cost. Further, ionic conductivity would be beneficial to the extension of active reaction sites for fuel oxidation from the TPBs to anode surfaces. Some other desirable, but not yet readily achieved, properties include tolerance to carbon deposition, sulfur poisoning, and reoxidation, which are vital to achieving direct utilization of readily available fuels or renewable fuels in SOFCs.

While these general requirements are well accepted, the specific values for each requirement may vary in a wide range, depending on the geometry of the electrodes and the design of the cells (such as the thickness and the length of the current path from the anode to the current collector). For example, the suggested anode electronic conductivity varies from 1 to 100 S/cm [8]. As explained elsewhere [8], the relaxed requirement of 1 S/cm for the effective electrical conductivity of a porous anode is valid only if a current collector layer is electrically well connected to the anode. Typically, the effective conductivity of a porous anode decreases rapidly as pores or an ionic conductor (such as YSZ) are introduced into the anode structure. For instance, the effective conductivity would be 10^2 to 10^3 S/cm for a conventional Ni-YSZ cermet anode with a composition of 50 vol% Ni and 50 vol% YSZ and a porosity of ~20 to 40% at 1000°C in a reducing atmosphere [12–16], whereas the electrical conductivity for pure Ni is ~2×10^4 S/cm [17].

2.1.2 ANODE MATERIALS

Anode materials studied in the early stage of SOFC development include graphite, platinum, iron, cobalt, and nickel. Ni was chosen as an anode material because of its low cost, good chemical stability, and excellent catalytic activity toward hydrogen oxidation and reforming of hydrocarbon fuels. However, pure nickel suffers from considerable mismatch in thermal expansion with YSZ, significant coarsening during operation, and poor binding to the electrolyte. The development of Ni-YSZ cermet anodes by Spacil [18] was groundbreaking in that the Ni-YSZ cermet readily meets most of the basic requirements for SOFC anodes. In a porous Ni-YSZ cermet anode, the Ni metal phase provides the required electronic conductivity and catalytic activity, whereas the YSZ ceramic phase lowers the coefficient of thermal expansion for the anode to match that of the electrolyte, prevents the Ni phase from coarsening, and offers a conduction path for oxide ions and thus may extend the active zones for anode reactions [1, 2]. Perhaps the most important function of the YSZ phase is its effectiveness in preventing Ni from aggregation. To date, porous Ni-YSZ cermet anodes are still widely used in SOFCs, although many different types of materials have been evaluated as the alternative anode materials. The idea of a metal-ceramic cermet anode is also used in the development of new SOFC systems based on alternative electrolyte materials such as doped ceria, $La_{1-x}Sr_xGa_{1-y}Mg_yO_3$ (LSGM), and proton conducting electrolytes (e.g., doped $BaCeO_3$-$BaZrO_3$). For each SOFC system based on a new electrolyte, nickel is mixed with the corresponding electrolyte to form a Ni-electrolyte cermet anode. The concept has also been adopted in searching for alternative anode materials with better tolerance to carbon deposition and sulfur poisoning. For example, Cu-ceria cermet anodes were used to enhance the tolerance to coking [19].

2.2   NI-YSZ CERMET ANODE

2.2.1   STARTING MATERIALS AND FABRICATION METHOD

The microstructure, properties, and performance of Ni-YSZ anodes depend sensitively on the microscopic characteristics of the raw materials (e.g., particles size and morphology of NiO and YSZ powders). The particle sizes of the starting YSZ powders vary usually from 0.2 to 0.3 μm, whereas those for the NiO powders are ~1 μm. The Ni to YSZ volume ratio usually varies from 35:65 to 55:45. For example, the reported Ni to YSZ volume ratios include 34:66 [20, 21], 40:60 [24], 43:57 [22], and 55:45 [23]. For a bilayer anode, the functional anode layer in contact with the electrolyte contains ~45 to 50 vol% Ni, whereas the anode support layer has ~35 to 40 vol% Ni [25, 26]. A pore former is usually added to tailor the shrinkage (for the cofiring) and to achieve sufficient porosity (>30 vol%) in the anode or the anode support layer.

For tubular SOFCs, Siemens Westinghouse originally developed an electrochemical vapor deposition (EVD) process, which was later replaced by a slurry coating or plasma spraying process [27, 28]. For planar SOFCs with an electrolyte-supported structure, the anode is usually fabricated by screen printing or tape casting [28]. For planar SOFCs with an anode-supported structure, tape casting is the predominant method used by industrial developers (e.g., Sulzer Hexis, ECN/InDec, FZJ, Risø, Global Thermoelectric, Allied Signal, Ceramic Fuel Cell Limited). Other methods used in anode formation include tape calendaring (Allied Signal), warm pressing (FZJ), screen printing (for anode functional layer in the case of two-layer anode at ECN/InDec), and vacuum slip casting (FZJ) [28]. The anode and electrolyte bilayer is usually fired at 1350 to 1400°C for 2 to 4 h, followed by the application and firing of the cathode. The cell is then heated up, while the anode is exposed to an inert gas (e.g., nitrogen), to the testing temperature (e.g., 750°C) before the inert gas is switched to a fuel gas containing hydrogen in order to reduce NiO to Ni.

In the following sections, the electrical conductivity, electrochemical activity toward hydrogen oxidation, and the sulfur poisoning behavior of Ni-YSZ cermet anodes will be discussed in detail, together with the effects of various processing procedures and testing conditions.

2.2.2   ELECTRICAL CONDUCTIVITY OF NI-YSZ CERMET ANODE

The electrical conductivity of a Ni-YSZ cermet anode depends on the composition (i.e., Ni to YSZ volume ratio), the microscopic features of the starting materials (e.g., particle size and distribution of NiO and YSZ powders), and the sintering and reduction conditions (e.g., temperature and atmosphere), as will be discussed in detail in the following sections.

2.2.2.1   Anode Composition (or Ni to YSZ volume ratio)

Since the conductivity of Ni is more than 5 orders of magnitude greater than that of YSZ under the fuel cell operating conditions, the electrical conductivity of a *porous* Ni-YSZ cermet anode changes several orders of magnitude, usually from ~0.1 S/cm to the range of ~10^3, as the Ni to YSZ volume ratio varies across the percolation threshold, which depends on the morphology, particle size, and distribution of each phase: Ni, YSZ, and pores. Shown in Figure 2.1 are the measured electrical conductivities versus the nickel volume content of the total solid for two types of cermets fabricated from a fine NiO powder (specific surface area of 3.5 m²/g, corresponding to primary particle size of 0.25 μm) and two types of YSZ powders (primary particle size of 0.1 and 0.3 μm, respectively) [12]. The anode conductivity versus Ni volume content followed an S-shaped curve, as predicted by the percolation theory; the electrical conductivity of the reduced anode increased from less than 0.1 S/cm to more than 10^2 S/cm as the Ni content increased from ~25 to ~35 vol%. Similar results were also obtained in other studies [13, 15].

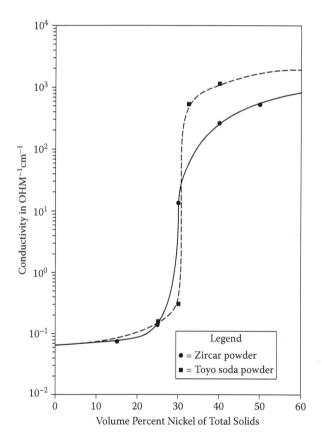

FIGURE 2.1 Change of electrical conductivity with respect to Ni content in the Ni-YSZ cermets at 1,000°C. Two types of YSZ were used to prepare the cermets: one was Toyo Soda powder with a specific surface area of 23 m²/g and an agglomerate size of ~0.3 μm, and the other was Zircar powder with a specific surface area of 47 m²/g and an agglomerate size of ~0.1 μm; the NiO used has a specific surface area of 3.5 m²/g. (From Dees, D.W. et al., *J. Electrochem. Soc.*, 134:2141–2146, 1987. Reproduced by permission of ECS-The Electrochemical Society.)

2.2.2.2   Effect of NiO and YSZ Particle Size

Percolation Threshold

The percolation threshold for randomly distributed spheres in a three-dimensional space is about 33% (i.e., one third of the volume fraction). It is important for the Ni and pore phases to exceed the threshold in order to ensure full access to fuel gas molecules (through the continuous pores) and electrons (through the continuous Ni phase) down at the dense electrolyte interface, where the anode is most active for fuel oxidation. This would leave little room for variation of the YSZ phase to achieve better performance. However, the percolation threshold can be significantly reduced by manipulating the distribution of each phase, which is determined by the shape and the particle size of Ni and YSZ. Results suggest that the percolation threshold for the Ni phase decreases (i.e., less volume fraction of nickel is required to form continuous nickel phase) as the particles size of NiO powder is reduced and the particle size of YSZ powder is increased. Shown in Figure 2.2 are the conductivities of three Ni-YSZ cermet anodes derived from an YSZ powder (Tosoh TZ-8Y, primary particle size of 0.1 to 0.2 μm) and NiO powders with three different particles sizes. Clearly, the percolation threshold was ~38 vol% Ni for the sample derived from a commercial NiO (particle size of ~1 to 20 μm) and the YSZ, ~33 vol% Ni for the sample derived from the commercial NiO ball-milled for 138 h (particle size of ~1 to 5 μm) and the YSZ, and ~23 vol% Ni for the sample derived from NiO prepared by the glycine nitrate process (GNP) (particle size of ~1 μm) and the YSZ [29]. Further, as shown in Figure 2.3, the conductivities of the Ni-YSZ cermet anode increased from less than 1 S/cm to ~10^3 S/cm when the particle size of YSZ was changed from 0.6 (i.e., c = 0%) to 27 μm (i.e., c = 100%), indicating that the percolation threshold decreased with YSZ particle size [14].The effect of particle size on the phase distribution and on the percolation threshold is insignificant when the difference in the particle size is relatively small (e.g., less than one order of magnitude). As shown in Figure 2.1, the percolation thresholds are similar for Ni-YSZ cermets derived from zirconia powders with primary particle size of 0.1 and 0.3 μm, respectively [12].

Effective Conductivity

Similar to the percolation threshold, the effective electrical conductivity of a porous Ni-YSZ cermet anode depends on the morphology, particle size, and distribution of the starting materials as well. In general, the effective conductivity increases as the NiO particle size is reduced when other parameters are kept constant. As shown in Figure 2.4 (samples 1 and 2), the cermet conductivity increased from ~10 S/cm to 10^3 S/cm as the NiO particle size was decreased from 16 to 1.8 μm while using the same YSZ powder (primary particle size of ~0.3 μm) and the same Ni to YSZ volume fraction [30].

Similarly, other studies concluded that the anode effective electrical conductivity increases with the YSZ particle size when other parameters (Ni to YSZ volume ratio and the particle size of NiO) remain constant, as can be seen in Figures 2.5 [13], 2.4 (compare samples 4, 5, and 6) [30], 2.3 [14], and 2.1 [12]. This is because it greatly influences the tendency of NiO clustering and downshifts the percolation threshold.

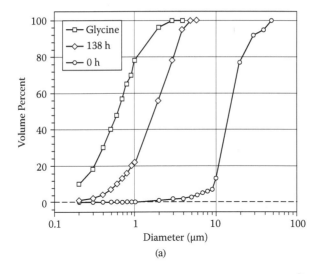

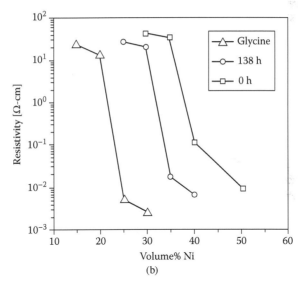

FIGURE 2.2 (a) Particle size distribution for three different NiO powders used in the Ni-YSZ cermets: the circle is for a commercial NiO powder with 0 h milling, the diamond is for the commercial NiO milled for 138 h, and the square is for the NiO powder prepared by the GNP process, and (b) resistivity versus Ni volume percent for the Ni-YSZ cermets made with the three different NiO powders at 1,000°C in reducing atmosphere. The YSZ powder has grain size of 0.1 to 0.2 µm, and the cermets were all sintered at 1,400°C for 2 h. (From Huebner, W. et al., *Proceedings of the Sixth International Symposium on Solid Oxide Fuel Cells,* 95(1):696–705, 1995. Reproduced by permission of ECS-The Electrochemical Society.)

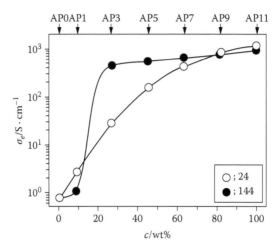

FIGURE 2.3 Electrical conductivity, σ_e, at 1000°C versus weight fraction of coarse YSZ, c/wt%, for 40 vol% Ni-60 vol% YSZ cermets made with a three-component mixture of NiO and fine YSZ (particle size of 0.6 μm) and coarse YSZ (particle size of 27 μm). The numbers 24 and 144 mean that the sintered cermets were reduced for 24 and 144 h, respectively. (From Itoh, H. et al., *J. Electrochem. Soc.*, 144, 641–646, 1997. Reproduced by permission of ECS-The Electrochemical Society.)

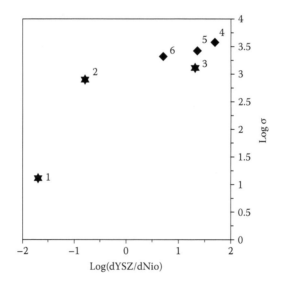

FIGURE 2.4 Electrical conductivity, σ (in S/cm), versus the YSZ to NiO particle size ratio (d_{YSZ}/d_{NiO}) for six Ni-YSZ cermets (1 to 6) made with different NiO and YSZ starting materials. The corresponding (YSZ, NiO) particle sizes are: 1 ($d_{YSZ} = 0.3$ μm, $d_{NiO} = 16$ μm), 2 ($d_{YSZ} = 0.3$ μm, $d_{NiO} = 1.8$ μm), 3 ($d_{YSZ} = 37$ μm, $d_{NiO} = 1.8$ μm), 4 ($d_{YSZ} = 50$ μm, $d_{NiO} = 1$ μm), 5 ($d_{YSZ} = 23$ μm, $d_{NiO} = 1$ μm), and 6 ($d_{YSZ} = 5$ μm, $d_{NiO} = 1$ μm). Samples 1 to 3 have Ni:YSZ volume ratio of 55:45 and samples 4 to 6 have Ni:YSZ volume ratio of 60:40. (From Tintinelli, A. et al., *Proceedings of the First European Solid Oxide Fuel Cells Forum,* 455–464, 1994. Reproduced by permission of ECS-The Electrochemical Society.)

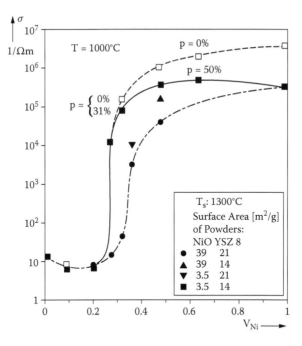

FIGURE 2.5 Electrical conductivity versus Ni volume fraction for Ni-YSZ cermets fabricated using different starting materials, as labeled in the graph. (From Ivers-Tiffée, E. et al., *Berichte der Bunsen-Gesellschaft für Physikalische Chemie*, 94:978–981, 1990. Permission pending.)

Effect of Porosity on Electrical Conductivity

The bulk conductivity of a fully *dense* cermet, σ_b, can be estimated from the effective conductivity, σ_e, and the porosity, f_p, of a *porous* cermet anode using the Bruggeman equation [12]

$$\sigma_e = \sigma_b (1 - f_p)^{3/2} \tag{2.1}$$

Clearly, the effective conductivity of a porous cermet electrode with a given composition (Ni to YSZ ratio) and phase distribution changes with porosity. When NiO particle size was reduced, the porosity of the cermet would be decreased as well [31,32]; Equation (2.1) suggests that the effective conductivity increases as the porosity is reduced [15].

Factors Influencing Electrical Conductivity

As stated, the particle size also influences the distribution of phases and the percolation threshold. In general, the small particles tend to cluster around the large particles to form a continuous path (lower percolation threshold for the smaller particles). Thus, if NiO particles are smaller than YSZ particles, we would expect high electrical conductivity. In contrast, if the YSZ particles are smaller, electrical conductivity would be lower because the small YSZ particles tend to cluster around the larger Ni particle, making them electrically isolated.

It should be noted that the mixing for Ni-YSZ cermets is determined by the sizes of the agglomerates *not* by the sizes of the primary particles that comprise the agglomerates. For example, in Figure 2.5, the electrical conductivity for Ni-YSZ cermets *increased* 1 to 4 times when the Brunauer-Emmett-Teller (BET)-specific surface area for the starting nickel oxide powder was *decreased* from 39 m^2/g (equivalent average primary particle size of 0.02 μm) to 3.5 m^2/g (equivalent average primary particle size of 0.25 μm). This appears to contradict the trend stated earlier, which would predict that NiO with larger surface area (and hence smaller particle size) should have higher electrical conductivity. This is because the BET-specific surface area may not reflect the degree of agglomeration; aggregates of small particles may still have high surface area, but behave as larger particles in terms of mixing with other particles. Ultrafine powders with particle size on the order of 0.02 μm will normally form agglomerates that may be hard to break in a normal mixing process. As a result, the actual size of NiO particles in the mixing process may well be on the order of submicron or even larger, which would explain the apparent contradiction observed in Figure 2.5. Another example can be found in the study of Tietz and co-workers [34] on Ni-YSZ cermets with Ni to YSZ volume ratio of 39:61, in which they showed that for two types of NiO with similar BET-specific surface areas of 55.7 and 47 m^2/g, the actual average particle size measured by the laser diffraction method was 0.6 and 14.7 μm, respectively. Accordingly, the former NiO with smaller average particle size resulted in a cermet with conductivity as high as 3,990 S/cm, while the latter with a much larger agglomerate size resulted in a cermet with conductivity of only 78 S/cm even though their specific surface area differs by only ~18% [34]. Thus, it is important to characterize the average particle size or the particle size distribution for the raw materials instead of the BET surface area since the distribution of particles during mixing is determined more critically by the particle size (distribution).

It is for this reason that ultrafine (~0.02 μm scale) starting materials, especially for NiO, do not necessarily lead to low percolation threshold, high electrical conductivity, and improved fuel cell performance as hoped, although finer starting materials may lead to lower sintering temperatures and, possibly, finer microstructures. In fact, it is interesting to note that several research groups reported excellent fuel cell performance (i.e., > ~1 W/cm^2 at 800°C) using commercial NiO powder, such as that from J.T. Baker [20, 21, 24], which has a BET-specific surface area of ~4 m^2/g, an average primary particle size of 0.2 μm, and an agglomerate size of ~0.8 μm [34].

In the pursuit of higher electrical conductivity, researchers should also be aware of the consequence of using very coarse YSZ powders (e.g., larger than 10 μm) which is poor mechanical strength [14, 30, 32]. According to the study by Yu et al. [32], the mechanical strength of the anode made from coarse YSZ (particle size of ~8 μm) and fine NiO (particle size of ~0.8 μm) was on the order of 20 MPa, which was only one third to one fifth of the anode made with fine YSZ powders (particle size of ~0.8 μm) with similar porosity (Figure 2.6) [32]. In addition, using very coarse YSZ powders may also increase the fabrication complexity by adding an extra step of particle coarsening via heat treatment of the commercial fine YSZ [30, 33] since uniformly shaped and sized coarse YSZ may not be cheaper and more available compared with fine grained YSZ. The effectiveness of YSZ on preventing Ni particles from coarsening would be reduced as well.

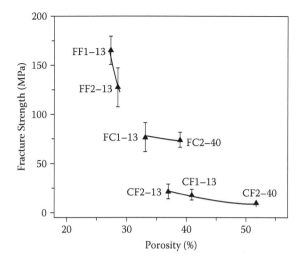

FIGURE 2.6 Fracture strength of Ni-YSZ cermets as a function of porosity. Standard deviation is superimposed on each average value. The starting NiO and YSZ particle sizes for FF1-13 and FF2-13 are both 0.8 μm, for FC1-13 and FC2-40 they are 0.8 and ~6 μm, for CF1-13, CF2-13, and CF2-40 they are 8 and 0.8 μm. The suffixes 13 and 40 represent the volume fraction of carbon black pore former added. (From Yu, J.H. et al., *J. Power Sources*, 163:926–932, 2007. Copyright by Elsevier, reproduced with permission.)

Some researchers explored Ni-YSZ cermets that used both fine (average size of 0.6 μm) and coarse (average size of 27.0 μm) YSZ as the starting materials. It was hoped that the coarse YSZ particles would form a frame that keeps the total volume unchanged while the fine YSZ particles would sustain the network of Ni and pore, providing good electrical conductivity and microstructural stability [14]. However, it is not clear how such an anode compares with conventional anodes made from both fine NiO and YSZ in terms of strength, electrical conductivity, and electrochemical activity.

Many studies have already demonstrated that the percolation threshold can be shifted to lower Ni content by various processing steps such as using NiO powders of particle size below ~1 μm (a percolation threshold of ~25 vol% Ni as in reference [29]), or increasing YSZ particle size to as large as 37 μm (a percolation threshold of ~26 vol% as in reference [30]), or using Ni-coated graphite pore former in the anode preparation (a percolation threshold of only ~15 vol% as in reference [35]). However, few researchers actually employ cermets with Ni to YSZ volume ratio less than 34:66 for SOFCs despite the facts that (a) lower Ni content may offer a closer match in thermal expansion coefficient with the electrolyte, and (b) the YSZ phases may be better connected to offer greater ionic conductivity and broader extension of the reaction sites. This is due probably to the uncertainty associated with the long-term conductivity of the anode of low Ni content. Furthermore, advantages such as higher electrical conductivity gained through those processing (e.g., coarsening of YSZ or coating of Ni with graphite) could be offset by the higher cost of materials and the degradation of the anode.

2.2.2.3 Influence of Processing Conditions
on Anode Electrical Conductivity

In addition to composition and starting materials particle size, the conductivity of the Ni-YSZ cermet anode is strongly influenced by processing procedures including the sintering and the reduction conditions of the cermet, which will be discussed in detail in this section.

Anode Sintering Condition

The exact influence of sintering condition on the obtained Ni-YSZ cermet bulk conductivity, σ_b, has not yet been fully revealed since most studies only report the effective conductivity of the porous Ni-YSZ cermet σ_e; see Equation (2.1) for definitions. Nevertheless, the influence of sintering condition, such as temperature and time, on the measured effective conductivity of the anode is fairly clear. It was found that the effective conductivity of Ni-YSZ anodes was higher with higher sintering temperatures and longer sintering time in the range of 1200 to 1350°C (see Figure 2.7) [15]. The difference in effective conductivity with respect to sintering temperature and time was linked to the porosity of the cermet: higher sintering temperature or longer sintering time led to lower porosity, which resulted in higher effective conductivity, as shown in Figure 2.8 [15]. It is not clear whether the electrical conductivity for a fully dense material is influenced in the same way by the anode sintering condition.

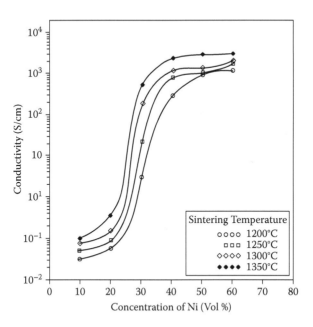

FIGURE 2.7 Change of electrical conductivity at 1000°C with respect to Ni volume content for Ni-YSZ cermets sintered for 2 h at 1200, 1250, 1300, and 1350°C, respectively. (From Pratihar, S.K. et al., *Proceedings of the Sixth International Symposium on Solid Oxide Fuel Cells*, 99(19):513–521. Reproduced by permission of ECS-The Electrochemical Society.)

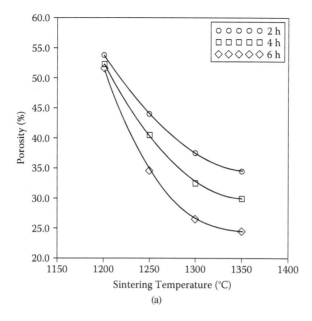

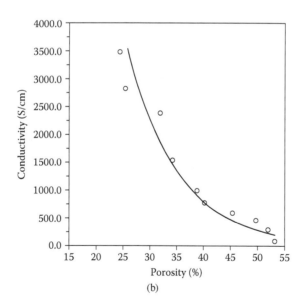

FIGURE 2.8 (a) Porosity versus sintering temperatures for Ni-YSZ cermets sintered for different times (2, 4, and 6 h); (b) Anode conductivity versus porosity for the Ni-YSZ cermets with Ni to YSZ volume ratio of 40:60. (From Pratihar, S.K. et al., *Proceedings of the Sixth International Symposium on Solid Oxide Fuel Cells*, 99(19):513–521. Reproduced by permission of ECS-The Electrochemical Society.)

Anode Reduction Condition

To the authors' surprise, the anode reduction condition, especially the reduction temperature, was found to be a factor that critically influences the anode electrical conductivity. In fact, according to the authors' study, the anode reduction condition determines whether an anode with Ni to YSZ volume ratio in the range of 35:65 to 70:30 is conductive or insulating, even though the anode is expected to be conductive according to the percolation theory. To the authors' knowledge, only Grahl-Madsen and co-workers [36, 37] have ever openly reported results on the influence of anode reduction temperature on the obtained anode electrical conductivity. They showed that the high-temperature electrical conductivity for cermets reduced at 1000°C was ~2 to 4 times higher than those reduced at 800°C—see Figure 2.9(a)—while for the same sintered NiO-YSZ composite, the room temperature electrical conductivity decreased almost linearly from 6000 to 1000 S/cm as the reduction temperature decreased from 1000 to 650°C; see Figure 2.9(b) [37]. Grahl-Madsen et al. [37] have also observed that the morphology of the anode was very different when the anode was reduced at different temperatures. The cermets reduced at 800°C showed many crevices between the Ni and the YSZ phase, while the cermets reduced at 1000°C showed close contact between Ni and YSZ.

Our study found the effects of reduction temperature and atmosphere on the anode conductivity to be even more drastic. If hydrogen was introduced at a high temperature (e.g., 750°C) after the sample had been heated up in air or N_2 to that temperature, the resulting NiO-YSZ composite had extremely high conductivity (sheet resistance of ~0.1 to 0.3 ohm). By contrast, if hydrogen (e.g., ~50 vol%) was introduced into the anode chamber from room temperature during the heating process to the high temperature (i.e., 750°C) at a slow ramp rate (e.g., 3°C/min or slower), the reduction was achieved at temperatures in the range of ~300 to 400°C, and the resulting Ni-YSZ cermet was almost *insulating* at room temperature (sheet resistance >10^6 ohm) even though all of the nickel oxide had been reduced into pure Ni, as confirmed by XRD. The estimated effective electrical conductivity at 750°C in H_2 for cermets reduced in the second method was on the order of ~2×10^{-3} S/cm (0.002 S/cm), which is only about one tenth of the ionic conductivity of YSZ at that temperature, indicating that Ni did not form an interconnected network.

Figure 2.10 compares the measured cell impedance for anode-supported cells reduced in these two different methods: the cell bulk resistance increased almost 100 times when a fuel mixture of 97 vol% H_2/3 vol% H_2O was introduced from room temperature instead of at 750°C. Figure 2.11(a) shows the scanning electron microscope (SEM) images of a Ni-YSZ composite reduced when hydrogen was introduced at 800°C, whereas Figure 2.11(b) shows the SEM images of a Ni-YSZ composite reduced when hydrogen was introduced from room temperature (so that the cermet was reduced at lower temperatures of ~300 to 400°C during the slow heating process). Significant differences in microstructures are apparent: the nickel phase in the cermet reduced 800°C formed a continuous network, while the nickel phase in the cermet reduced during the heating process formed isolated, spherical Ni grains.

The complete reduction of an anode at a lower temperature (e.g., ~200 to 300°C) is catastrophic in our experience. This is because once an anode forms the structure as in Figure 2.11(b) and develops low conductivity due to reduction at a low temperature, it will not improve after subsequent heat treatments at high temperature in reducing atmosphere.

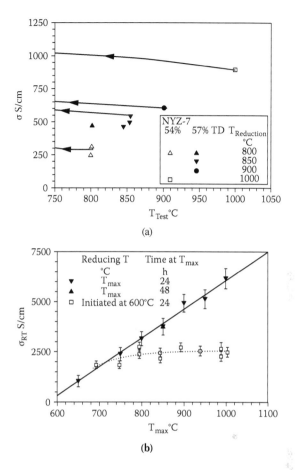

FIGURE 2.9 (a) Change of electrical conductivity with testing temperature for Ni-YSZ cermets reduced at temperatures of 800, 850, 900, and 1000°C. The percentages 54 and 57% represent the cermet relative density. The samples were reduced at the highest temperature and the conductivity measurements were carried as the sample was cooled down. (b) Change of room temperature conductivity, σ_{RT}, versus the maximum temperature the sample experienced, T_{max}, for cermets reduced at the fixed temperature of T_{max} (closed symbol) or reduced during the heating process from 650°C to T_{max}. (From Grahl-Madsen, L. et al., *J. Mater. Sci.*, 41:1097–1107, 2006. With kind permission from Springer Science and Business Media.)

This is supported by the study of Grahl-Madsen et al. as shown by the dotted line in Figure 2.9(b). It appears that the reduction reaction finished at ~750°C for samples reduced during the heating process from 650°C, so the conductivity did not increase further during soaking in H_2 at higher temperatures up to 1000°C; the resulting room temperature conductivity of the samples remained the same as that for samples reduced at a fixed temperature of ~750°C and was much lower than that for samples reduced at a higher fixed temperature (e.g., 900 or 1000°C).

In addition, cell open circuit voltage is usually much lower than the theoretical value if the anode is fully reduced at a lower temperature (i.e., ~200 to 300°C).

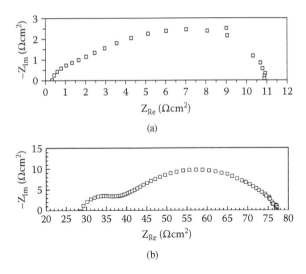

FIGURE 2.10 Impedance spectra measured at 750°C for anode-supported cells, for which a fuel mixture of 97% H_2/3% H_2O was introduced into the anode chamber (a) at 800°C after the cell was heated up to 800°C with the anode exposed to air, and (b) from room temperature during the heating process to 800°C.

In the authors' study, whenever the anode was reduced at a lower temperature, the cell open-circuit voltage (OCV) was only ~0.8 to 0.9 V in humidified hydrogen even though no gas leakage was detected.

A third problem associated with reducing the anode at a lower temperature is that the strength of the cermet anode would be significantly lower. Wang et al. [38] showed that the strength of the Ni-YSZ cermet reduced in a method in which 5% H_2/95% N_2 was introduced from room temperature during the heating process was 30% lower than that for cermets reduced in a method in which 5% H_2/95% N_2 was introduced at a fixed high temperature (e.g., 800°C), even though both types of samples had the same reduced fraction of NiO and porosity.

As expected from the above discussion, if the reducing gas (mixture) is introduced during the heating process, the actual anode reduction temperature is also influenced by two other factors, which are the concentration of the reducing agent (e.g., hydrogen) and the heating rate. This is because the anode reduction process needs a significant amount of hydrogen and time for completion. Therefore, if the concentration of the reducing agent (e.g., hydrogen) is low, for example 5% H_2/95% N_2 as in Wang et al.'s study, even though NiO reduction starts at ~300°C, it would not finish until the sample temperature reaches relative high temperatures (i.e., ~600 to 800°C). In that case, a significant portion of the NiO is reduced at high temperature, and the resulting anode is still somewhat conductive. Some researchers adopted this method for the reduction of anode (i.e., introducing a low concentration of hydrogen during the slow heating process) and did not seem to meet the problem of insulating anodes [23, 30]. However, the authors did find that the surface of the anode reduced in this way (i.e., 5% H_2 introduced from room temperature) has higher sheet resistance than

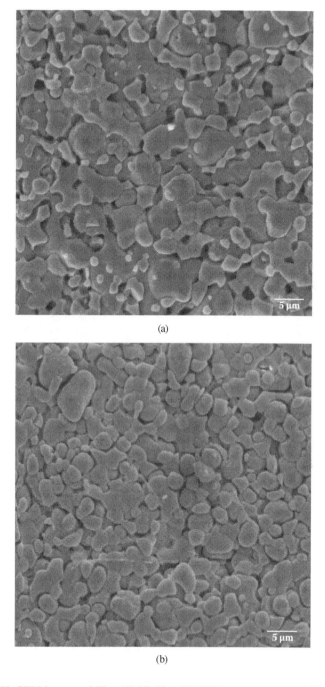

(a)

(b)

FIGURE 2.11 SEM images of 55 vol% Ni-45 vol% YSZ cermets prepared by sintering at 1475°C for 5 h and reduction in different ways: (a) hydrogen was introduced at 800°C after the sample was heated up to 800°C in air or nitrogen and (b) hydrogen was introduced from room temperature during the heating process.

the inside of the anode because the surface was reduced first at lower temperatures. Similarly, if the heating rate is high, a significant portion of the anode is also reduced at elevated temperatures and the anode would also be conductive.

It is surprising that the problem of low anode conductivity and low strength for anodes reduced at lower temperatures (e.g., ~200 to 400°C) is seldom mentioned in the literature. While most researchers use a general description such as the "anode was reduced *in situ*," many researchers such as de Souza et al. [20], Primdahl and Mogensen [39], Chung et al. [40], and Mai et al. [41] did specify clearly in their studies the anode was reduced *in situ* in a particular way like this: the cermet anode was heated up in inert gas or air to a designated temperature, usually at 750 to 900°C, then and only then, the inert gas or air was changed to hydrogen (in the case of heating up in air, an extra purge step using inert gas was required before switching to hydrogen). It is not clear whether those authors adopted this specific sequence because they have met the same problem.

Nevertheless, because the phenomena of low conductivity and strength when the anode is reduced at lower temperatures are rarely mentioned in the literature, the fundamental reasons for them are not clear at this time. Grahl-Madsen et al. [37] speculated that the formation of continuous Ni network at high temperatures (e.g., at 1000°C) is due to the redistribution of freshly formed metal when its surface is still active, while the failure to achieve such an effect at lower temperatures (e.g., at 650°C) is due to slow reduction kinetics and the failure to achieve Ni connectivity in subsequent high-temperature treatment is due to the loss of surface activity of the nickel. Wang et al. [38] attributed the lower strength and poor microstructure connectivity for Ni-YSZ cermets reduced at lower temperatures to the damaging effect of nickel on the YSZ framework. They argued that when NiO was reduced to Ni at lower temperatures, it would introduce microcracks in the YSZ network during the subsequent heating process since Ni has higher coefficient of thermal expansion than YSZ. Further studies are required to validate these hypotheses and completely reveal the fundamental reasons for such effects.

2.2.3 ELECTROCHEMICAL PERFORMANCE OF NI-YSZ CERMET ANODE

One of the fundamental requirements for a high performance anode is to have excellent catalytic activity toward the electrochemical oxidation of the fuel (e.g., hydrogen). This is reflected as low anode polarization or interfacial resistance. This area has seen intensive research for quite some time and is covered very well by the reviews of McEvoy [2], Zhu and Deevi [3], and Jiang and Chan [4]. In this section, the focus will still be on revealing the influences of processing and testing parameters on the obtained anode electrochemical performance.

2.2.3.1 Influence of Anode Composition on Anode Electrochemical Performance

Like electrical conductivity, the anode composition (i.e., Ni to YSZ volume ratio) also influences the anode activity or polarization. The lowest anode interfacial resistance is usually obtained when the Ni to YSZ volume ratio is ~40:60. For example, Kawada et al. [42] found that anode interfacial resistance reached a minimum when the Ni content was 40 vol%, as shown in Figure 2.12. This was verified by several other independent studies [25, 31, 43, 44]. For example, Koide [25] found that the

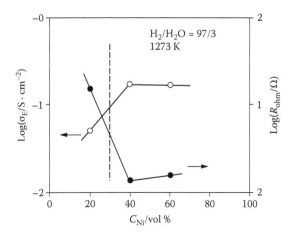

FIGURE 2.12 Effect of Ni content on the electrode/electrolyte interface conductivity (σ_E) and the ohmic resistance (R_{ohm}) for Ni-YSZ cermet anodes prepared using NiO-YSZ powder mixtures precalcinated at 1400°C and sintered at 1500°C. (From Kawada, T. et al., *J. Electrochem. Soc.*, 137:3042–3047, 1990. Reproduced by permission of ECS-The Electrochemical Society.)

anode polarization resistance (at a constant current density of 300 mA/cm²) reached a minimum at ~40 vol% Ni, as shown in Figure 2.13.

However, there seems to be some disagreements on the exact influence of anode composition on the cell/anode bulk resistance. While Kawada et al. [42] indicated that cell bulk resistance also reached minimum when the Ni content was 40 vol% (Figure 2.12), Koide et al. [25] showed that as the Ni content increased, cell bulk resistance decreased continuously, and cell maximum power output also increased (Figure 2.13).

In addition to anodes with a uniform composition, double-layer anodes with two different compositions are also investigated. The layer next to the electrolyte is called "anode functional layer (AFL)," and the layer outside is called "anode support layer (ASL)." Generally, the AFL has slightly higher nickel volume content than the ASL. For example, Koide et al. [25] proposed a double-layer anode structure for electrolyte-supported cells: the interfacial layer right next to the electrolyte was nickel rich (Ni:YSZ = 61:39 by vol.) while the bulk layer was YSZ rich (Ni:YSZ = 40:60 by vol.). They believed that (a) anode ohmic resistance (labeled as "IR resistance" by Koide et al. [25]) mainly comes from contact resistance between Ni and YSZ electrode across the interface and not from the anode body, and (b) the electrochemical reaction can occur at a certain distance away from the electrolyte. Therefore, using a Ni-rich cermet as the interfacial layer would be effective in reducing ohmic resistance, while a higher YSZ content bulk layer will "reduce polarization resistance." According to the authors, the ohmic resistance of the double-layer anode was similar to that for a single-layer anode with a Ni content of 87 vol%, while the interfacial resistance is similar to that for a single-layer anode with a Ni content of 40 vol%. A 50% increase in the total cell performance was observed over cells with a uniform anode of 40 vol% Ni to 60 vol% YSZ composition, and a cell life of 8,000 h was achieved [25]. Similar bilayer anode structures for anode-supported cells were

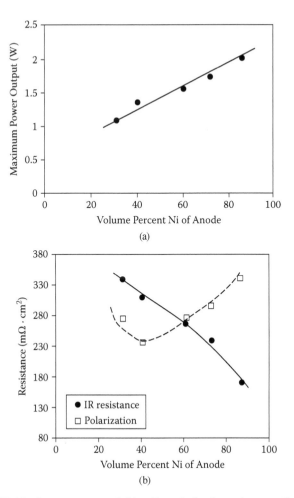

FIGURE 2.13 (a) Maximum power and (b) cell total ohmic resistance (labeled as "IR resistance") and interfacial resistance (labeled as "polarization") at constant current density of 0.3 A/cm² versus the volume percent of Ni in the Ni-YSZ cermet for electrolyte-supported cells with an active area of 2 cm² operated at 1000°C. (From Koide, H. et al., *Solid State Ionics*, 132:253–260, 2000. Copyright by Elsevier, reproduced with permission.)

adopted by the groups at FZJ [45, 46] and PNNL [26, 47], etc. with the exception that the nickel content in the anode functional layer was usually lower (i.e., 45 to 50 vol% Ni), while the anode support layer still contained ~40 vol% Ni.

2.2.3.2   Influence of Starting Materials Particle Size on Anode Electrochemical Performance

Effect of Starting Materials Particle Size

Studies show that anode electrochemical performance increases as the starting materials—NiO and YSZ—particle size decreases. For example, Huebner et al. [31] found that as the starting NiO particle size decreased, the anode overpotential decreased significantly and also became more stable. Similar results were obtained by Jiang et al. [44] using NiO and 3 mol% Y_2O_3-ZrO_2 (3YSZ). They found that both the anode ohmic resistance and the polarization resistance decreased as the NiO particle size decreased, as shown in Figure 2.14(a). On the other hand, Hikita et al. [48] found that the electrochemical performance of the Ni-8YSZ cermet anode reached minimum overpotential when the starting 8YSZ/NiO particle size ratio equaled ~0.01.

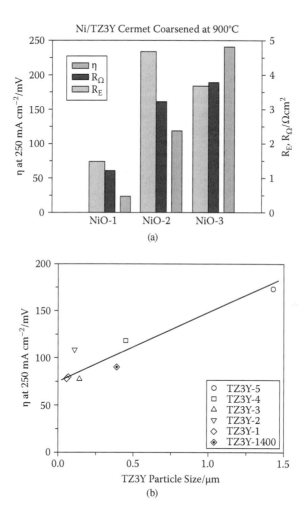

FIGURE 2.14 (a) Influence of NiO particle size on the anode overpotential η (at a constant current density of 250 mA/cm^2), anode ohmic resistance R_Ω, and anode interfacial resistance, R_E, for cermets made from YSZ and three different types of NiO: NiO-1, NiO-2, and NiO-3, of which the particle size is 1, 5, and 10 µm, respectively. (b) Influence of TZ3Y particle size on the anode overpotential η (at a constant current density of 250 mA/cm^2) for cermets made from 600°C-calcinated NiO-1 and six different TZ3Y powders with different particle sizes. (From Jiang, S.P. et al., *Solid State Ionics*, 132:1–14, 2000. Copyright by Elsevier, reproduced with permission.)

Jiang et al. [44] also found that the anode overpotential decreased as the YSZ particle size decreased in the range of ~0.1 to 1.5 μm, as shown in Figure 2.14(b).

Effect of Precalcination of the NiO-YSZ Powder Mixture

For electrolyte-supported cells, many researchers observed that precalcination of the NiO-YSZ powder mixture together significantly enhanced anode performance. For example, Kawada et al. [42] found that precalcination of NiO-YSZ powder mixtures at temperatures above 1200°C before sintering of the anode for electrolyte-supported cells led to higher anode polarization conductance, which is the inverse of the polarization resistance, as shown in Figure 2.15. Huebner et al. [31] also observed that precalcination of the NiO-YSZ powder mixture at 1400°C led to cermets with lower anode overpotential. The effect of starting materials precalcination on the anode performance was also studied by Jiang, who found that the powder mixture calcinated at 1300°C before sintering at 1400°C provided the best performance for cermet with a Ni:8 mol% Y_2O_3-ZrO_2 (8YSZ) volume ratio of 50:50 [49]. Jiang also observed that both anode bulk resistance and interfacial resistance decreased when the precalcination temperature increased as long as the precalcination temperature was lower than the anode sintering temperature [49]. Another benefit of precalcination of the NiO-YSZ powder mixture, according to Kawada [42], is that less degradation was observed, which was supported by Huebner's study [31].

However, despite those positive reports, the authors would not recommend precalcination of the starting NiO-YSZ powder mixture as a necessary step in the processing of solid oxide fuel cells for the following reasons. First, the mechanism for the enhanced electrochemical performance for anodes when the NiO-YSZ precalcinated together has not been explained clearly. The phenomenon is actually counterintuitive because it has been shown that coarsening of NiO or YSZ alone leads to lower anode

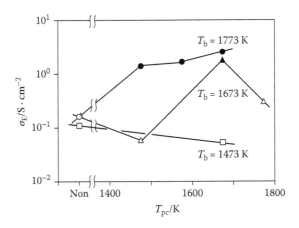

FIGURE 2.15 Effect of NiO-YSZ powder mixture precalcination temperature, T_{pc}, and anode sintering temperature, T_b, on the anode interface conductance, σ_E. "Non" in the T_{pc} axis represents the electrode prepared without precalcination treatment. (From Kawada, T. et al., *J. Electrochem. Soc.*, 137:3042–3047, 1990. Reproduced by permission of ECS-The Electrochemical Society.)

performance [44, 49]. Second, some other researchers obtained high-performance electrolyte-supported cells using similar NiO and YSZ powders but without the precalcination of the NiO-YSZ powder mixture; see the studies by Primdahl and Mogensen [39] and Koide et al. [25]. Third, it is noted that for all the studies involving precalcination of the NiO-YSZ powder mixture, the NiO-YSZ powder mixture after the high-temperature precalcination treatment were not used directly but were ground down and mixed again by ball-milling for a significant period of time [49]. Considering the fact the NiO and YSZ had already been mixed before the precalcination step, it is likely that the additional steps of precalcination and ball milling again in the anode processing may serve the role of helping to achieve a more thorough mixing of NiO and YSZ powders, which would lead to better anode performance. Last, the applicability of precalcination of the NiO-YSZ mixture seems to be limited because most studies involving precalcination of anode powder mixtures were carried out on electrolyte-supported cells; for anode-supported cells, to the authors' knowledge, no studies have identified precalcination of the starting powder mixture as a beneficial step up until now.

2.2.3.3 Influence of Sintering Temperature on Anode Electrochemical Performance

The influence of sintering temperature on the anode performance is fairly clear: most studies suggest that increasing the anode-sintering temperature leads to increase in the anode performance by decreasing both the anode bulk resistance and interfacial resistance. For example, Kawada et al. [42] studied the influence of the anode-sintering temperature on the anode polarization resistance for electrolyte-supported cells (Figure 2.16). They found that as the anode-sintering temperature increased, both anode ohmic and interfacial resistances decreased dramatically. The results were confirmed by later studies of Huebner et al. [31], Jiang et al. [44, 49], and Primdahl et al. [50] with the exception that further temperature increase from 1400 to 1500°C led to negligible improvement or even a small drop in cell performance in those later studies. Jiang [49] claimed that a high-sintering temperature has no effect on the conductivity of the pure Ni anode, but is essential to the formation of good bonding with YSZ, leading to the formation of a rigid structure that supports the Ni phase yet limits Ni coarsening.

2.2.3.4 Influence of Testing Atmosphere on Anode Electrochemical Performance

The influence of the atmosphere on anode electrochemical activity has been studied. The major parameters investigated include hydrogen concentration and water concentration. For the influence of H_2 concentration on the electrode kinetics, Mizusaki et al. [51] used Ni pattern electrode and found that the anode interfacial conductance did not change significantly with changing partial pressure of H_2 when the H_2O concentration was kept constant, and if the water concentration was low, there seemed to be a tendency for anode interfacial conductance to increase with increasing pH_2; see Figure 2.17. The result was confirmed by later studies [52, 53]. Nakagawa et al. [52] found that by changing the partial pressure of H_2 from ~0.03 to 1 while keeping the pH_2O on the order of ~0.001, anode polarization did not change much, while Wen et al. [53] found that changing the fuel from pure H_2 to 20% H_2/80% Ar does not change the anode overpotential significantly. Primdahl and Mogensen [39] found that anode

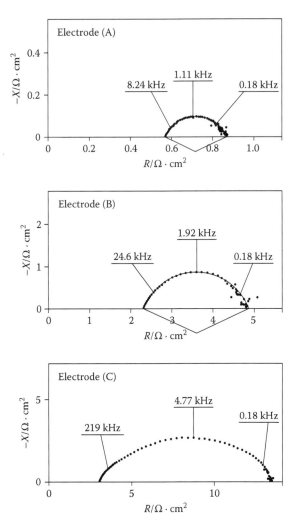

FIGURE 2.16 Impedance spectra for Ni-YSZ cermet anodes prepared at different NiO-YSZ precalcination temperature (T_{pc}) and sintering temperature (T_b): Electrode A: $T_{pc} = 1400°C$, $T_b = 1500°C$; Electrode B: $T_{pc} = 1400°C$, $T_b = 1400°C$, Electrode C: $T_{pc} = 1400°C$, $T_b = 1200°C$. (From Kawada, T. et al., *J. Electrochem. Soc.*, 137:3042–3047, 1990. Reproduced by permission of ECS-The Electrochemical Society.)

interfacial resistance could be separated into three portions. The anode conductance associated with the high-frequency process increased slowly as pH_2 increased slowly, while the polarization for intermediate- and low-frequency processes increased first slowly with pH_2, then at $pH_2 \approx 0.5$, it increased significantly with pH_2. Similar observation was made by Jiang and Badwal [54] who found that as the water content was fixed, the anode conductance increased sharply when $pH_2 > {\sim}30$ vol%.

The effect of moisture on the anode activity has also been widely studied. It is generally accepted that the addition of H_2O accelerates the anode kinetics. Dees et al.

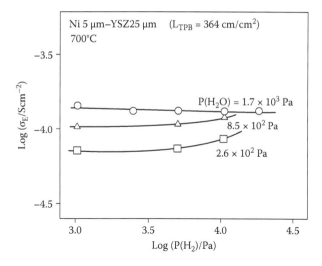

FIGURE 2.17 Change of electrode interfacial conductance, σ_E, with respect to partial pressure of hydrogen for Ni-YSZ pattern electrode in fuels with different H_2O partial pressure. (From Mizusaki, J. et al., *Solid State Ionics*, 70(71):52–58, 1994. Copyright by Elsevier, reproduced with permission.)

[12] found that the interfacial resistance for an anode symmetrical cell in a uniform atmosphere decreased significantly as the H_2O partial pressure increased up to ~3 vol%. However, the experiment is not completely conclusive since under open-circuit condition, when the moisture content increased from 10^{-3} to 10^{-2}, the problem of H_2O depletion would be less significant as the anodes work cathodically to reduce water and produce H_2, which would also display reduced polarization resistance in the impedance spectra. Nevertheless, Nakagawa et al. [52] studied the influence of H_2O on the cell voltage-current curve and found that as the H_2O content increased, the OCV for the cell dropped accordingly, but the initial slope for the voltage-current curve decreased significantly as water was added, indicating that the addition of moisture decreased the anode overpotential and the anode interfacial resistance (Figure 2.18). Jiang and Badwal [54] also showed in their study that as the water content increases, the anode overpotential becomes lower at lower current density condition (Figure 2.19). Primdahl and Mogensen [39] found that with the same H_2 concentration, the anode conductance increased with increasing moisture content up to 25 vol% H_2O. Later, Wen et al. [53] confirmed the study by Nakagawa et al. and indicated that as the fuel was changed from pure H_2 to 80% H_2/20% H_2O, the anode overpotential decreased, and the difference was very obvious at lower current densities.

Even though the effect of moisture on the anode kinetics is well known, interpretation of experimental results on the effect of moisture can be tricky. As Nakagawa et al. [52] pointed out, the measurement of the total cell impedance under the OCV condition is not convincing since the reduction of polarization could as well be due to the availability of H_2O for the cathodic reaction. In addition, the measurement of cell performance under the constant voltage or constant current conditions may also lead to wrong conclusions about the effect of water, because the addition of H_2O will

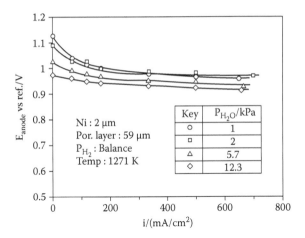

FIGURE 2.18 Potential-current density curves for the Ni electrodes in hydrogen fuels containing different concentration of H_2O. (From Nakagawa, N. et al., *J. Electrochem. Soc.*, 142:3474–3479, 1995. Reproduced by permission of ECS-The Electrochemical Society.)

decrease cell open-circuit voltage. For example, if a full cell without a reference electrode is tested, then under the galvanostatic condition, one will observe a decrease in cell terminal voltage after water is introduced into the anode chamber. However, this is due to the decrease in cell open-circuit voltage. The effect of H_2O on OCV must be deducted to discern the real effect of moisture on the anode overpotential. In addition, the fundamental reason for the drop in anode polarization due to the presence of water is still not clear at this stage.

2.2.3.5 Influence of Current/Voltage on Anode Electrochemical Performance

For electrolyte-supported cells, many studies indicate that anode resistance decreases significantly as anodic current passes through the anode. For example, van Herle et al. [55] found that anode resistance decreased dramatically from 2.4 to 0.5 and to 0.1 Ω when the cell current increased from 0 to 95 and then to 567 mA (Figure 2.20). Similarly, Primdahl and Mogensen [39] studied the effect of anode overpotential on the anode interfacial conductance and found that the anode interfacial resistance decreased significantly as the anode overpotential increased, which was also verified by Jiang and Badwal [43].

2.2.3.6 Influence of Porosity on Anode Electrochemical Performance

The influence of porosity on the electrochemical activity has not been studied much for electrolyte-supported cells because anode pastes for electrolyte-supported cells are made for screen printing, and thus contain significant amounts of organics, which almost guarantees sufficient porosity. In addition, since the anode thickness for electrolyte-supported cells is only on the order of ~50 μm, the concentration polarization itself becomes much less of an issue. In fact, Jiang et al. [44] showed that anode overpotential for cermet anodes prepared with extra graphite pore formers

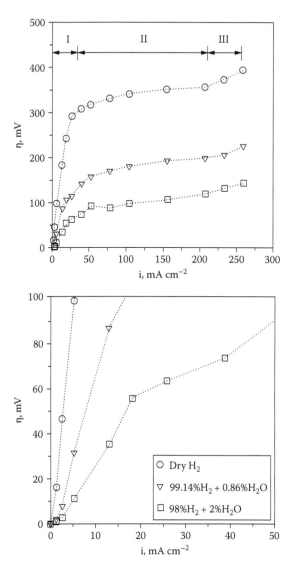

FIGURE 2.19 Anode overpotential versus current density in hydrogen with different concentration of H_2O. The plot at the bottom is the enlarged part for the polarization in region I. (From Jiang, S.P. and Badwal, S.P.S., J. *Electrochem. Soc.*, 144:3777–3784, 1997. Reproduced by permission of ECS-The Electrochemical Society.)

was worse than those without when the electrode thickness was less than 80 μm. However, the positive effect of pore formers to anode activity started to show up as the anode thickness increased above ~100 μm, as shown in Figure 2.21.

For anode-supported cells, the addition of pore former is a must, especially when fine starting materials are used. It was found that the anode porosity is very low when the starting materials are very fine. For example, Huebner et al. [31] found

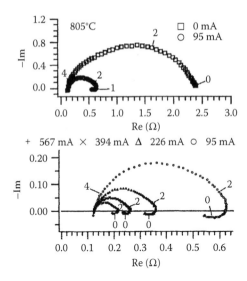

FIGURE 2.20 Change of anode/electrolyte impedance spectra with respect to change in anodic current. (From van Herle, J. et al., *Proceedings of the Fifth International Symposium on Solid Oxide Fuel Cells*, 97(40): 565–574, 1997. Reproduced by permission of ECS-The Electrochemical Society.)

that the porosity for a cermet (target Ni:YSZ volume ratio of 30:70) sintered from GNP NiO and fine YSZ at 1400°C was only 4%. Yu et al. [32] studied the relative density of Ni-YSZ cermets (55:45 by volume) and found that even with the addition of 40 vol% graphite, the resulting porosity was only ~15%, which is still too

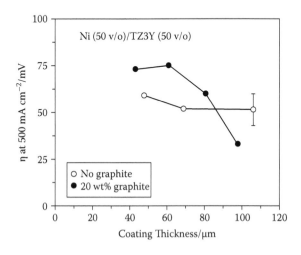

FIGURE 2.21 Change of anode overpotential versus anode thickness for anodes made with and without 20 wt% graphite as pore former in electrolyte-supported cells. (From Jiang, S.P. et al., *Solid State Ionics*, 132:1–14, 2000. Copyright by Elsevier, reproduced with permission.)

low. Low porosity in the anode for anode-supported cells will significantly limit the mass transfer process and lead to large anode overpotential. For example, Primdahl et al. [56] found that the anode interfacial resistance for anode prepared without pore former was almost twice as much as that made with 40 vol% added pore former; see Figure 2.22(a). Zhao and Virkar [23] also studied the influence of porosity on the performance of anode-supported cells and found that when the anode porosity increased from 32 to 57%, the maximum cell power density increased from 0.72 to 1.55 W/cm²—see Figure 2.22(b)—which is quite striking since the fuel utilization is rather low (up to 6 A of current for a H₂ flow of 300 cc/min, which led to fuel utilization of <30%). Therefore, the porosity of anode-supported cells is usually fairly high to ensure high fuel cell performance. For example, Energy Research Centre of the Netherlands (ECN) cells even have a porosity of ~55% [57].

2.2.4 Sulfur Poisoning of Ni-YSZ Cermet Anodes

The sulfur poisoning of SOFCs with Ni-YSZ cermet anodes has been studied extensively and, in this section, the previous studies will be briefly summarized.

2.2.4.1 Short-Term Sulfur Poisoning Behavior

Quick Poisoning Behavior

The short-term sulfur poisoning behavior is rather clear now. The cell performance decreases quickly in an almost linear way as soon as sulfur poison (usually in the form of hydrogen sulfide or H_2S) is introduced into the fuel stream. Then, after some time, which depends on the H_2S concentration and the cell structure, the rate of cell performance degradation will decrease dramatically or even stop completely. The earliest report in the open literature about the sulfur poisoning of solid oxide fuel cells is probably that by Feduska and Isenberg [58] at Westinghouse who studied the effect of 50 ppm H_2S on the performance of a 7-cell stack operated at 1000°C for 800 h on a fuel mixture of 5% H_2/10% CO/85% CO_2 at a constant current density of 150 mA/cm² (Figure 2.23). They reported that the only effect caused by sulfur impurity was a quick drop of the operating voltage (U) from 0.62 to 0.59 V (a ~5% drop) per cell. The cell voltage recovered to the original level when the H_2S impurity was removed from the fuel stream. Similar studies on poisoning by 1 to 100 ppm H_2S of cell stacks at temperatures of 900 to 1000°C were carried out by Singhal et al. [59], Ray [60], and Maskalick and Ray [61] at Westinghouse, Stolten et al. [62] at Daimer-Zenz/Dorknier in Germany, Iritani et al. [63] at Electrical Power Development Co. and Mitsubishi in Japan, Batawi et al. [64] at Sulzer Hexis Ltd. in Switzerland, and Waldbillig et al. [65] at Versa Power in Canada. Depending on H_2S concentration, temperature, and cell structure, the relative drop in stack power output is usually 1 to 15%.

Change in Impedance Spectrum

Dees et al. [66, 67] reported that the sulfur poisoning was due to a large increase in anode interfacial polarization resistance (R_p). They found that total R_p for an Ni-YSZ cermet anode/electrolyte/anode symmetrical cell in 97% H_2/3% H_2 increased from 0.27 to 0.45 Ω/cm² (an ~67% increase) when 100 ppm H_2S was introduced into the

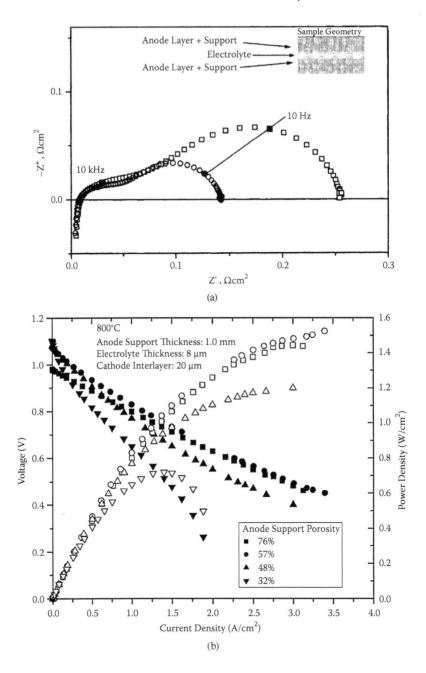

FIGURE 2.22 (a) Impedance spectra for symmetrical cells prepared without (square) and with (circle) 40 vol% corn starch as pore former in 3% H_2O/H_2 at 850°C. (From Primdahl, S. et al., *Proceedings of the Sixth International Symposium on Solid Oxide Fuel Cells*, 99(19): 793–802, 1999. Reproduced by permission of ECS-The Electrochemical Society.) (b) Influence of anode support porosity on the performance of cells at 800°C. (From Zhao, F. and Virkar, A.V., *J. Power Sources*, 141:79–95, 2005. Copyright by Elsevier, reproduced with permission.)

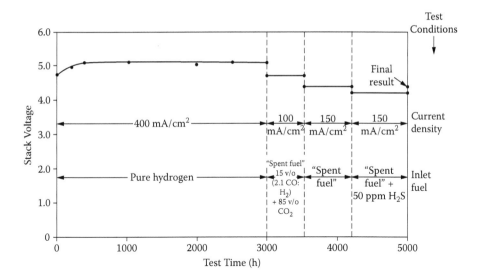

FIGURE 2.23 Stability of 7-cell SOFC stack tested at 1000°C in pure hydrogen and spent fuel (10 vol% CO + 5 vol% H$_2$ + 85 vol% CO$_2$) without and with 50 ppm H$_2$S. (From Feduska, W. and Isenberg, A.O., *J. Power Sources*, 10:89–102, 1983. Copyright by Elsevier, reproduced with permission.)

fuel flow at 1000°C while the bulk resistance (R_b) remained largely unchanged. Similar results were obtained by Geyer et al. [68] and Primdahl and Mogensen [69].

Dependence on Sulfur Concentration and Temperature

Systematic studies on the influence of operating parameters such as sulfur concentration and temperature indicate that sulfur poisoning generally becomes more severe as the temperature decreases or as the pH_2S/pH_2 increases. The first such study was probably the one carried out by Singhal et al. [59] who found that the relative power output drop due to sulfur poisoning for a full cell became more severe with increasing H$_2$S concentration or decreasing temperature. The sulfur poisoning behavior with respect to H$_2$S concentration and temperature was verified by Matsuzaki and Yasuda [70, 71] in a wide temperature range from 750 to 1000°C and a wide H$_2$S concentration range from 0.02 to 15 ppm using impedance spectroscopy measurements on half cells. They found that (a) the critical H$_2$S concentration (i.e., above which the sulfur poisoning effect became significant) decreased rapidly from 2 to 0.5 and then to 0.05 ppm as the cell temperature decreased from 1000 to 900 and then to 750°C; (b) the degradation became more severe as H$_2$S concentration increased (Figure 2.24), which were confirmed by Zha et al. [72] under functional full cells, as shown in Figure 2.25. Matsuzaki and Yasuda also observed that the rates of sulfur poisoning and recovery increased with increasing temperature and that at lower temperatures, the recovery was much slower compared with the poisoning while at higher temperatures the recovery was almost as fast as poisoning.

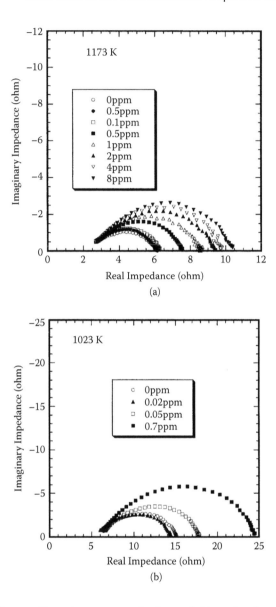

FIGURE 2.24 Change of impedance spectra for the anode/electrolyte interface with respect to change in H_2S concentration at (a) 900°C, and (b) 750°C. (From Matsuzaki, Y. and Yasuda, I., *Solid State Ionics*, 132:261–269, 2000. Copyright by Elsevier, reproduced with permission.)

    Matsuzaki and Yasuda did not observe an obvious dependence of the time needed for the influence of the sulfur impurity to saturate with respect to sulfur concentration when H_2S concentration was in the range of 2 to 15 ppm at 1000°C. Such dependence, however, was observed at lower temperatures (e.g., 750°C) in fuels with 0.1 to 10 ppm H_2S by Waldbillig et al. [65] and Sprenkle et al. [73] on anode-supported cells, as shown in Figure 2.26.

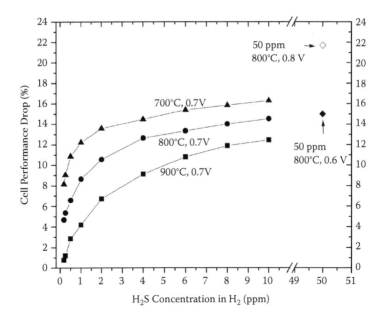

FIGURE 2.25 Relative cell performance drop, $\Delta P = \Delta (UI)$, versus H_2S concentration in H_2 for electrolyte-supported cells at different temperatures. (From Zha, S. et al., *J. Electrochem. Soc.*, 154:B201–B206, 2007. Reproduced by permission of ECS-The Electrochemical Society.)

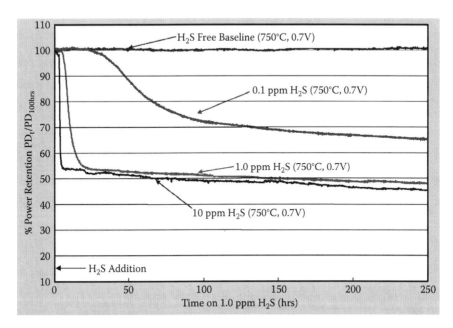

FIGURE 2.26 Change of cell performance versus time for anode-supported cells subject to different concentrations of H_2S poison. (From Sprenkle, V. et al., Sulfur Poisoning Studies on the Delphi-Battelle SECA Program, 2007. Permission pending.)

Dependence on Cell Current Density and Voltage

There are some differences in the observed dependence of sulfur poisoning behavior on cell current or voltage. For cell testing carried out under the galvanostatic condition, Singhal et al. [59] reported that the relative power output drop caused by exposure to 10 ppm H_2S increased from 10.3 to 15.6% when the cell current density increased from 160 to 250 mA/cm^2 at 1000°C. Similarly, Waldbillig et al. [65] also reported that when a hydrogen fuel with 1 ppm H_2S was used, the relative drop in cell power output was 6.5, 9.8, and 11.8% for a constant cell current density of 250, 500, and 990 mA/cm^2, respectively at 750°C. Xia and Birss [74] indicated that the relative cell power output drop caused by 10 ppm H_2S increased from 19 to 56% when the current density increased from 130 to 400 mA/cm^2 at 800°C.

However, under potentiostatic conditions, Zha et al. [72] showed that the relative drop in cell power output due to exposure to 50 ppm H_2S was 32.3, 18.5, and 15.5% when cell voltage was 0.8, 0.6, and 0.3 V, respectively (the corresponding current densities before sulfur poisoning were 115, 252, and 513 mA/cm^2). Therefore, on one hand, under galvanostatic conditions, the relative power output drop due to sulfur poisoning *increases* with increasing current density while on the other hand, under potentiostatic condition, the relative power output drop due to sulfur poisoning *decreases* with increasing current density.

Such an "apparent contradiction" was resolved by Cheng et al. [75] using equivalent circuit analyses and verified using impedance measurement. As shown in Figure 2.27, the relative increase in total cell resistance is always *smaller* at *higher* cell current density and lower cell terminal voltage no matter whether the cell was tested under constant current or constant voltage condition. The contradiction was attributed to the changing resistance for the external circuit to maintain constant current or constant voltage under those two modes of testing (i.e., galvanostatic and potentiostatic).

However, now there is still the uncertainty as to the relative increase in anode/electrolyte interfacial resistance under different current densities. Primdahl and Mogensen [39] found that the relative increase in anode interfacial resistance due to sulfur poisoning is independent of temperature and cell current density (up to 100 mA/cm^2) when the anode was subject to 35 ppm H_2S at 1000°C. Whether this is also the case when the cell temperature is lower (i.e., at 750°C), the H_2S concentration is lower (i.e., ~1 ppm), and the current density is higher (i.e., up to 1 A/cm^2) is not clear at the current stage.

In addition, even though it had been shown that the relative increase in total cell resistance is always smaller when the cell current density is higher or voltage is lower, alleviating sulfur poisoning by passing a large current through the cell is not very practical due to the fact that overall cell efficiency would decrease dramatically when the cell voltage is low and the current density is high.

Dependence on CO to H₂ Ratio

The influence of CO to H_2 ratio on the short-term sulfur poisoning behavior was studied by Sasaki et al. [76, 77] who found that the ratio CO to H_2 would not influence much unless the CO concentration is extremely high (e.g., CO:H_2O = 90:10 or higher), as shown in Figure 2.28. This was consistent with the observation of similar

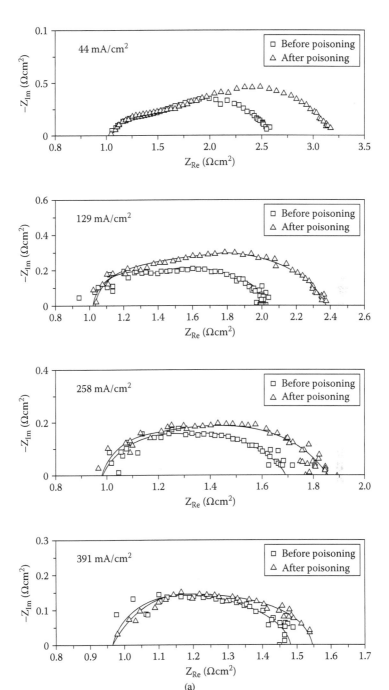

FIGURE 2.27 Change of total impedance spectra for an electrolyte-supported cell when it was poisoned by 10 ppm H_2S under (a) different constant current densities, and (b) different constant cell voltage. (From Cheng, Z. et al., *J. Power Sources*, 172:688–693, 2007. Copyright by Elsevier, reproduced with permission.)

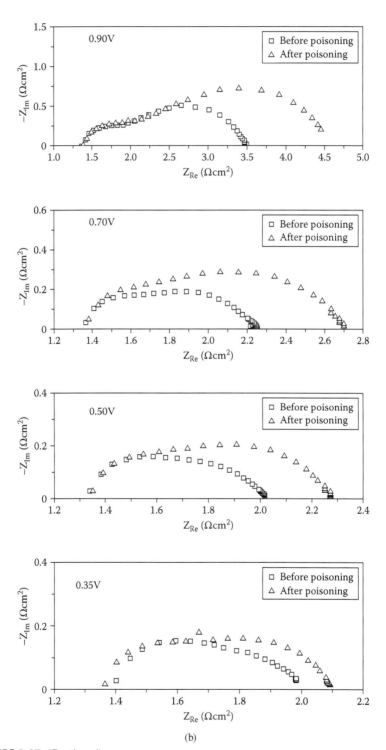

(b)

FIGURE 2.27 (Continued).

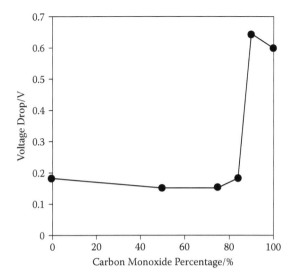

FIGURE 2.28 Voltage drop due to sulfur poisoning versus CO concentration for an SOFC button cell operated on a H_2-CO fuel mixture containing 5 ppm H_2S at 1000°C with current density of 200 mA/cm². (From Sasaki, K. et al., *Proceedings of the Ninth International Symposium on Solid Oxide Fuel Cells*, 2005-07: 1267–1275. Reproduced by permission of ECS-The Electrochemical Society.)

sulfur poisoning behavior for SOFCs operating on fuels with significant amounts of CO (CO:H_2O of up to 2:1) [58, 60, 61, 78].

2.2.4.2 Long-Term Sulfur Poisoning Behavior

There are some uncertainties about the long-term behavior of solid oxide fuel cells subject to sulfur contaminants in the fuel. Some studies showed the cell performance reached a steady state right after the quick poisoning stage. For example, in the study of Feduska and Isenberg [58], even though initial pH_2S/pH_2 value was as high as 1000 ppm, after the quick poisoning finished in a few hours, no further performance degradation was observed in the next 800 h (see Figure 2.23). A similar phenomenon, quick saturation of sulfur poisoning, was also observed by Maskalick and Ray [61], Iritani et al. [63], and Batawi et al. [64] with much lower H_2S concentration of 0.1 to 1 ppm.

However, there have also been studies showing that, after the initial quick poisoning, the cell performance continued to degrade at a much slower, but still significant, rate (usually in an almost linear way) for hundreds of hours (see Figures 2.26, 2.29, and 2.30). Zha et al. [72] and Sprenkle et al. [73] named this continued degradation "2nd stage sulfur poisoning." Table 2.1 summarizes the studies that involved long-term sulfur poisoning tests for solid oxide fuel cells.

It is still too early to conclude on the nature of this 2nd stage sulfur poisoning for several reasons. As shown in Table 2.1, all the studies that did not show 2nd stage sulfur poisoning were those carried out on relatively large fuel cell stacks [58, 61, 63, 64], while those that showed the 2nd stage degradation were carried out on small systems, which usually included only a single cell [59, 61, 72, 73, 78]. Due to the

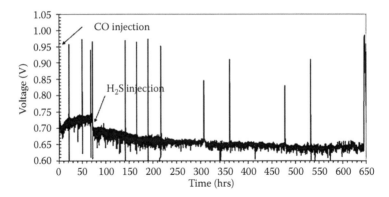

FIGURE 2.29 Change of cell voltage versus time for an electrolyte-supported cell stack at 850°C when 207 ppm H_2S was introduced into a fuel mixture 24.8% H_2/40% CO/35.7% N_2. (From Trembly, J.P. et al., *J. Power Sources*, 158: 263–273, 2006. Copyright by Elsevier, reproduced with permission.)

limited time scale explored, whether this 2nd stage sulfur poisoning will eventually manifest in SOFC stacks after longer operation in sulfur-containing fuels (e.g., thousands of hours) is not clear. Nevertheless, such a difference in the sulfur poisoning behavior between single cells and stacks is not likely to be related to the inferior stability of single cells as compared with stacks because stable performance in

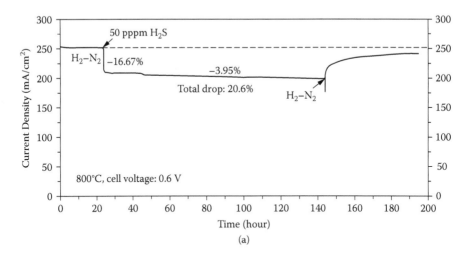

FIGURE 2.30 Sulfur poisoning and regeneration for solid oxide fuel cells showing (a) incomplete recovery. (From Zha, S. et al., *J. Electrochem. Soc.*, 154: B201–B206, 2007. Reproduced by permission of ECS–The Electrochemical Society.) (b) Sulfur poisoning and regeneration for solid oxide fuel cells showing complete recovery. (From Singhal, S.C. et al., Anode Development for Solid Oxide Fuel Cells, 1986. Permission pending.) (c) Sulfur poisoning and regeneration for solid oxide fuel cells showing cell failure. (From Sasaki, K. et al., *Proceedings of the Ninth International Symposium on Solid Oxide Fuel Cells*, 2005-07: 1267–1275. Reproduced by permission of ECS-The Electrochemical Society.)

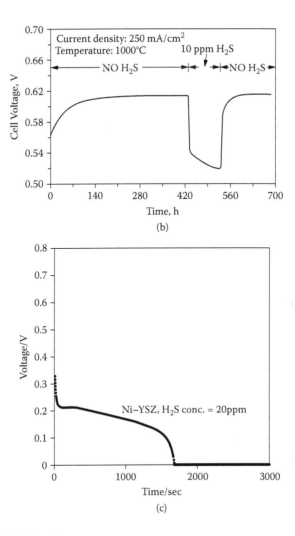

FIGURE 2.30 (Continued).

H₂S-free fuels was demonstrated in many of those studies showing this 2nd stage sulfur poisoning using single cells [59, 60, 61, 72, 73]. From the study of Maskalick and Ray [61], it also seems that for a similar cell structure and temperature, the 2nd stage sulfur poisoning will show up if the H₂S concentration is 1 or 5 ppm instead of 0.1 ppm. However, if this is the case, it would be difficult to understand why this 2nd stage sulfur poisoning was not observed in some other studies in which the H₂S concentration was much higher (e.g., >100 ppm as in the studies of Feduska and Isenberg [58] and Batawi et al. [64]).

In addition, among all the studies showing the 2nd stage sulfur poisoning, only Trembly et al. [78] demonstrated a saturation-like behavior after ~450 h of operation (Figure 2.29). Whether this continued 2nd stage sulfur poisoning will saturate is not

TABLE 2.1
Summary of the Studies Involving Long-Term Sulfur Poisoning Tests for SOFCs

Studies	Cell/Stack Structure	T (°C)	Fuel Utilization (%)	j (mA/cm^2)	pH_2S/pH_2 (ppm)	Observations About the 2nd Stage Slow Poisoning	
						Time (h)	Reached Saturation
Feduska and Isenberg [58]	7 cathode-supported tubular cell stack	1000	N/A	150	333–1000	Not observed for 800 h	
Iritani et al. [63]	22 cathode-supported tubular cell stack	900	60	200	1	Not observed for 530 h	
Batawi et al. [64]	5-cell stack	950	N/A	200	100	Not observed for 450 h	
Ray [60], Maskalick, and Ray [61]	Single cathode-supported tubular cell	1000	85	350	0.1	Not observed for 500 h	
		1000	85	350	1		
		1025	N/A	350	5	1500	No
		1000				450	No
Singhal et al. [59]	Cathode-supported	1000	N/A	250	10	80	No
Trembly et al. [78]	Single electrolyte-supported cell	850	40[a]	200	837	450	Yes
Sprenkle et al. [73]	Single anode-supported button cell	750	N/A	1100[b]	0.1, 1, 10	200	No
Zha et al. [72]	Single electrolyte-supported button cell	800	<5[a]	250[b]	2	24	No
		800	<5[a]	250[b]	100	120	No

Note: N/A:

[a]　Estimated fuel utilization value.

[b]　Initial current density under constant voltage condition.

very clear. If there is saturation, it would be difficult to understand why in the study of Maskalick and Ray [61] that even after 1,500 h of operation in only 1 ppm H_2S, the cell still did not reach saturation. On the other hand, if there is no saturation, it would be hard to explain the sulfur-anode interaction mechanism, as bulk reaction has been considered thermodynamically forbidden.

2.2.4.3 Reversibility of Sulfur Poisoning

As to the reversibility of the sulfur poisoning of solid oxide fuel cells, complete recovery of poisoned cells has been observed by Feduska and Isenberg (Figure 2.23) [58], Singhal et al. (Figure 2.30(b)) [59], Maskalic and Ray [61], Matsuzaki and Yasuda [70, 71], Sasaki et al. [77], and Zha et al. [72], while incomplete recovery has been observed by Iritani et al. [63], Waldbillig et al. [65], Xia and Birss [74], Trembly et al. [78], and Zha et al. (Figure 2.30(a)) [72]. Cell failure was rare, but still observed by Sasaki et al. (Figure 2.30(c)) [76, 77] and Kurokawa et al. [79]. The general trend regarding reversibility is that sulfur poisoning tends to be more reversible if the temperature is higher, the H_2S concentration is lower, or the time of exposure to H_2S is shorter, while it tends to be less reversible if the temperature is lower, the H_2S concentration is higher, or the time of exposure is longer (Matsuzaki and Yasuda [70, 71] and Zha et al. [72]).

2.2.4.4 Sulfur Poisoning Mechanism

For the sulfur poisoning mechanism for solid oxide fuel cells with Ni-YSZ cermet anodes, some researchers proposed that it might be due to the adsorption of sulfur on nickel surface [65, 70, 76], while some other researchers proposed that it might be due to the formation of conventional multilayer nickel sulfides such as Ni_3S_2 or NiS [74, 78, 81, 82]. Recently, the formation of conventional multilayer nickel sulfides has been definitively ruled out as the mechanism for sulfur poisoning by thermodynamic considerations [76, 77] as well as *ex situ* and *in situ* Raman experiments [80]. The conventional multilayer nickel sulfides observed in some earlier studies [78, 81, 82] on samples subject to sulfur poisoning were attributed to the artifact induced during the cooling of samples from elevated temperatures to room temperature in sulfur-containing fuels [80]. Specifically, as the temperature decreases, the equilibrium concentration to form conventional multilayer nickel sulfide shifts to much lower H_2S concentration. For example, the equilibrium pH_2S/pH_2 value decreases from 3×10^{-3} at 1000 K to only 1×10^{-5} at 600 K for Ni_3S_2 and from 4×10^{-2} at 1,000 K to only 4×10^{-5} at 500 K for NiS. Therefore, take the example of a fuel mixture with pH_2S/pH_2 of only 10^{-4} (100 ppm), even though the formation of conventional multilayer nickel sulfides (e.g., Ni_3S_2 and NiS) is thermodynamically forbidden at high temperatures (e.g., 1000 K), it will be energetically favorable as the temperature drops to 600 K for Ni_3S_2 and 500 K for NiS in the same fuel during the cooling process [80].

For the adsorption mechanism of sulfur poisoning, up until now, only Waldbillig et al. [65] claimed the detection of adsorption-level sulfur (i.e., monolayer to submonolayer) on the nickel surface, and *in situ* detection of sulfur on the nickel surface is not yet achieved [80]. Nevertheless, the adsorption mechanism is unambiguously supported by many theoretical studies [83–88]. Among them, the recent work by Wang and Liu [86] is of particular interest as it includes an updated Ni-S phase

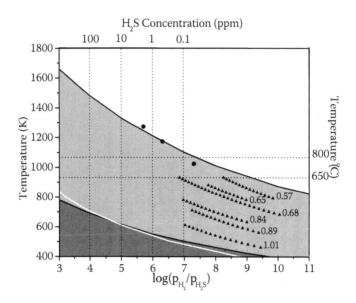

FIGURE 2.31 Calculated S–Ni phase diagram. The white, light gray, and dark gray regions represent the phase of clean Ni surface, adsorbed sulfur atoms on Ni surfaces, and Ni_3S_2 bulk phase, respectively. The white line is taken from a phase diagram based on classical thermodynamics for bulk phases. The triangle symbols are from experimental results of chemisorbed sulfur with different "area coverage," and the number adjacent to each set of data points represents the value of "fractional sulfur-saturation coverage." The circle symbols are critical H_2S concentrations determined for Ni-YSZ cermet anode, above which the cell performance degraded significantly due to sulfur poisoning. (Adapted with modifications from Wang, J.H. and Liu, M., *Electrochem Comm*, 9:2212–2217, 2007. Copyright by Elsevier, reproduced with permission.)

diagram calculated using density function theory (DFT) (see Figure 2.31). Different from traditional phase diagrams, which only have nickel and bulk nickel sulfides, this new phase diagram includes a transition region between the nickel and nickel sulfide (i.e., $Ni_3\ S_2$) regions. This transition region, as shown in light gray in Figure 2.31, represents the equilibrium conditions under which sulfur would only adsorb onto the nickel surface with different surface coverage from submonolayer to a monolayer. The typical sulfur poisoning condition—for example, 800°C with $pH_2S/pH_2 = 1$ ppm—will fall right into this transition region for sulfur adsorption on nickel surface, i.e., the formation 2-dimensional (2D) sulfides.

The adsorption mechanism for sulfur poisoning is also supported by several other considerations in addition to the energetic consideration. First, the rate of sulfur poisoning is very fast, which is consistent with the adsorption mechanism. Second, the relative extent of cell performance degradation, especially the relative increase in anode interfacial resistance, is consistent with the estimated surface coverage of sulfur on nickel. For example, in the study of Matsuzaki and Yasuda [70], the anode interfacial resistance increased by over 140% when 0.7 ppm H_2S was present (see Figure 2.24(b)), which was roughly in line with an estimated equilibrium coverage of slightly larger than 0.6 (see Figure 2.31). Third, the boundary between sulfur poisoning and non-poisoning determined experimentally is consistent with calculated boundary between a clean nickel surface and a surface with a lower limit coverage (that corresponds to the maximum adsorption energy 1/16 coverage in Figure 2.31) is well below that for conventional multilayer nickel sulfide formation. In addition, the effect that higher cell current leads to less relative cell resistance increase, as discussed previously is attributed to the oxidation of adsorbed sulfur species on the Ni on anode surface by the oxygen ions from the electrolyte [75].

2.3 ALTERNATE ANODES FOR SOFCS

In addition to traditional nickel-YSZ cermet anodes, many alternative materials have been studied as potential anodes for SOFCs. Such studies arise primarily from the limitations of the nickel-YSZ cermet anode when practical hydrocarbon fuels are used directly. In fact, most people believe one major advantage of SOFCs over other types of fuel cells is the potential for the direct utilization, or through internal reforming, of a wide variety of fuels including fossil, biomass, and other renewable fuels. Yet it is exactly this very application that proves to be most problematic for the current Ni-YSZ cermet anode due to the severe carbon formation on the anode and sulfur poisoning. These two problems, together with the change of NiO and Ni during redox condition, pose serious obstacles to the successful commercialization of SOFCs.

In recent years, there have been numerous studies on alternate anode materials. The areas of interest include carbon-tolerant anode materials, sulfur-tolerant anode materials, and redox-stable anode materials. The idea is that by developing alternative anode materials and structure, the reforming and the desulfurization unit could be eliminated, which would reduce the system complexity and cost dramatically. In this section, the studies into these new, alternative anode materials will be briefly touched upon. Because the number of candidate materials studied is quite large, the amount of study on any individual candidate anode material is rather small, and not much work has been done to reproduce the results reported. Therefore, it is not possible to fully evaluate the real potentials of those new materials proposed by different groups of researchers. Therefore, the focus would be on the fundamental issues for these alternative materials, instead of on the processing and properties of a specific candidate material.

2.3.1 CARBON-TOLERANT ANODES

As stated, one of the fundamental problems encountered in the direct oxidation of hydrocarbon fuels in SOFCs is carbon deposition on the anode, which quickly deactivates the anode and degrades cell performance. The possible buildup of carbon can lead to failure of the fuel-cell operation. Applying excess steam or oxidant reagents to regenerate anode materials would incur significant cost to SOFC operation. The development of carbon tolerant anode materials was summarized very well in several previous reviews and are not repeated here [7–9]. In this section, the focus will be on theoretical studies directed toward understanding the carbon deposition processes in the gas-surface interfacial reactions, which is critical to the development of better carbon-tolerant anode materials and the design of cost-effective and long-life SOFCs.

The carbon coking process involves complicated heterogeneous reactions, which consist of a series of elementary steps, including the adsorption of carbon-containing species on the surfaces, decomposition of adsorbates on the surface, and, eventually, the formation of graphite or other carbon species from atomically adsorbed carbon on anode surfaces. The potential energy surface (PES) of the complicated heterogeneous coking processes has been resolved computationally by rationally combining the elementary steps in the mechanism study [89, 90].

From the mechanism study, two key factors are considered in the development of carbon-tolerant anode materials: loosen or isolate the atomic carbon adsorption, as illustrated in Figure 2.32. For the first one as shown in Figure 2.32(a), the simplest way to identify potential carbon-tolerant anode materials is to compute the atomic carbon adsorption energy directly. The metals with stronger adsorption energy are more likely to be poisoned from carbon deposition. The adsorption energies of atomic carbon adsorbed on metal surfaces correspond to the overlap between carbon p orbital and metal d bands, which are related to the electronic and geometrical effects of the electrode materials. For the electronic effect, the highest adsorption energies of the common transition metal and alloys of Ni, Co, Fe, FeAl, and Fe_3Si [91, 92]

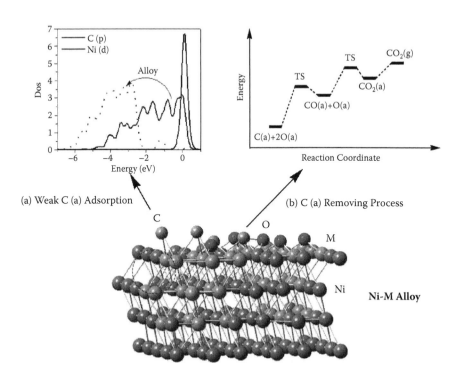

FIGURE 2.32 Two key factors (a) atomic carbon adsorption and (b) removing processes, for better carbon-tolerant materials.

TABLE 2.2
The Adsorption Energies (eV) of Atomic Carbon Adsorption (a) on Different Transition Metals and Their Alloys and (b) on Different Adsorption Sites of the Ni Surfaces

(a)

Metal	Energy
Ni(111)[a]	6.76
Co(0001)[a]	6.58
Fe(110)[b]	7.80
FeAl(110)[b]	8.20
Fe$_3$Si(110)[b]	7.70
Fe(100)[b]	8.20
FeAl(100)[b]	6.80
Fe$_3$Si(100)[b]	6.70

(b)

Surface					
Ni(111)[c]	FCC	HCP	BRI	ATOP	
	6.76	6.68	6.27	4.39	
Ni(110)[c]	LB	4FH	SB	ATOP	
	7.09	6.81	5.98	4.42	
Ni(210)[c]	e-BRI	3FT	BRI-T	sb-BRI	ATOP
	5.81	5.27	5.18	5.14	4.10
Ni(531)[c]	sb-BRI	ne-BRI	sbt	ne-FCC	ne-HCP
	7.65	7.46	7.08	6.93	6.91

Note: 3FT: threefold on terrace site; 4FH: fourfold hollow site; BRI: bridge site; BRI-T: bridge on terrace site; e-BRI: edge bridge site; ne-BRI: near edge bridge site; ne-FCC: near edge FCC; ne-HCP: near edge HCP; SB: short bridge site; sb-BRI: step bottom bridge site; sbt: step bottom top site; LB: long bridge site.

[a] From Klinke, II D.J. et al., *J. Catal.*, 178:540–554, 1998.

[b] From Jiang, D.E. and Carter, E.A. Jr., *Phys. Chem. B*, 109:20469–20478, 2005.

[c] From Li, T. et al., *J. Chem. Phys.* 121:10241–10249, 2004.

are summarized in Table 2.2(a) and for the geometrical effect, surface structures and adsorption sites have been tested on the Ni surfaces, as listed in Table 2.2(b) [93].

The calculation of carbon adsorption energy shows that atomic carbon can adsorb on the transition metals tightly, even stronger than atomic sulfur adsorption, which indicates that atomic carbon is a severe poison species as well. As long as the atomic carbon is formed during the cell operation process, it will immediately poison metal surfaces and degrade cell performance. The development of a better carbon-tolerant anode with weak carbon-metal bonds can be achieved by examining the density of state (DOS) of the designed alloy, the d band of proper alloy will have smaller over-lap with p orbital with atomic carbon, as illustrated in Figure 2.32(a).

The other key factor for removing atomic carbon from the surface can be achieved by isolating atomic carbon adsorption. The isolated atomic carbon adsorption is easier to remove from surfaces than carbon clusters. By examining the mechanism of the removing process, the appropriate alloy will reduce reaction barriers in the key process, as illustrated in Figure 2.32(b).

An alternative way to solve the carbon-cracking problem is to apply oxidant reagents to react with adsorbed carbon forming carbon mono- or dioxide and desorbing from the surface. For example, Wang et al. [94] has computed the reaction mechanism of CO_2 reforming of CH_4 on Ni (111) surface. Furthermore, the carbon adsorbates could be removed by reacting with oxygen ions from cathode through electrolyte to form CO_2. This has been proven to be feasible in Cu/GDC systems experimentally [95–99]. However, the reaction mechanism of this process is rather complicated and has not been examined in detail.

In addition, carbon coking on anode surfaces can also be avoided by operating the fuel cell under proper conditions. Atomic adsorbed carbon can desorb from the surface when the operation temperature and partial pressures of carbon-containing species are changed. The related thermodynamic prediction of carbon adsorption on Ni (111) surface has been examined by Kalibaeva et al. [100]. Their finite temperature analysis showed that adsorbed carbon tends to form clusters, in which the nickel surface is likely to divide into graphite-adsorbed and clean-nickel surfaces at finite temperature. As temperature increases, the required pressure for the transition from clean to full graphite covered surface becomes higher. However, no detailed adsorption phase diagrams of carbon-metal systems have been predicted.

2.3.2 SULFUR-TOLERANT ANODES

Although sulfur poisoning for SOFCs has been known for some time, it is interesting to note that most earlier studies on sulfur-tolerant anodes were directed toward the utilization of fuels containing significant amount of H_2S (usually on the percentage level). Only recently did studies on low concentration sulfur-containing fuels start to show up in the literature. Table 2.3 provides a brief summary of both types of studies—studies that use exclusively platinum as the anode material for SOFC are excluded since they do not have practical significance. Clearly, for any of the candidate materials, the number of studies is quite small—often by only one group. For materials that have been studied by multiple groups, the results could be totally conflicting. Take the example of strontium-doped lanthanum titanate (LST), while some researchers reported sulfur poisoning by H_2S in the range of 26 to 1000 ppm [106, 107], some others reported no poisoning effect in 1000 ppm H_2S and an enhancement effect in 5000 ppm [108].

Due to the problems mentioned above, only limited evaluation of the real potential for any of these materials is possible. For example, as shown in the table, metal sulfides are functional as anodes in fuels with very high concentrations (up to 100%) of H_2S [101–105]. However, these sulfide materials have limited potential because they tend to decompose back into metal phase in a typical reformed fuel containing only ppm-level sulfur poison based on thermodynamic considerations [120]. In addition, for materials such as $La_{0.7}Sr_{0.3}VO_3$ (LSV) and $Gd_2Ti_{1.4}Mo_{0.6}O_7$ (GTMO), even though they demonstrate sulfur tolerance (or even sulfur enhancement effect), consideration from many other aspects would greatly limit, if not eliminate, their practical applications. For the example of the LSV material, such considerations include (a) its electrical conductivity is only 1/10 to 1/200 of that for pure nickel, (b) it is made from V_2O_5, which is highly poisonous, (c) it can be oxidized into $La_{0.7}Sr_{0.3}VO_4$ and lose its property and expand significantly, and (d) it is difficult to integrate this material into current SOFC fabrication procedure. Nevertheless, studies into new materials are still of scientific importance since they may shed some light into future directions for anode studies for SOFCs. For example, Choi et al. [119] recently demonstrated that the enhanced sulfur tolerance (in 100 ppm H_2S) for a dense Ni-YSZ anode sputtered with niobium oxide was due to the transformation of the metal surface from NbO_2 to niobium sulfides (Figures 2.33 and 2.34). Although the cell performance is still

TABLE 2.3
Summary of Previous Studies on Potential Sulfur-Tolerant Anode Materials for Solid Oxide Fuel Cells

Materials	Year First Reported	Major Observations
$CuCo_2S_4$, $CuFe_2S_4$, $CuNi_2S_4$, $NiFe_2S_4$, $NiCo_2S_4$, WS_2 [101, 102]	1987	P_{max} of up to 10 mW/cm^2 at 900°C in 100% H_2S for $NiFe_2S_4$. Anode exchange current density decreases in the sequence of: $NiFe_2S_4 > WS_2 > CuCo_2S_4 > CuFe_2S_4 > NiCo_2S_4 > CuNi_2S_4$
$CoS_{1.035}$, WS_2, $Li_2S/CoS_{1.035}$ [103]	1999	P_{max} of up to ~400 mW/cm^2 (after IR correction) at 770°C in 15% H_2S. Pt mesh current collector.
Co-Mo-S, Fe-Mo-S, Ni-Mo-S, and MoS_2 with or without Ag and YSZ [104, 105]	2003	P_{max} of up to ~220 mW/cm^2 (after IR correction) at 850°C in 100% H_2S. Pt current collector layer.
$La_{0.35}Sr_{0.65}TiO_3$ (LST) $-Ce_{1-x}La_xO_{2-x}$ composite [106, 107]	2003	P_{max} of ~400 mW/cm^2 at 850°C in H_2 Poisoned by 26–1,000 ppm H_2S. Pt or Au current collector layer.
$La_{0.4}Sr_{0.6}TiO_3$ (LST) [108]	2004	P_{max} of ~200 mW/cm^2 at 1000°C in H_2. Performance not influenced by up to 1000 ppm H_2S and enhanced by 5,000 ppm H_2S. Pt current collector layer.
$La_{0.7}Sr_{0.3}VO_3$ (LSV) [109–112]	2004	P_{max} of ~280 mW/cm^2 at 950°C in 5% H_2S/ 95% CH_4. Better performance in fuels with 5% H_2S than in pure H_2. Pt current collector layer.
Cu-CeO_2 cermet [113]	2005	P_{max} of ~320 mW/cm^2 at 800°C, in H_2/H_2O. Sulfur resistance up to 450 ppm. Au current collector layer P_{max} of ~350 mW/cm$_2$ at 950°C in H_2.
$La_{0.75}Sr_{0.25}Cr_{0.5}Mn_{0.5}O_3$ (LSCM) [114]	2005	Cell poisoned in the presence of 10% H_2S. Pt current collector layer.
$Gd_2Ti_{1.4}Mo_{0.6}O_7$ (GTMO) [115]	2005	P_{max} of ~340 mW/cm^2 at 950°C in 10%H_2S/90% H_2. Better performance in fuels with 10% H_2S than in pure H_2. Pt current collector layer. P_{max} of ~300 mW/cm^2 at 750°C in H_2.

(Continued)

TABLE 2.3 (CONTINUED)
Summary of Previous Studies on Potential Sulfur-Tolerant Anode Materials for Solid Oxide Fuel Cells

Materials	Year First Reported	Major Observations
Ni-YSZ cermet infiltrated by Mo or W precursors [116]	2005	Still poisoned by sulfur (C_4H_4S) but to a lesser extent and showed gradual recovery. No Pt current collector layer. P_{max} of ~300 mW/cm^2 at 750°C in H_2.
$Sr_2MgMoO_{6-\delta}$ and $Sr_2MnMoO_{6-\delta}$ [117, 118]	2006	P_{max} of ~800 mW/cm^2 at 800°C in H_2. Still slightly poisoned by 5–50 ppm H_2S. Pt current collector layer.
Ni-YSZ infiltrated with ceria [79]	2007	P_{max} of ~300 mW/cm^2 at 700°C in H_2. Still poisoned by 40 ppm H_2S but to a lesser extent
Ni-YSZ sputtered with Nb_2O_5 [119]	2008	P_{max} of 50 mW/cm^2 at 700°C in H_2 Not poisoned by 50 ppm H_2S

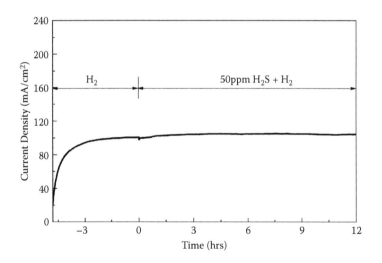

FIGURE 2.33 Change of current density as function of time for a cell with a structure of Pt/ YSZ/dense Ni-YSZ with Nb_2O_5 deposition in dry H_2 and 50 ppm H_2S balanced with H_2 at 700°C after 0.5 V is applied. (From Choi, S. et al., *J. Electrochem. Soc.*, 155:B449–B454, 2008. Reproduced by permission of ECS-The Electrochemical Society.)

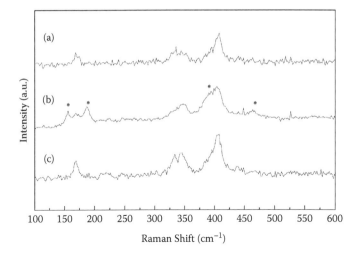

FIGURE 2.34 Raman spectra of (a) NbO_2 powder after reduction in H_2 at 800°C for 24 h, (b) NbO_2 powder after being exposed to 100 ppm H_2S balanced with H_2 at 700°C for 15 h showing NbS2 peaks marked by asterisk, and (c) Nb_2O_5 powder after regeneration in H_2. (From Choi, S. et al., *J. Electrochem. Soc.*, 155:B449–B454, 2008. Reproduced by permission of ECS-The Electrochemical Society.)

low, the study together with the studies by Smith and McEvoy [116] and Kurokawa et al. [79] may help to set the path for modifying the Ni-YSZ cermet anodes to achieve enhanced sulfur tolerance.

2.4 SUMMARY

Ni-YSZ cermet anodes satisfy most of the basic requirements for SOFC anodes. The effective conductivity of a Ni-YSZ cermet anode increases with the Ni to YSZ volume ratio, relative density, and decreasing the particle size ratio of NiO to YSZ. While coarse YSZ powders may result in poor mechanical strength and low stability, coarse NiO powders may lead to poor effective conductivity. The effective conductivity increases with the temperature at which the NiO is reduced to Ni metal in a reducing atmosphere. Further, very low reduction temperatures (e.g., below ~400°C) may result in not only low electrical conductivity, but also poor mechanical strength.

Catalytic activity and electrochemical performance generally increase as the NiO and YSZ particle sizes are reduced. However, ultrafine powders are prone to agglomeration during the milling and mixing process; the distributions of the phases (and hence the percolation threshold and many other important properties) are determined by the agglomeration size, not by the primary particle size.

The presence of a small amount of water vapor (up to $pH_2O/pH_2 = $ ~0.03) in fuel reduces anode overpotential. For anode-supported cells, the use of pore formers is important to tailor the shrinkage during cofiring and to create adequate porosity for better performance. The difference in cell power output could differ by as much as 100% for cells as porosity changes from ~30 to ~50%.

A small amount of sulfur in the fuel dramatically degrades the performance of Ni-YSZ anodes due to the adsorption of sulfur on Ni surfaces. The extent of sulfur poisoning, as measured by the relative increase in cell resistance, always increases with H_2S concentration in the fuel, but decreases with cell operating temperature and cell current density. Sulfur poisoning of Ni-based anode is generally more reversible as the cell temperature increases and as H_2S concentration or exposure time is reduced.

The development of new, alternative anode materials has recently attracted considerable interest. Several new materials show improved tolerance to sulfur poisoning and carbon deposition. However, critical issues associated with each candidate material are yet to be overcome. The traditional Ni-YSZ cermet anode still offers the best performance when clean hydrogen is used as the fuel and will continue to play an important role in SOFCs.

ACKNOWLEDGMENTS

This work was supported by the U.S. Department of Energy Office of Basic Energy Sciences, grant DE-FG02-06ER15837. The authors would like to thank Dr. Shaowu Zha for valuable comments and suggestions.

SYMBOLS AND ABBREVIATIONS

σ_b	Bulk conductivity of a dense cermet (S/cm)
σ_e	Effective conductivity (S/cm)
f_p	Porosity, unitless
pH_2	Partial pressure of hydrogen (Pa)
pH_2S	Partial pressure of hydrogen sulfide (Pa)
T	Temperature (°C)
U	Voltage (V)
J	Current density, A/cm^2 or mA/cm^2
CTE	Coefficient of thermal expansion
DOS	Density of state
GNP	Glycine nitrate process
LSGM	$La_{1-x}Sr_xGa_{1-y}Mg_yO_3$
LST	La1-xSrxTiO3
LSV	$La_{1-x}Sr_xVO_3$
OCV	Open-circuit voltage
SOFC	Solid oxide fuel cell
TPB	Triple-phase boundary
YSZ	Yttria-stabilized zirconia

REFERENCES

1. Minh NQ and Takahashi T. *Science and Technology of Ceramic Fuel Cells*. The Netherlands: Elsevier Science B. V., 1995.
2. McEvoy A. Anodes. In: Singhal SC, Kendall K, editors. *High Temperature Solid Oxide Fuel Cells: Fundamentals, Design, and Applications*. Oxford, UK, Elsevier, 2003; 140–171.

3. Zhu WZ and Deevi SC. A review on the status of anode materials for solid oxide fuel cells. *Mater Sci Eng* 2003; A362: 228–239.

4. Jiang SP and Chan SH. A review on anode materials development in solid oxide fuel cells. *J Mater Sci* 2004; 39: 4405–4439.

5. Sun C and Stimming U. Recent anode advances in solid oxide fuel cells. *J Power Sources* 2007; 171: 247–260.

6. Fergus JW. Oxide anode materials for solid oxide fuel cells. *Solid State Ionics* 2006; 177: 1529–1541.

7. Mogensen M and Kammer K. Conversion of hydrocarbons in solid oxide fuel cells. *Annu Rev Mater Res* 2003; 33: 321–331.

8. Atikinson A, Barnett S, Gorte RJ, Irvine JTS, McEvoy AJ, Mogensen M, Singhal SC, and Vohs J. Advanced anodes for high-temperature fuel cells. *Nature Mater* 2004; 3: 17–27.

9. Gorte RJ, Vohs JM, and McIntosh S. Recent developments on anodes for direct fuel utilization in SOFC. *Solid State Ionics* 2004; 175: 1–6.

10. Gong M, Liu X, Trembly J, and Johnson C. Sulfur-tolerant anode materials for solid oxide fuel cell application. *J Power Sources* 2007; 168: 289–298.

11. Sarantaridis D and Atkinson A. Redox cycling of Ni-based solid oxide fuel cell anodes: A review. *Fuel Cells* 2007; 7: 246–258.

12. Dees DW, Claar TD, Ealser TE, Fee DC, and Marzek FC. Conductivity of porous Ni/ ZrO_2-Y_2O_3 cermets. *J Electrochem Soc* 1987; 134: 2141–2146.

13. Ivers-Tiffée E, Wersing W, Schießl M, and Greiner H. Ceramic and Metallic Components for a Planar SOFC. *Berichte der Bunsen-Gesellschaft für Physikalische Chemie* 1990; 94: 978–981.

14. Itoh H, Yamamoto T, and Mori M. Configurational and electrical behavior of Ni-YSZ cermet with novel microstructure for solid oxide fuel cell anodes. *J Electrochem Soc* 1997; 144: 641–646.

15. Pratihar SK, Baus RN, Mazumder S, and Maiti HS. Electrical conductivity and micro-structure of Ni-YSZ anode prepared by liquid dispersion method. In: Singhal SC, Dokiya M, editors. *Proceedings of the Sixth International Symposium on Solid Oxide Fuel cells (SOFC-VI)*, Pennington, NJ: The Electrochemical Society, 1999; 99(19): 513–521.

16. Lee J-H, Moon H, Lee H-W, Kim J, Kim J-D, and Yoon K-H. Quantitative analysis of microstructure and its related electrical property of SOFC anode, Ni-YSZ cermet. *Solid State Ionics* 2002; 148: 15–26.

17. Lide DR. *CRC Handbook of Chemistry and Physics, 82nd edition*, Boca Raton, FL: CRC Press, 2001; 12–46.

18. Spacil HS, U.S. Patent 1970; 3,503,809.

19. Gorte RJ, Vohs JM, and McIntosh S. Recent developments on anodes for direct fuel utilization in SOFC. *Solid State Ionics* 2004; 175: 1–6.

20. de Souza S, Visco SJ, and De Jonghe LC. Thin-film solid oxide fuel cell with high performance at low temperature. *Solid State Ionics* 1997; 98: 57–61.

21. Liu J and Barnett SA. Thin yttrium-stabilized zirconia electrolyte solid oxide fuel cells by centrifugal casting. *J Am Ceram Soc* 2002; 85: 3096–3098.

22. Kim J-W, Virkar AV, Fung K-Z, Mahta K, and Singhal SC. Polarization effects in intermediate temperature, anode-supported solid oxide fuel cells. *J Electrochem Soc* 1999; 146: 69–78.

23. Zhao F, Virkar AV. Dependence of polarization in anode-supported solid oxide fuel cells on various cell parameters. *J Power Sources* 2005; 141: 79–95.

24. Leng YJ, Chan SH, Khor KA, and Jiang SP. Performance evaluation of anode-supported solid oxide fuel cells with thin-film YSZ electrolyte. *Int J Hydrogen Energy* 2004; 29: 1025–1033.

25. Koide H, Someya Y, Yoshida T, and Maruyama T. Properties of Ni/YSZ cermet as anode for SOFC. *Solid State Ionics* 2000; 132: 253–260.
26. Simner SP, Stevenson JW, Meinhardt KD, and Canfield NL. Development of fabrication techniques and electrodes for solid oxide fuel cells. In: Yokokawa H, Singhal SC, editors. *Proceedings of the Seventh International Symposium on Solid Oxide Fuel cells (SOFC-VII)*, Pennington, NJ: The Electrochemical Society, 2001; 2001(16): 1051–1060.
27. Singhal SC. Progress in tubular solid oxide fuel cell technology. In: Singhal SC, Dokiya M, editors. *Proceedings of the Sixth International Symposium on Solid Oxide Fuel cells (SOFC-VI)*, Pennington, NJ: The Electrochemical Society, 1999; 99(19): 39–51.
28. Tietz F, Buchkremer H-P, and Stöver D. Components manufacturing for solid oxide fuel cells. *Solid State Ionics* 2002; 152-153: 373–381.
29. Huebner W, Anderson HU, Reed DM, Sehlin S, and Deng X. Microstructure-property relationships of Ni:ZrO$_2$ anodes. In: Dokiya M, Yamamoto O, Tagawa H, Singhal SC, editors. *Proceedings of the Fourth International Symposium on Solid Oxide Fuel Cells (SOFC-IV)*, Pennington, NJ: The Electrochemical Society, 1995; 95(1): 696-705.
30. Tintinelli A, Rizzo C, and Giunta G. Ni-YSZ porous cermets: microstructure and electrical conductivity. In Bossel U, editor. *Proceedings of the first European Solid Oxide Fuel Cells Forum*. Lucerne, Switzerland: European Fuel Cell Forum, 1994; 455–464.
31. Huebner W, Reed DM, and Anderson HU. Solid oxide fuel cell performance studies: anode development. In: Singhal SC, Dokiya M, editors. *Proceedings of the Sixth International Symposium on Solid Oxide Fuel cells (SOFC-VI)*, Pennington, NJ: The Electrochemical Society, 1999; 99(19): 503–512.
32. Yu JH, Park GW, Lee S, and Woo SK. Microstructural effects on the electrical and mechanical properties of Ni-YSZ cermet for SOFC anode. *J Power Sources* 2007; 163: 926–932.
33. van Berkel FPF, van Heuveln FH, and Huijsmans JPP. Status of SOFC component development at ECN. In: Singhal SC, Iwahara H, editors. *Proceedings of the Third International Symposium on Solid Oxide Fuel Cells (SOFC-III)*, Pennington, NJ: The Electrochemical Society, 1993; 93(4): 744–751.
34. Tietz F, Dias FJ, Simwonis D, and Stöver D. Evaluation of commercial nickel oxide powders for components in solid oxide fuel cells. *J Eur Ceram Soc* 2000; 20: 1023–1034.
35. Corbin SF and Qiao X. Development of solid oxide fuel cell anodes using metal-coated pore-forming agents. *J Am Ceram Soc* 2003; 86: 401–406.
36. Grahl-Madsen L, Larsen PH, Bonanos N, Engell J, and Linderoth S. Manufacture and properties of Ni-YSZ anode supports and current collectors. In: Huijsmans J, editor. *Proceedings of the Fifth European SOFC Forum*. Lucerne, Switzerland: European Fuel Cell Forum, 2002; 82.
37. Grahl-Madsen L, Larsen PH, Bonanos N, Engell J, and Linderoth S. Mechanical strength and electrical conductivity of Ni-YSZ cermets fabricated by viscous processing. *J Mater Sci* 2006; 41: 1097-1107.
38. Wang Y, Walter ME, Sabolsky K, and Seabaugh MM. Effects of powder sizes and reduction temperature on the strength of Ni-YSZ anodes. *Solid State Ionics* 2006; 177: 1517-1527.
39. Primdahl S and Mogensen M. Oxidation of hydrogen on Ni/yttria-stabilized zirconia cermet anodes. *J Electrochem Soc* 1997; 144: 3409–3419.
40. Chung BW, Pham A-Q, Haslam JJ, and Glass RS. Influence of electrode configuration on the performance of electrode-supported solid oxide fuel cells. *J Electrochem Soc* 2002; 149: A325–A330.

41. Mai A, Haanappel VAC, Uhlenbruck S, Tietz F, and Stöver D. Ferrite-based perovskites as cathode materials for anode-supported solid oxide fuel cells, Part I. Variation of composition. *Solid State Ionics* 2006; 176: 1341–1350.

42. Kawada T, Sakai N, Yokokawa H, Dokiya M, Mori M, and Iwata T. Characteristics of slurry-coated nickel zirconia cermet anodes for solid oxide fuel cells. *J Electrochem Soc* 1990; 137: 3042–3047.

43. Jiang SP and Badwal SPS. An electrode kinetics study of H_2 oxidation on Ni/Y_2O_3-ZrO_2 cermet electrode of the solid oxide fuel cell. *Solid State Ionic* 1999; 123: 209–224.

44. Jiang SP, Callus PJ, and Badwal SPS. Fabrication and performance of $Ni/3\%$ $mol\%$ Y_2O_3-ZrO_2 cermet anodes for solid oxide fuel cells. *Solid State Ionics* 2000; 132: 1–14.

45. Buchkremer HP, Diekmann U, de Haart LGJ, Kabs H, Stimming U, and Stöver D. Advances in the anode supported planar SOFC technology. In: Stimming U, Singhal SC, Tagawa H, Lehnert W, editors. *Proceedings of the Fifth International Symposium on Solid Oxide Fuel cells (SOFC-V)*, Pennington, NJ: The Electrochemical Society, 1997; 97(40): 160–170.

46. Basu RN, Blaß G, Buchkremer HP, Stöver D, Tietz F, Wessel E, et al. Fabrication of simplified anode supported planar SOFCs—a recent attempt. In: Yokokawa H, Singhal SC, editors. *Proceedings of the Seventh International Symposium on Solid Oxide Fuel Cells (SOFC-VII)*, Pennington, NJ: The Electrochemical Society, 2001; 2001(16): 995–1001.

47. Simner SP, Anderson MD, Pederson LR, and Stevenson JW. Performance Variability of $La(Sr)FeO_3$ SOFC Cathode with Pt, Ag, and Au Current Collectors. *J Electrochem Soc* 2005; 152: A1851–A1859.

48. Hikita T. Research and development of planar solid oxide fuel cells at Tokyo Gas. In: Badwal SPS, Bannister MJ, Hannink RHJ, editors. *Science and Technology of Zirconia V.*, Lancaster, PA: Technical Publishing Company, 1993; 674–681.

49. Jiang SP. A comparative study of fabrication and performance of $Ni/3$ mol $\%$ Y_2O_3-ZrO_2 and $Ni/8$ mol $\%$ Y_2O_3-ZrO_2 cermet electrodes. *J Electrochem Soc* 2003; 150: E548–E559.

50. Primdahl S, Sørensen BF, and Mogensen M. Effect of nickel oxide/yttria-stabilized zirconia anode precursos sintering temperature on the properties of solid oxide fuel cells. *J Am Ceram Soc* 2000; 83: 489–494.

51. Mizusaki J, Tagawa H, Saito T, Yamamura T, Kamitani K, Hirano K, et al. Kinetic studies of the reaction at the nickel pattern electrode on YSZ in H_2-H_2O atmospheres. *Solid State Ionics* 1994; 70(71): 52–58.

52. Nakagawa N, Sakurai H, Kondo K, Morimoto T, Hatanaka K, and Kato K. Evaluation of the effective reaction zone at $Ni(NiO)/zirconia$ anode by using an electrode with a novel structure. *J Electrochem Soc* 1995; 142: 3474–3479.

53. Wen C, Kato R, Fukunaga H, Ishitani H, and Yamada K. The overpotential of nickel/yttria-stabilized zirconia cermet anodes used in solid oxide fuel cells. *J Electrochem Soc* 2000; 147: 2076–2080.

54. Jiang SP and Badwal SPS. Hydrogen oxidation at the nickel and platinum electrodes on yttria-tetragonal zirconia electrolyte. *J Electrochem Soc* 1997; 144: 3777-3784.

55. van Herle J, Ihringer R, and McEvoy AJ. Development of a standard Ni-YSZ cermet anode for intermediate temperatures. In: Stimming U, Singhal SC, Tagawa H, Lehnert W, editors. *Proceedings of the Fifth International Symposium on Solid Oxide Fuel Cells (SOFC-V)*, Pennington, NJ: The Electrochemical Society, 1997; 97(40): 565–574.

56. Primdahl S, Jørgensen MJ, Bagger C, and Kindl B. Thin anode supported SOFC. In: Singhal SC, Dokiya M, editors. *Proceedings of the Sixth International Symposium on Solid Oxide Fuel cells (SOFC-VI)*. Pennington, NJ: The Electrochemical Society, 1999; 99(19): 793–802.

57. Ouweltjes JP, van Berkel FPF, Nammensma P, and Christie GM. Development of 2nd generation, supported electrolyte, flat plate SOFC components at ECN. In: Singhal SC, Dokiya M, editors. *Proceedings of the Sixth International Symposium on Solid Oxide Fuel Cells (SOFC-VI)*. Pennington, NJ: The Electrochemical Society, 1999; 99(19): 803–811.

58. Feduska W and Isenberg AO. High temperature solid oxide fuel cell—technical results. *J Power Sources* 1983; 10: 89–102.

59. Singhal SC, Ruka RJ, Bauerle JE, and Spengler CJ. *Anode Development for Solid Oxide Fuel Cells*. Washington, DC: U.S. Department of Energy, 1986; Report No. DOE/MC/22046-2371 (DE87011136).

60. Ray ER. Contaminant Effects in Solid Oxide Fuel Cells. In: Hubner WJ, editor. *Proceedings of the Third Annual Fuel Cells Contractors Review Meeting*, Washington, DC: U.S. Department of Energy, 1992; document No. DOE/METC—91/6120 (DE91002085): 108–116.

61. Maskalick NJ and Ray ER. *Contaminant Effects in Solid Oxide Fuel Cells*. Washington, DC: U.S. Department of Energy, 1992; Report No. DOE/MC/26355—92/C0062 (DE92040255).

62. Stolten D, Späh R, and Schamm R. Status of SOFC development at Daimler-Benz/ Dornier. In: Stimming U, Singhal SC, Tagawa H, Lehnert W, editors. *Proceedings of the Fifth International Symposium on Solid Oxide Fuel Cells (SOFC-V)*, Pennington, NJ: The Electrochemical Society, 1997; 97(40): 88–93.

63. Iritani J, Kougami K, Komiyama N, Nagata K, Ikeda K, and Tomida K. Pressurized 10 kW class module SOFC. In: Yokokawa H, Singhal SC, editors. *Proceedings of the Seventh International Symposium on Solid Oxide Fuel Cells (SOFC-VII)*, Pennington, NJ: The Electrochemical Society, 2001; 2001(16): 63–71.

64. Batawi E, Weissen U, Schuler A, Keller M, and Voisard C. Cell Manufacturing Processes at Sulzer Hexis. In: Yokokawa H, Singhal SC, editors. *Proceedings of the Seventh International Symposium on Solid Oxide Fuel Cells (SOFC-VII)*, Pennington, NJ: The Electrochemical Society, 2001; 2001(16): 140–147.

65. Waldbillig D, Ivey DG, and Wood A. The poisoning effect of H_2S fuel impurities on SOFC anodes. In: Fuel Cell and Hydrogen Technologies—*Proceedings of the International Symposium on Fuel Cell and Hydrogen Technologies*, 1st. Montreal, Canada: Canadian Institute of Mining, Metallurgy, and Petroleum, 2005; 237–249.

66. Dees DW, Balachandran U, Dorris SE, Heiberger JJ, McPheeters CC, and Picciolo JJ. Interfacial effects in monolithic solid oxide fuel cells. In: Singhal SC, editor. *Proceedings of the First International Symposium on Solid Oxide Fuel Cells*, Pennington, NJ: The Electrochemical Society, 1989; 89(11): 317–321.

67. Dees DW, Balachandran U, Dorris SE, Heiberger JJ, McPheeters CC, and Picciolo JJ. Electrode development in monolithic solid oxide fuel cells. In: White RE, Appleby AI, editors. *Proceedings of the Symposium on Fuel Cells*, November 6-7, 1989, San Francisco, CA, Pennington, NJ: The Electrochemical Society, 1989; 89(14): 130–136.

68. Geyer J, Kohlmüller H, Landes H, and Stübner R. Investigation into the kinetics of the Ni-YSZ-cermet-anode of a solid oxide fuel cell. In: edited by Stimming U, Singhal SC, Tagawa H, Lehnert W, editors. *Proceedings of the Fifth International Symposium on Solid Oxide Fuel Cells (SOFC-V)*, Pennington, NJ: The Electrochemical Society, 1997; 97(40): 585–594.

69. Primdahl S and Mogensen M. Limitations in the hydrogen oxidation rate on Ni/YSZ anodes. In: Singhal SC, Dokiya M, editors. *Proceedings of the Sixth International Symposium on Solid Oxide Fuel Cells (SOFC-VI)*, Pennington, NJ: The Electrochemical Society, 1999; 99(19): 530–540.

70. Matsuzaki Y and Yasuda I. The poisoning effect of sulfur-containing impurity gas on a SOFC anode: part I. dependence on temperature, time, and impurity concentration. *Solid State Ionics* 2000; 132: 261–269.
71. Matsuzaki Y and Yasuda I. Effect of a sulfur-containing impurity on electrochemical properties of a Ni-YSZ cermet electrode. In: Yokokawa H, Singhal SC, editors. *Proceedings of the Seventh International Symposium on Solid Oxide Fuel Cells (SOFC-VII)*, Pennington, NJ: The Electrochemical Society, 2001; 2001(16): 769-779.
72. Zha S, Cheng Z, and Liu M. Sulfur poisoning and regeneration in Ni-based anodes in solid oxide fuel cells. *J Electrochem Soc* 2007; 154: B201–B206.
73. Sprenkle V, Kim JY, Meinhardt K, Lu C, Chick L, Canfield N et al. Sulfur poisoning studies on the Delphi-Battelle SECA program. Presented at The 31st International Cocoa Beach Conference and Exposition on Advanced Ceramics and Composites, 2007; Daytona Beach, FL.
74. Xia SJ and Birss VI. Deactivation and recovery of Ni-YSZ anode in H_2 fuel containing H_2S. In: Singhal SC, Mizusaki J, editors. *Proceedings of the Ninth International Symposium on Solid Oxide Fuel Cells (SOFC-IX)*, Pennington, NJ: The Electrochemical Society, 2005; 2005(7): 1275-1283.
75. Cheng Z, Zha S, and Liu M. Influence of cell voltage and current on sulfur poisoning behavior of solid oxide fuel cells. *J Power Sources* 2007; 172: 688–693.
76. Sasaki K, Susuki K, Iyoshi A, Uchimura M, Imamura N, Kusaba H et al. Sulfur tolerance of solid oxide fuel cells. In: Singhal SC, Mizusaki J, editors. *Proceedings of the Ninth International Symposium on Solid Oxide Fuel Cells (SOFC-IX)*, Pennington, NJ: The Electrochemical Society, 2005; 2005(7): 1267–1275.
77. Sasaki K, Susuki K, Iyoshi A, Uchimura M, Imamura N, Kusaba H et al. H_2S poisoning of solid oxide fuel cells. *J Electrochem Soc* 2006; 153: A2023–A2029.
78. Trembly JP, Marquez AI, Ohrn TR, and Bayless DJ. Effects of coal syngas and H_2S on the performance of solid oxide fuel cells: single-cell tests. *J Power Sources* 2006; 158: 263–273.
79. Kurokawa H, Sholkalapper TZ, Jacobson CP, De Johghe LC, and Visco SJ. Ceria nanocoating for sulfur tolerant Ni-based anodes of solid oxide fuel cells. *Electrochem Solid-State Lett* 2007; 10: B135–B138.
80. Cheng Z and Liu M. Characterization of sulfur poisoning of Ni–YSZ anodes for solid oxide fuel cells using in situ Raman microspectroscopy. *Solid State Ionics* 2007; 178: 925–935.
81. Dong J, Zha S, and Liu M. Study of sulfur-nickel interaction using Raman spectroscopy. In: Singhal SC, Mizusaki J, editors. *Proceedings of the Ninth International Symposium on Solid Oxide Fuel Cells (SOFC-IX)*, Pennington, NJ: The Electrochemical Society, 2005; 2005(07): 1284–1293.
82. Dong J, Cheng Z, Zha S, and Liu M. Identification of nickel sulfides on Ni-YSZ cermet exposed to H_2 fuel containing H_2S using Raman spectroscopy. *J Power Sources* 2006; 156: 461–465.
83. Marquez AI, Abreu YD, and Botte GG. Theoretical investigation of NiYSZ in the presence of H_2S. *Electrochem Solid-State Lett* 2006; 9:A163–A166.
84. Choi YM, Compson C, Lin MC, and Liu M. A mechanistic study of H_2S decomposition on Ni- and Cu-based anode surfaces in a solid oxide fuel cell. *Chem Phys Lett* 2006; 421: 179–183.
85. Choi YM, Compson C, Lin MC, and Liu M. Ab initio analysis of sulfur tolerance of Ni, Cu, and Ni-Cu alloys for solid oxide fuel cells. *J Alloys Compd* 2007; 427: 25–29.
86. Wang JH and Liu M. Computational study of sulfur–nickel interactions: A new S–Ni phase diagram. *Electrochem Commun* 2007; 9: 2212–2217.

87. Wang JH and Liu M. Surface regeneration of sulfur-poisoned Ni surfaces under SOFC operation conditions predicted by first-principles based thermodynamic calculations. *J Power Sources* 2008; 176: 23–30.

88. Galea NM, Kadantsev ES, and Ziegler T. Studying reduction in solid oxide fuel cell activity with density functional theory-effects of hydrogen sulfide adsorption on nickel anode surface. *J Phys Chem C* 2007; 111: 14457–14468.

89. Besenbacher F, Chorkendorff I, Clausen BS, Hammer B, Molenbroek AM, Norskov JK et al. Design of a surface alloy catalyst for steam reforming. *Science* 1998; 279: 1913–1915.

90. Nikolla E, Holewinski A, Schwank J, and Linic S. Controlling carbon surface chemistry by alloying: carbon tolerant reforming catalyst. *J Am Chem Soc* 2006; 128: 11354–11355.

91. Klinke II DJ, Wilke S, and Broadbelt LJ. A theoretical study of carbon chemisorption on Ni (111) and Co (0001) surfaces. *J Catal* 1998; 178: 540–554.

92. Jiang DE and Carter EA Jr. Effects of alloying on the chemistry of CO and H_2S on Fe surfaces. *Phys Chem B* 2005; 109: 20469-20478.

93. Li T, Bhatia B, and Sholl DS. First-principles study of C adsorption, O adsorption, and CO dissociation on flat and stepped Ni surfaces. *J Chem Phys* 2004; 121: 10241–10249.

94. Wang SG, Cao DB, Li YW, Wang J, and Jiao H. CO_2 reforming of CH_4 on Ni(111): A density functional theory calculation. *J Phys Chem B* 2006; 110: 9976–9983.

95. Steele BCH, Kelly I, Middleton PH, and Rudkin R. Oxidation of methane in solid state electrochemical reactors. *Solid State Ionics* 1988; 28: 1547–1552.

96. Putna ES, Stubenrauch J, Vohs JM, and Gorte RJ. Ceria-based anodes for the direct oxidation of methane in solid oxide fuel cells. *Langmuir* 1995; 11: 4832–4837.

97. Park S, Craciun R, Vohs JM, and Gorte RJ. Direct oxidation of hydrocarbons in a solid oxide fuel cell: I. methane oxidation. *J Electrochem Soc* 1999; 146: 3603-3605.

98. Park S, Vohs JM, and Gorte RJ. Direct oxidation of hydrocarbons in a solid-oxide fuel cell. *Nature* 2000; 404: 265–267.

99. Kim H, Vohs JM, and Gorte RJ. Direct oxidation of sulfur-containing fuels in a solid oxide fuel cell. *Chem Commun* 2001; 2334–2335.

100. Kalibaeva G, Vuilleumier R, Meloni S, Alavi A, Ciccotti G, and Rosei R. Ab initio simulation of carbon clustering on an Ni (111) Surface: A model of the poisoning of nickel-based catalysts. *J Phys Chem B* 2006; 110: 3638–3646.

101. Pujare NU, Semkow KW, and Sammells AF. A Direct H_2S/Air Solid Oxide Fuel Cell. *J Electrochem Soc* 1987; 134: 2639-2640.

102. Pujare NU, Tsai KJ, and Sammells AF. An electrochemical claus process for sulfur recovery. *J Electrochem Soc* 1989; 136: 3662–3678.

103. Yates C and Winnick J. Anode materials for a hydrogen sulfide solid oxide fuel cell. *J Electrochem Soc* 1999; 146: 2841–2844.

104. Liu M, Wei G, Luo JL, Sanger AR, and Chuang KT. Use of metal sulfides as anode catalysts in H_2S-Air SOFCs. *J Electrochem Soc* 2003; 150: A1025–A1029.

105. Wei GL, Luo JL, Sanger AR, and Chuang KT. High-performance anode for H_2S-Air SOFCs. *J Electrochem Soc* 2004; 151: A232–A237.

106. Marina O and Stevenson J. Development of ceramic composites as SOFC anodes. Presented at the SECA core technology program review meeting, Albany NY, September 2003, see the link at DOE SECA program website: http://www.netl.doe.gov/publications/proceedings/03/seca-review/marina.pdf

107. Marina O and Stevenson J. SOFC Anode Materials Development at PNNL. In: 2004 Office of Fossil Energy Fuel Cell Program Annual Report, Office of Fossil Energy, U.S. Department of Energy, 2004; 90–92.

108. Mukundan R, Brosha EL, and Garzon FH. Sulfur tolerant anodes for SOFCs. *Electrochem Solid-State Lett* 2004; 7: A5–A7.

109. Aguilar L, Zha S, Li S, Winnick J, and Liu M. Sulfur-tolerant materials for the hydrogen sulfide SOFC. *Electrochem Solid-State Lett* 2004; 7: A324–A326.

110. Aguilar L, Zha S, Cheng Z, Winnick J, and Liu M. A solid oxide fuel cell operating on hydrogen sulfide (H$_2$S) and sulfur-containing fuels. *J Power Sources* 2004; 135: 17–24.

111. Cheng Z, Zha S, Aguilar L, and Liu M. Chemical, electrical, and thermal properties of strontium doped lanthanum vanadate. *Solid State Ionics* 2005; 176: 1921-1928.

112. Cheng Z, Zha S, Aguilar L, Wang D, Winnick J, and Liu M. A solid oxide fuel cell running on H$_2$S/CH$_4$ fuel mixtures. *Electrochem Solid-State Lett* 2006; 9: A31-A33.

113. He H, Gorte RJ, and Vohs JM. Highly sulfur tolerant Cu-ceria anodes for SOFCs. *Electrochem Solid-State Lett* 2005; 8: A279–A280.

114. Zha S, Tsang P, Cheng Z, and Liu M. Electrical Properties and Sulfur Tolerance of La$_{0.75}$Sr$_{0.25}$Cr$_{1-x}$Mn$_x$O$_3$ under Anodic Conditions. J Solid State Chem 2005; 178: 1844–1850.

115. Zha S, Cheng Z, and Liu M. A sulfur-tolerant anode materials for SOFCs: Gd$_2$Ti$_{1.4}$Mo$_{0.6}$O$_7$. *Electrochem Solid-State Lett* 2005; 8: A406–A408.

116. Smith M and McEvoy AJ. Sulfur-tolerant cermet anodes. In: Singhal SC, Mizusaki J, editors. *Proceedings of the Ninth International Symposium on Solid Oxide Fuel Cells (SOFC IX)*, Pennington, NJ: The Electrochemical Society, 2005; 2005(7):1437–1444.

117. Huang Y-H, Dass RI, Xing Z-L, and Goodenough JB. Double perovskites as anode materials for solid oxide fuel cells. *Science* 2006; 312: 254–256.

118. Huang Y-H, Dass RI, Denyszyn JC, and Goodenough JB. Synthesis and characterization of Sr$_2$MgMoO$_{6-\delta}$ an anode material for the solid oxide fuel cell. *J Electrochem Soc* 2006; 153: A1266–A1272.

119. Choi S, Wang J, Cheng Z, and Liu M. Surface modification of Ni-YSZ using niobium oxide for sulfur-tolerant anodes in solid oxide fuel cells. *J Electrochem Soc* 2008; 155: B449–B454.

120. Cheng, Z, Zha, S, and Liu, M. Stability of materials as candidates for sulfur-resistant anode of solid oxide fuel cells, *J Electrochem Soc*, 2006; 153: A1302–A1309.

3 Cathodes

San Ping Jiang and Jian Li

CONTENTS

3.1   INTRODUCTION

In solid oxide fuel cells (SOFCs), the cathode is the material where pure oxygen or oxygen from air is reduced through the following electrochemical reaction [1]

$$O_2 + 2V_O^{\bullet\bullet} + 4e^- = 2O_O^x \tag{3.1}$$

where, in Kröger–Vink notation, $V_O^{\bullet\bullet}$ is a vacant oxygen site, and O_O^x is an oxygen ion on a regular oxygen site in the Y_2O_3-ZrO_2 (YSZ) lattice. As illustrated by Equation (3.1), the oxygen reduction process requires the presence of oxygen and electrons as well as the possibility for generated oxide ions to be transported away from the reaction site into the bulk of the electrolyte. When the electrode material and the electrolyte material possess only electronic and ionic conductivity, respectively, such as Sr-doped $LaMnO_3$ (LSM) electrode and YSZ electrolyte, these criteria are fulfilled in the vicinity of the triple-phase boundary (TPB) between the electrode, electrolyte, and oxidant gas. If the electrode material possesses mixed electronic and ionic conductivity, for example, Sr-doped $LaCoO_3$ (LSC), oxygen species may be transported through the bulk of the electrode and the reaction zone could be extended to the electrode surface away from the electrode–electrolyte interface. Under certain conditions, electronic conductivity may be induced in the electrolyte surface close to the TPB, which will also expand the reaction zone around the TPB. Despite the extensive studies there still exist considerable disagreements on the exact reaction sites for the O_2 reduction even on the most common cathode such as LSM [1].

There has been significant progress in reducing the operation temperature of SOFC from traditional 1000°C to intermediate temperature range of 600 to 800°C. Decreasing the operation temperature greatly improves the long-term stability of the materials and the system cost can be reduced by using less costly metal alloys as the interconnect [2] and for manifold components. On the other hand, reduction in the operation temperature results in increasing the electrolyte and electrode resistivity and the polarization losses. To compensate for the increase in ohmic losses at lower temperatures, electrolytes with higher ionic conductivity or thinner films are used [3]. In SOFCs based on thin electrolyte films, the overall losses of the cell are generally dominated by the polarization losses for the O_2 reduction on the cathode [4]. The cathodic polarization loss can be as high as 65% of the total voltage loss in the intermediate temperature SOFCs or immediate-temperature solid oxide fuel cells (ITSOFCs) [5]. This is partly due to the high-activation energy and slow-reaction kinetics for the O_2 reduction reactions when compared with the hydrogen oxidation reactions. Consequently the development of cathodes with high performance and high stability becomes increasingly critical for the development of ITSOFC technologies.

In this chapter the technological development in cathode materials, particularly the advances being made in the material's composition, fabrication, microstructure optimization, electrocatalytic activity, and stability of perovskite-based cathodes will be reviewed. The emphasis will be on the defect structure, conductivity, thermal expansion coefficient, and electrocatalytic activity of the extensively studied manganite-, cobaltite-, and ferrite-based perovskites. Alterative mixed ionic and electronic conducting perovskite-related oxides are discussed in relation to their potential application as cathodes for ITSOFCs. The interfacial reaction and compatibility of the perovskite-based cathode materials with electrolyte and metallic interconnect is also examined. Finally the degradation and performance stability of cathodes under SOFC operating conditions are described.

3.2 LANTHANUM MANGANITE-BASED PEROVSKITES

3.2.1 STRUCTURE, OXYGEN NONSTOICHIOMETRY, AND DEFECT MODEL

Undoped $LaMnO_3$ is orthorhombic at room temperature and shows an orthorhombic/rhombohedral crystallographic transformation at ~600°C [6]. This transformation has been attributed to the oxidation of some Mn^{3+} to Mn^{4+} ions. Thus, the orthorhombic/

rhombohedral transition temperature is dependent on the Mn^{4+} content and sensitive to the stoichiometry of the material. Doping lower-valence cations, such as Sr^{2+} and Ca^{2+}, for the La sites increases the Mn^{4+} concentration in $LaMnO_3$, thus affecting the transformation temperature. Depending on the doping level, $La_{1-x}Sr_xMnO_{3-\delta}$ can display three lattice types: rhombohedral ($0 \leq x \leq 0.5$), tetragonal ($x = 0.5$), and cubic ($x = 0.7$) [7]. Zhang et al. [8] also observed three different perovskite-type phases at room temperature: orthorhombic for $0 < x < 0.15$, hexagonal for $0.15 < x < 0.45$, and cubic for $x > 0.45$. The unit cell volume of the LSM decreased with increasing Sr concentration.

Zheng and Pederson [9] systematically studied the phase behavior of $La_{1-x}Sr_xMnO_3$ with various Sr doping content and A/B cation ratio. As the Sr content is increased to $x = 0.2$, the perovskite structure adopts an orthorhombic distortion. This changes to a monoclinic or hexagonal structure for $0.2 \leq x \leq 0.3$. When the Sr content increases to $x = 0.3$, the structure reverts to orthorhombic symmetry. The influence of the A/B cation ratio on the changes in the lattice constants and cell volume of the perovskite phase is minor, as compared to the effect of Sr content. However, A/B cation ratio has significant effect on the minor phases formed. In the case of A/B < 1, the main minor phases observed are Mn_3O_4. When A/B $= 1$ and A/B > 1, the secondary phase La_2O_3 and its hydrated product $La(OH)_3$ are observed. $La(OH)_3$ is not a desirable phase in SOFC cathode materials due to its low melting point ($\sim 252°C$) and its tendency to swell and degrade the strength of the cathode. Excess La_2O_3 is also known to react with YSZ to form a highly resistive lanthanum zirconate phase at the LSM/YSZ interface [10]. LSM with A-site nonstoichiometric compositions (e.g., A/B $= 0.9$) is generally a preferred choice for the cathodes of ceramic fuel cells at high temperatures.

$LaMnO_3$-based oxides can have the oxygen-excess nonstoichiometry as well as the oxygen-deficient nonstoichiometry. This is generally denoted by $La_{1-x}A_xMnO_{3\pm\delta}$ (A is divalent cation, such as Sr^{2+} and Ca^{2+}, "+" means oxygen excess, and "–" means oxygen deficiency). Mizusaki et al. [11–14] and Miyoshi et al. [15] investigated in detail the oxygen nonstoichiometry of $La_{1-x}Sr_xMnO_{3\pm\delta}$ as a function of the oxygen partial pressure, P_{O_2}, operating temperature, and the composition. Figure 3.1 shows the oxygen nonstoichiometry of $La_{1-x}Sr_xMnO_{3\pm\delta}$ as a function of oxygen partial pressure and temperatures [13]. For $LaMnO_3$ and $La_{1-x}Sr_xMnO_{3\pm\delta}$ with $x < 0.4$, the oxygen content exhibits two plateaus in its oxygen partial pressure dependence, one is around the oxygen excess ($3 + \delta$) at high oxygen partial pressure and the other around the stoichiometric composition ($\delta = 0$) at intermediate oxygen partial pressures (10^{-5} to 10^{-10} Pa); see Figure 3.1(a-c). At lower oxygen partial pressures, the oxides become oxygen deficient ($3 - \delta$) and the charge compensation of the positive effective charges of $V_O^{\cdot\cdot}$ is maintained by Mn reduction. The oxygen excess also disappears for the oxide with $x \geq 0.5$; see Figure 3.1 (d). $La_{1-x}Sr_xMnO_{3\pm\delta}$ appears to be an oxygen-excess nonstoichiometry region under normal fuel cell operation conditions and becomes oxygen deficient only at very low partial pressure of oxygen, e.g., $< 10^{-10}$ Pa at 900°C for $x = 0.2$. The oxygen-excess nonstoichiometry of $La_{1-x}Sr_xMnO_{3+\delta}$ is most interesting as it is rarely observed in other perovskite-type oxides.

Various defect models have been proposed to explain the defect structure of the doped $LaMnO_3$ oxides particularly in the oxygen-excess region. At high oxygen

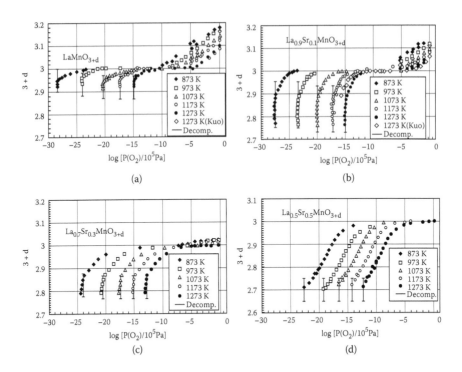

FIGURE 3.1 Oxygen nonstoichiometry of (a) LaMnO$_{3+\delta}$, (b) La$_{0.9}$Sr$_{0.1}$MnO$_{3+\delta}$, (c) La$_{0.7}$Sr$_{0.3}$ MnO$_{3+\delta}$, and (d) La$_{0.5}$Sr$_{0.5}$MnO$_{3+\delta}$ as a function of oxygen partial pressure and temperature. (After Mizusaki, J. et al., *Solid State Ionics*, 129:163–177, 2000. With permission.)

partial pressure the electrical charge due to oxidation of Mn can be compensated for by formation of two possible of defects: interstitial oxide ions or cation vacancies. Neutron powder diffraction and high-resolution transmission electron microscopy (HRTEM) results indicate that cation site vacancies, rather than oxygen interstitials, are responsible for the oxygen-excess nonstoichiometry [16, 17]. This is consistent with the close-packed nature of perovskite lattice, which could not accommodate an excess of oxygen as an interstitial oxygen ion. Thus apparent oxygen excess can only be considered in terms of a Schottky-type disorder, which involves the formation of cation vacancies [18]. To explain the upper limit of δ, van Roosmalen and Cordfunke [19–21] considered the charge disproportionation reaction of Mn^{3+} into Mn^{2+} and Mn^{4+} with a parameter ξ, i.e., the amount of Mn^{2+} and Mn^{4+} formed by the disproportion of Mn^{3+}. Since the La/Mn ratio is kept constant on incorporation of excess oxygen, the La site vacancy model inevitably requires occupation of the La ions on the Mn site according to [19–21]:

$$LaMnO_{3+\delta} + \frac{\delta}{2}O_2 = \frac{3+\delta}{3}\left(La^{3+}_{1-2\delta/(3+\delta)} V_{2\delta/(3+\delta)}\right)$$

$$\times \left(Mn^{3+}_{3(1-2\delta)/(3+\delta)} Mn^{4+}_{3.2\delta/(3+\delta)} La^{3+}_{\delta/(3+\delta)}\right)O^{2-}_3 \qquad (3.2)$$

Upon dissolution of 3/2 oxygen molecules, 6 Mn^{4+} ions and 2 vacancies, i.e., 1 each at La and Mn site, are formed:

$$6Mn_{Mn}^x + 3/2O_2 \Leftrightarrow 6Mn_{Mn}^{\bullet} + 3O_O^x + V_{La}''' + V_{Mn}''' \tag{3.3}$$

where $Mn_{Mn}^{\bullet}$ and Mn_{Mn}^x are Mn^{4+} and Mn^{3+} ions.

Satisfactory results have been reported for the model in the quantitatively explaining the relation between the Sr doping level, excess oxygen δ, and oxygen partial pressure [22]. Nevertheless, the model can only apply to the perovskite system in which charge disproportionation would occur. Mizusaki et al. [13] proposed a vacancy exclusion model to explain the upper limit of δ, in which a sphere of a vacancy exclusion space consisting of nine unit cells (nine La ions) and only a single La ion inside this sphere can enter into an Mn site to create a La site vacancy. This yields the upper limit of δ, $\delta_{max} = 3/17 = 0.1764$ [13]. The maximum oxygen excess can be determined from the maximum number of vacancy excluding spaces available in the lattice. This explains the observed disappearance of the oxygen excess region for LSM at x > 0.4 as for x > 0.4, there is no room for the vacancy excluding space around the cation vacancies. A modified vacancy exclusion model was proposed by Nakamura and Ogawa [23] in which the first 6 and second 12 nearest neighboring La (Mn) sites around a La (Mn) site vacancy are excluded from the additional formation of vacancies. However, the assumption of constant [Mn^{4+}] in the LSM lattice in both models is questioned by Yokokawa et al. [24].

Alonso et al. [25, 26] prepared $LaMnO_{3+\delta}$ with high δ values ($0.11 \leq \delta \leq 0.29$) by thermal decomposition of metal citrates by annealing at high oxidizing conditions of low temperature (800 to 1100°C) and high oxygen pressure (200 bar). The neutron powder diffraction results revealed substantially higher proportion of Mn vacancies as compared to the La vacancies. This indicates that the ratio of the La/Mn vacancies would depend strongly on the preparative oxidation conditions.

In the oxygen-deficient region, the predominant ionic defect is the oxygen vacancy, $V_O^{\bullet\bullet}$. The charge neutrality in the solid is maintained by reduction of transition element in B-site to the lower valence state. This can be represented as [13]:

$$\frac{1}{2}O_2 + V_O^{\bullet\bullet} + 2Mn_{Mn}^x \Leftrightarrow 2Mn_{Mn}^{\bullet} + O_O^x \tag{3.4}$$

$$\frac{1}{2}O_2 + V_O^{\bullet\bullet} + 2Mn_{Mn}' \Leftrightarrow 2Mn_{Mn}^x + O_O^x \tag{3.5}$$

where Mn_{Mn}' is Mn^{2+} ion. Since the electronic conduction in $La_{1-x}Sr_xMnO_{3\pm\delta}$ is hopping-type and p-type irrespective of the oxygen content ($\delta < 0$ and $\delta > 0$), disproportion of Mn ion into Mn^{2+}, Mn^{3+}, and Mn^{4+} could occur. The reduction of the transition element in B-site to maintain charge neutrality also results in the increase of the average ionic radius of the cation and the consequent lattice expansion [27]. The isothermal expansion has an almost linear relationship with oxygen-deficit nonstoichiometry.

The defect chemistry of $La_{1-x}Sr_xMnO_{3\pm\delta}$ under cathodic polarization conditions has been subjected to intensive studies due to its importance in the fundamental understanding of the O_2 reduction mechanism and electrocatalytic activity. Yasumoto et al. [28, 29] studied the effect of oxygen nonstoichiometry on the electrocatalytic activity of a LSM cathode under polarization using impedance spectroscopy. The results suggest that oxygen nonstoichiometry affects the cathode reactivity under polarized and nonpolarized states through the exchange current density

$$\sigma_E = \beta \left(\frac{F}{RT} \right) \times P_{O_2}^{1/2} \tag{3.6}$$

where σ_E is the electrode polarization conductivity or activity, P_{O_2} is the oxygen partial pressure, and β ($mAcm^{-2}$) is proportional to the exchange current density and is a function of δ, T, and overpotential. β was found to increase linearly with decreasing δ at a constant polarization except under a large anodic overpotentials [29].

Lee et al. [30] used *in situ* X-ray photo electron spectroscopy (XPS) measurement on $La_{0.9}Sr_{0.1}MnO_3$ as a function of cathodic polarization. The XPS results showed the peaks of Mn 2p spectra were shifted to the lower binding energy as the applied potential became more cathodic, indicating the reduction of Mn ions. The oxygen reduction and the concomitant formation of Mn^{2+} ions and oxygen vacancies are proposed as:

$$2Mn_{Mn}^{\bullet} + O_{O,LSM}^{x} + V_{O,YSZ}^{\bullet\bullet} + 2e' \rightarrow 2Mn_{Mn}^{x} + V_{O,LSM}^{\bullet\bullet} + O_{O,YSZ}^{x} \tag{3.7}$$

$$2Mn_{Mn}^{x} + O_{O,LSM}^{x} + V_{O,YSZ}^{\bullet\bullet} + 2e' \rightarrow 2Mn_{Mn}' + V_{O,LSM}^{\bullet\bullet} + O_{O,YSZ}^{x} \tag{3.8}$$

For the oxygen reduction on LSM, there exists a significant hysteresis as shown by thermogravimetry studies of Hammouche et al. [31]. Jiang et al. [32, 33] studied the electrode behavior of LSM in air by cyclic voltammetric technique. At 950°C, LSM exhibited significant hysteresis and this hysteresis increases with the decreasing scanning rate. Chen et al. [34] investigated the electrode behavior of $La_{0.85}Sr_{0.15}MnO_3$ by cyclic voltammetry at a low oxygen partial pressure (100 Pa). Figure 3.2 is the cyclic voltammograms of the LSM electrode as a function of reverse potential at 900°C and oxygen partial pressure of 100 Pa [34]. The current peaks, $p_{c,1}$ and $p_{a,1}$ and $p_{c,2}$ and $p_{a,2}$, correspond to the reduction and oxidation processes of electrochemical species. Based on the CV and potential responses in current step experiments, they considered that at the initial stage of the cathodic polarization, the reduction of Mn^{4+} to Mn^{3+} is accompanied by the removal of the metal vacancies.

$$V_{La}''' + V_{Mn}''' + 6Mn_{Mn}^{\bullet} + 3O_{O,LSM}^{x} + 3V_{O,YSZ}^{\bullet\bullet} + 6e' \rightarrow 6Mn_{Mn}^{x} + 3O_{O,YSZ}^{x} \tag{3.9}$$

Further reduction of manganese ions and the generation of oxygen vacancies could take place only after all metal vacancies are consumed under cathodic polarization. However, the oxygen vacancy formation and the removal of cation vacancies may not be the only explanation for the hysteresis behavior as shown recently by Wang and Jiang [35].

The surface segregation of Sr is of particular interest as SrO affects the surface reactivity and the activation behavior of the LSM electrode. Jiang and Love [36] studied the activation behavior of $La_{0.72}Sr_{0.18}MnO_3$ cathode after treatment of the LSM coating with diluted hydrochloric acid (HCl) solution. The etched solution

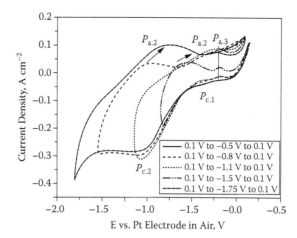

FIGURE 3.2 Cyclic voltammograms of an LSM electrode as a function of reverse potential at 900°C and oxygen partial pressure of 100 Pa. The scan rate is 200 mVs^{-1}. (From Chen, X.J. et al., *Electrochem. Solid-State Lett.*, 7: A144–A147, 2004. With permission.)

was analyzed by ICP-AES method. The concentration ratio of La/Sr/Mn based on the La content was 0.74/8.4/15.8, significantly different from the measured ratio of 0.74/0.15/1 of the bulk composition of the LSM electrode coating. This indicates that the acid etching primarily dissolved the LSM surface layers rather than the whole LSM particles. The high Mn concentration is not surprising as the LSM composition studied was A-site substoichiometric with excess Mn. The high concentration of Sr in the etched solution indicates that Sr could be enriched on the LSM surface. The LSM electrode after the HCl treatment showed a much smaller activation effect as compared to that without the acid treatment [36]. The marked segregation of Sr has also been observed for related perovskites, such as La$_{1-x}$Sr$_x$VO$_3$ [37], La$_{1-x}$Sr$_x$CoO$_3$ [38] and for layered perovskites, such as La$_{2-x}$Sr$_x$NiO$_4$ [39]. The intrinsic segregation of Sr species in the manganite, cobaltite, or nickelate phase is apparently strain driven because the ionic radius of the Sr^{2+} dopant is larger than that of the host La^{3+}. The dopant produces less elastic strain when accommodated at the surface rather than at bulk sites. An enhanced concentration of Sr is therefore available at the surface of these materials which, as reviewed recently by Jiang [40], has a significant effect on the polarization and activation behavior of LSM cathodes.

3.2.2 Electronic Conductivity and Thermal Expansion Coefficient

LaMnO$_3$ is an intrinsic *p*-type conductor. Electronic conductivity is enhanced by substitution of the La^{3+} site with divalent ions such as strontium or calcium. Of the alkaline-earth dopants, Sr substitution is preferred for SOFC applications because the resultant perovskite forms stable compounds with high conductivity in the oxidizing atmosphere found at the cathode [41]. Extensive data show that La$_{1-x}$Sr$_x$MnO$_3$, where x = ~0.1 – 0.2, provides high conductivity while maintaining mechanical and chemical stability with YSZ [41, 42].

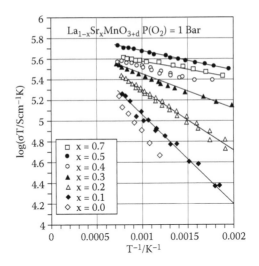

FIGURE 3.3 Temperature dependence of electronic conductivity (σ) of $La_{1-x}Sr_xMnO_{3+d}$ ($0 \leq x \leq 0.7$) at pure oxygen (P_{O_2} = 1bar). (From Mizusaki, J. et al., *Solid State Ionics*,132:167–180, 2000. With permission.)

Mizusaki [14] studied in detail the dependence of electronic conductivity and Seebeck coefficient of $La_{1-x}Sr_xMnO_3$ ($0 \leq x \leq 0.7$) on P_{O_2} (i.e., the nonstoichiometry). Electronic conductivity increases with increase in x and the maximum electronic conductivity was obtained for x = 0.5. Zhang et al. [8] and Li et al. [7] also found that the electrical conductivity showed a peak value of 200 to 485 S/cm at x = 0.5 at 1000°C. Figure 3.3 shows the plots of log σT against $1/T$ for $La_{1-x}Sr_xMnO_3$ ($0 \leq x \leq 0.7$) at pure oxygen (P_{O_2} = 1 bar) [14]. The plot for any composition falls on a straight line, suggesting that the conduction is by a small polaron hopping. The hopping conduction is generally expressed by [43]:

$$\sigma T = (\sigma T)^o \exp\left(-\frac{E_a}{kT}\right) = A\left(\frac{hv^o}{k}\right)c(1-c)\exp\left(\frac{E_a}{kT}\right) \qquad (3.10)$$

Here, $(\sigma T)^o$ and E_a are pre-exponential constant and activation energy, respectively, and c is the ratio of the carrier occupancy of the level on which the hopping conduction takes place. The conductivity is also a function of nonstoichiometry or partial pressure of oxygen. In the oxygen excess region (or high P_{O_2} > 10^{-5} bar), σ is constant, while it decreases sharply with the decrease in oxygen content in the oxygen deficient region (low P_{O_2} < 10^{-5} – 10^{-10} bar) [14].

The electronic conductivity depends significantly on the preparation and crystalline structure of the LSM specimen. Li et al. [7] prepared LSM samples by conventional sintering and plasma spraying. Depending on the composition, the electronic conductivity of the sintered sample is between 40 and 485 Scm⁻¹ at 1,000°C and that of plasma-sprayed coatings between 50 and 201 Scm⁻¹. The electronic conductivity of the plasma-sprayed samples is 50% lower than that of the sintered sample. On an

A-site deficient $La_{0.65}Sr_{0.3}MnO_3$ specimen sintered at 1400°C, the conductivity as low as 3.5 Scm^{-1} at 1000°C has been reported [44].

Similar to Sr^{2+} doping, Ca^{2+} doping also significantly enhances the electronic conductivity of $LaMnO_3$. The conductivity $La_{1-x}Ca_xMnO_3$ increases up to x = 0.5, but decreases with further Ca substitution. The conductivity of $La_{1-x}Ca_xMnO_3$ increases with increasing temperature with a metallic-type conduction transition at about 600 to 700°C, and has a conductivity as high as 90 Scm^{-1} at 1000°C [45]. Above x = 0.6, the conductivity of $La_{1-x}Ca_xMnO_3$ decreases with increasing temperature. The thermal expansion coefficient (TEC) is 11.84×10^{-6} K^{-1} for $La_{0.8}Ca_{0.2}MnO_3$ [45]. The thermal expansion properties of Ca-doped $LaMnO_3$ were also studied by Mori et al. [46].

For undoped $LaMnO_3$ the TEC is in the range of 11.33×10^{-6} to 12.4×10^{-6} K^{-1} [46, 47]. This is slightly higher than that of most commonly used YSZ electrolyte, which is approximately 10.3×10^{-6} K^{-1} in the temperature range from 50 to 1000°C in air or in H_2 atmosphere.

Ca-doped $YMnO_3$ system has also been investigated as a possible SOFC cathode material [48]. There is a miscibility gap in the $Y_{1-x}Ca_xMnO_3$ system for x < 0.25 as $CaMnO_3$ is cubic and $YMnO_3$ is a perovskite. However, single-phase perovskite structure can be obtained x > 0.25. Undoped $YMnO_3$ is difficult to densify, but $Y_{1-x}Ca_xMnO_3$ for $0.3 \leq x \leq 0.6$ densifies readily at 1400°C in air. The TEC of $Y_{1-x}Ca_xMnO_3$ increases with increasing calcium from 7×10^{-6}K^{-1} for x = 0.3 to 12×10^{-6}K^{-1} for x = 0.6, so the TEC for intermediate composition can match that of YSZ electrolyte (~10×10^{-6}K^{-1}). The conductivity of Ca-doped $YMnO_3$ also increases with increasing Ca content.

3.2.3 OXYGEN DIFFUSION AND SURFACE EXCHANGE COEFFICIENT

Oxygen diffusion and transport properties are very important for the oxygen reduction reaction at the electrode–electrolyte interface in SOFC. Among the elemental steps for the oxygen reduction reactions, there are at least two steps associated with oxygen diffusion: oxygen surface exchange reaction between the electrode and gaseous phase, and surface and bulk diffusion of oxygen species. The oxygen diffusion and incorporation properties of LSM materials can be characterized by two parameters: oxygen surface exchange coefficient (k) and oxygen chemical diffusion coefficient (D^*). The most common technique in the measurement of k and D^* is the ^{16}O/^{18}O isotope exchange measurement. In this method, the oxide sample is annealed in ^{18}O-enriched oxygen and the net isotope flux crossing an O_2/solid surface is directly proportional to the difference in isotope fractions between the gas and the solid. An oxygen tracer is introduced and the sample is quenched. The isotope concentration profile in the sample is measured by secondary ion mass spectrometry (SIMS), and k and D^* parameters are obtained by fitting the experimental data.

Carter et al. [49] measured the oxygen diffusion coefficient and oxygen surface exchange coefficient of Sr-doped $LaMnO_3$ by SIMS technique. A-site doping increases the concentration of oxygen vacancies and can also increase the oxygen self-diffusion coefficient, D^*. For example, at 900°C D^* is 3×10^{-12} cm^{2}s^{-1} for $La_{0.5}Sr_{0.5}MnO_{3-\delta}$, which is higher than 4×10^{-14} cm^{2}s^{-1} measured for $La_{0.65}Sr_{0.35}MnO_{3-\delta}$. The activation energy of the oxygen self-diffusion coefficient is 300 to 350 kJmol^{-1} for manganite-based perovskites [49]. This indicates that the association enthalpy, ΔH_a, resulting from the formation of complex defects, such as $[M'_{La} - V_O^{\bullet\bullet}]$, could be very high. The oxygen diffusion properties of $La_{0.8}Sr_{0.2}MnO_3$ have also been reported by De Souza et al. [50], and its D^* value is one order of magnitude lower than that of $La_{0.5}Sr_{0.5}MnO_{3-\delta}$ at the same temperatures.

Horita et al. [51] investigated the microstructure and oxygen diffusion at a $LaMnO_3$ electrode and YSZ electrolyte interface. A-site nonstoichiometric $La_{0.92}MnO_3$ dense films were prepared on a single crystal YSZ by RF-sputtering. The interface is characterized by the formation of a convex ring between LSM and YSZ with high concentration of manganese. The convex formation is due to the cation interdiffusion inside YSZ. The convex region shows high ^{18}O concentration, indicating the short oxygen diffusion path between LSM electrode and YSZ electrolyte. At 1000°C the D^* and k values of $LaMnO_3$ were found to be 2.45×10^{-13} cm²s⁻¹ and 7.45×10^{-8} cms⁻¹, respectively. For the $La_{1-x}Sr_xMnO_{3\pm\delta}$ ($x = 0.05, 0.10, 0.15$ and 0.20), the D^* is of the order of 10^{-12} to 10^{-11} cm²s⁻¹ and oxygen ionic conductivity is estimated to be 10^{-7} to 10^{-6} Scm⁻¹ at 1000°C [52]. The activation energy for the chemical diffusion of oxygen for LSM is in the range of 250 to 300 kJmol⁻¹ and close to the value for $La_{0.9}Sr_{0.1}CoO_{3\pm\delta}$ (270 kJmol⁻¹) in which oxide ions are transported by the vacancy diffusion mechanism [52]. This indicates the similar vacancy diffusion mechanism also applies to the LSM materials.

Ji et al. [53] investigated the electrical conductivity and oxygen diffusion properties of LSM/YSZ composites. The percolation threshold was identified at ~30 wt% (or 28 vol%) of LSM. The oxygen diffusion coefficients of the composites with 30 wt% and 40 wt% LSM were much higher than those of the LSM materials, but they were slightly lower than that of the YSZ at the same temperature. The most interesting observation is that these two composites showed an enhanced effective surface exchange coefficient, i.e., greater than either of the parent materials (see Figure 3.4).

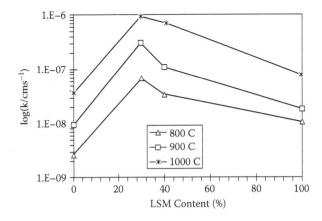

FIGURE 3.4 Measured effective surface oxygen exchange coefficient of the LSM/YSZ composites as a function of the LSM content. (From Ji, Y. et al., *Solid State Ionics*, 176:937–943, 2005. With permission.)

The result implies that both oxygen vacancies and electrons are important for the surface exchange process.

As the incorporation and bulk diffusion of oxygen inside LSM cannot be expected to occur to a significant degree, the TPBs become the reaction sites for the O_2 reduction. The very low oxygen ionic conductivity of LSM is considered to be the main factor in the high-polarization losses of LSM cathode for the O_2 reduction reaction in reduced SOFC operating temperatures [54]. Table 3.1 lists the oxygen diffusion coefficient, oxygen surface coefficient and oxygen ionic conductivity of undoped and doped $LaMnO_3$.

3.2.4 Polarization, Activation, and Microstructure Optimization

The impedance polarization performance of LSM electrode is closely related to the mechanism and kinetics of the oxygen reduction reactions. O_2 reduction at SOFC cathodes is the most heavily studied subject, and this subject is sufficiently broad and complex to warrant its own review. Interested readers should consult the recent excellent articles by Adler [1] and Fleig [55]. Here, only the polarization performance and its influencing factors are discussed.

The initial polarization behavior for the oxygen reduction on LSM-based cathodes is characterized by a well-known activation phenomenon, as shown in

TABLE 3.1
Oxygen Diffusion Coefficient (D^*), Oxygen Surface Exchange Coefficient (k), and Oxygen Ionic Conductivity (σ_i) of $LaMnO_3$-Based Oxides Measured by SIMS Technique

Composition	Temp. (°C)	$D^*(cm^2\ s^{-1})$	$K(cm\ s^{-1})$	$\sigma_i(S\ cm^{-1})$	Ref.
$La_{0.92}MnO_3$	1000	2.45×10^{-13}	7.45×10^{-8}		51
$La_{0.65}Sr_{0.35}MnO_3$	900	4×10^{-14}	5×10^{-8}		49
$La_{0.5}Sr_{0.5}MnO_3$	900	3×10^{-12}	9×10^{-8}		49
	800	8×10^{-14}	1×10^{-7}		49
	700	2×10^{-15}	1×10^{-8}		49
$La_{0.95}Sr_{0.05}MnO_3$	900	2.44×10^{-13}		1.10×10^{-7}	52
$La_{0.90}Sr_{0.10}MnO_3$	1000	4.78×10^{-12}		2.09×10^{-6}	52
$La_{0.80}Sr_{0.20}MnO_3$	1000	1.33×10^{-11}		5.76×10^{-6}	52
	900	1.27×10^{-12}		5.93×10^{-7}	52
$La_{0.80}Sr_{0.20}MnO_3$	1000	6.6×10^{-13}	5.62×10^{-8}		50
	900	1.6×10^{-13}	1.78×10^{-8}		50
	800	4.0×10^{-15}	5.62×10^{-9}		50
	700	3.1×10^{-16}	1.01×10^{-9}		50
YSZ-40 wt%LSM	900	6×10^{-9}	1×10^{-7}	3×10^{-3}	53
	800	1×10^{-9}	7×10^{-7}		53

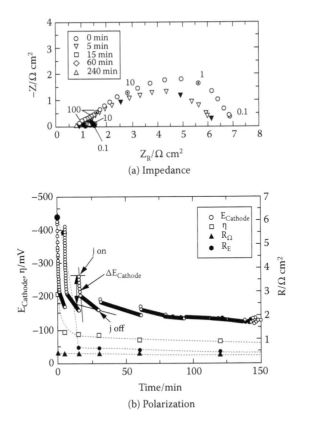

(a) Impedance

(b) Polarization

FIGURE 3.5 (a) Impedance and (b) polarization behavior of a freshly prepared $La_{0.72}Sr_{0.18}MnO_3$ electrode for O_2 reduction as a function of cathodic polarization time at $200\,mAcm^{-2}$ and $900°C$ in air. The impedance was measured at open circuit and the numbers are frequencies in hertz. (From Jiang, S.P. and Love, J.G., *Solid State Ionics*, 138:183–190, 2001. With permission.)

Figure 3.5 [36]. For the O_2 reduction reaction on freshly prepared LSM electrodes, the initial polarization losses are very high and decrease significantly with the cathodic polarization/current passage (see Figure 3.5b). Consistent with the polarization potential, the impedance responses at open circuit decrease rapidly with the application of the cathodic current passage. For example, the initial electrode polarization resistance, R_E, is 6.2 Ωcm^2 and after cathodic current treatment for 15 min R_E is reduced to 0.7 Ωcm^2; see Figure 3.5 (a). The reduction in the electrode polarization resistance is substantial. The analysis of the impedance responses as a function of the cathodic current passage indicates that the effect of the cathodic polarization is primarily on the reduction in the low-frequency impedance [10]. Such activation effect of cathodic polarization/current on the electrochemical activity of the cathodes was also reported on LSM/YSZ composite electrodes [56–58]. Nevertheless, the magnitude of the activation effect on the composite electrodes is relatively small.

The impedance behavior for the O_2 reduction on LSM-based cathodes is generally characterized by two depressed arcs and has been analyzed using equivalent circuits [59–61]. Jiang et al. [62, 63] studied in detail the electrode behavior of LSM cathodes over the temperature range of 850 to 1000°C and oxygen partial pressure of 1 to 21 kPa. The electrode process associated with the high frequency arc is essentially independent of O_2 partial pressure with an activation energy of ~74 kJmol^{-1} while that at the low frequency arc has an activation energy of ~202 kJmol^{-1} and has a reaction order with respect to O_2 partial pressure of ~0.5 at low temperatures and ~0.85 at high temperatures. The O_2 reduction reaction is limited by the diffusion of oxygen species on the LSM surface and the migration of oxygen ions into the YSZ electrolyte.

The cathodic overpotentials of the LSM-based perovskites are dependent on the A-site cations. Ishihara et al. [64] investigated the oxygen reduction activities of $Ln_{0.6}Sr_{0.4}MnO_3$ (Ln is La, Pr, Nd, Sm, Gd, Yb, and Y) at intermediate temperatures. The electrodes were prepared by slurry coating, followed by firing at 950°C for 10 min. With Ln = La, Pr, Nd, Sm, and Gd, the conductivity was about 100 to 240 Scm^{-1} at 800°C. For $Ln_{0.6}Sr_{0.4}MnO_3$ (Ln = Yb and Y), the electric conductivity was between 35 and 50 Scm^{-1} at 800°C. The overpotentials of the cathodes at 1000°C decreased in the following order:

$$Y > Yb > La > Gd > Nd > Sm > Pr$$

The electrochemical activity of the electrodes follows a similar order as the conductivity in respect to the Ln cations in $Ln_{0.6}Sr_{0.4}MnO_3$. This indicates the importance of the electronic conductivity for the cathodic activity of the materials. Further work established that Sr-doped $PrMnO_3$ is far superior to Sr-doped $LaMnO_3$ in terms of thermal expansion compatibility with YSZ electrolyte and electrochemical activity in the intermediate temperatures (Figure 3.6) [65]. Praseodymium oxides exhibit nonstoichiometry in air. The high activity of the $Pr_{0.6}Sr_{0.4}MnO_3$ could be attributed to the facile redox cycle of Pr ions. Huang et al. [66] also showed that A-site nonstoichiometry $Pr_{0.55}Sr_{0.4}MnO_4$ has significantly lower R_E than that of $La_{0.55}Sr_{0.4}MnO_4$ for the O_2 reduction at 850°C.

Different results were reported by Sasaki et al. [67]. They studied the electrical conductivity and polarization performance of $Ln_{1-x}Sr_xMnO_3$ (Ln = Pr, Nd, Sm, and Gd) and found that the influence of the lanthanide ions at the A-site on the electrical conductivity and electrocatalytic activity is not significant. For example, the electrical conductivity was ~200 Scm^{-1} at 1000°C for all $Ln_{0.7}Sr_{0.3}MnO_3$ and the polarization performance showed no significant dependence on the kind of lanthanide elements.

Due to the high activation energy of the O_2 reduction reaction and negligible oxygen ionic conductivity of LSM-based electrodes, the polarization resistance for the reaction on the LSM-based cathodes increases significantly with the decrease of the SOFC operating temperature. For example, for the O_2 reduction on a LSM electrode, R_E was 0.39 Ωcm^2 at 900°C and increased dramatically to 55.7 Ωcm^2 at 700°C [62].

Various strategies have been developed to improve the electrocatalytic activities of the LSM-based cathodes. Murray and Barnett [68, 69] showed that the addition of YSZ and gadolinia-doped ceria (GDC) phase to LSM significantly reduced the electrode polarization resistance. Figure 3.7 shows the electrode polarization resistance of the LSM/GDC tested in the air [69]. The electrode polarization resistance

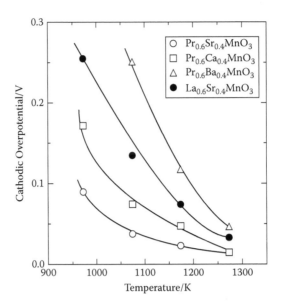

FIGURE 3.6 Cathodic overpotential of some Mn-based perovskites at a current density of 0.1 A cm^{-2} as a function of operating temperature. (From Ishihara, T. et al., *J. Electrochem. Soc.*, 142:1519–1524, 1995. With permission.)

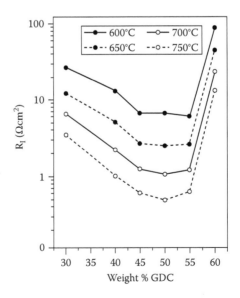

FIGURE 3.7 Plots of electrode polarization resistance versus the cathode composition at different temperatures. (From Murray, E.P. et al., *Solid State Ionics*, 143: 265–273, 2001. With permission.)

decreased as the GDC concentration increased to 50 wt%. The rapid increase of the electrode polarization resistance with further increase in GDC content above 50 wt% may be due to a decrease in electrical conductivity.

Xia et al. [70] studied LSM/GDC composite cathodes fabricated by the sol-gel process. The sol-gel derived composite cathodes showed very low electrode polarization resistance, 0.16 Ωcm^2 at 750°C, lower than that obtained by the slurry coating technique [71]. Using a functionally graded structure can also reduce the electrode polarization resistance of LSM-based cathode, as shown by Hart et al. [72] on the cathodes consisting of five prints of LSM/YSZ composite electrodes and five prints of $La_{0.8}Sr_{0.2}CoO_3$ electrodes. Recently, Liu et al. [73, 74] used combustion chemical vapor deposition (CCVD) to fabricate functionally graded LSM/LSC/GDC electrodes and achieved electrode polarization resistance of 0.43 Ωcm^2 at 700°C. The high electrocatalytic activity of the LSM-based cathode fabricated by CCVD has been attributed to the functionally graded and nanostructured interface.

The electrochemical performance of an LSM cathode can also be enhanced substantially by introducing catalytically active nanoparticles, such as doped CeO_2, into the LSM porous structure. Wet impregnation is one of the most effective methods to introduce the nano-sized particles into porous SOFC electrode [75]. The microstructure of the impregnated electrode is characterized by a uniform distribution of very fine and well-dispersed oxide particles in the porous electrode backbone. Jiang et al. [76, 77] studied in detail the microstructure and electrochemical behavior of $Gd_{0.1}Ce_{0.9}O_3$-impregnated LSM cathodes. For the pure LSM electrode, LSM grains were in the range of 0.7 to 1.2 μm. After wet impregnation with a $Gd_{0.1}Ce_{0.9}(NO_3)_x$ solution, very fine particles were formed around LSM particles and their particle size was in the range of 100 to 200 nm. The electrode polarization (interface) resistance (R_E) of the pure LSM electrode is 11.7 Ωcm^2 at 700°C. In the case of the 5.8 mg/cm^{-2} GDC-impregnated LSM electrode, R_E is 0.21 Ωcm^2 at 700°C, which is 56 times smaller than that of the pure LSM cathode at the same temperature (Figure 3.8).

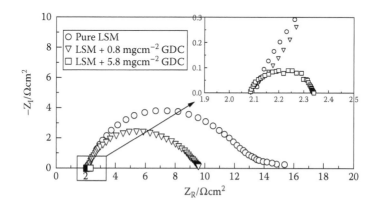

FIGURE 3.8 Impedance curves of pure LSM, 0.8 mgcm^{-2} GDC impregnated LSM and 5.8 mgcm^{-2} GDC impregnated LSM electrodes measured at 700°C in air. The impedance curves were measured after cathodic current treatment at 200 mAcm^{-2} for 120 min at the same temperature.

The electrode polarization resistance of the 5.8 mg/cm^{-2} GDC-impregnated LSM electrode is close to or lower than those of mixed ionic and electronic conducting cathodes, such as La$_{1-x}$Sr$_x$Co$_{1-y}$Fe$_y$ O$_{3-\delta}$ (LSCF) and Gd$_{0.8}$Sr$_{0.2}$CoO$_3$ [77]. This indicates that through the proper composition optimization and materials engineering of the electrode structure, LSM-based electrodes can be used for ITSOFCs.

The performance of LSM cathodes appears to be related to the ionic conductivity of the electrolyte or the interface. Tsai and Barnett [78] studied the addition of Y$_2$O$_3$-doped CeO$_2$ (YDC) interlayers on both sides of the YSZ electrolyte on the performance of the cell with LSM cathode and Ni/YSZ anode. Adding thin porous YDC layers on the YSZ electrolyte surface significantly reduced the interfacial resistance, leading to substantial improvement in the cell performance. The effect of a YDC interlayer on the cathode side was suggested to promote the oxygen surface exchange coefficient.

3.3 LANTHANUM COBALTITE AND FERRITE PEROVSKITES

LaMnO$_3$-based cathode materials show good performance at high temperatures. However, due to the negligible ionic conductivity, pure LSM materials are not suitable for operation at temperatures lower than 800°C. Development of new cathode materials is needed for intermediate temperature operation. The increasing interest recently upon LSCF perovskite materials as cathodes for SOFC mainly stems from their high electronic and ionic conductivities and excellent catalytic activity for oxygen reduction at reduced temperatures. A good comparison of the electrochemical activities of LSM and LSCF electrodes for the O$_2$ reduction has been provided by Jiang [54].

3.3.1 STRUCTURE, OXYGEN NONSTOICHIOMETRY, AND DEFECT MODEL

LaCoO$_3$ is rhombohedral from room temperature to 1000°C. For LSCF perovskite, the phase structure depends on the dopant level and temperature. In the La$_{0.8}$Sr$_{0.2}$Co$_{1-y}$Fe$_y$O$_{3-\delta}$ system, the room-temperature phases are rhombohedral and orthorhombic for $0 \leq y \leq 0.7$ and $0.8 \leq y \leq 1$, respectively [79]. For y = 0.8, i.e., La$_{1-}$$_xSr_xCo_{0.2}Fe_{0.8}O_{3-\delta}$, the orthorhombic/rhombohedral transition at room temperature occurs at x = 0.3. When Sr content is increased to x ≥ 0.6, the second phase was observed, while the main phases were rhombohedral for 0.6 < x < 0.8 and cubic for x > 0.8 [80].

Acceptor-doped cobaltite perovskite-type oxides are characterized by enhanced lattice oxygen vacancy formation, which results in substantial departure from oxygen stoichiometry at increased temperatures. A change in the vacancy concentration is accompanied by a corresponding change in the average valence of either the cobalt or oxygen ions. The high concentration of oxygen vacancies in conjunction with their relatively high mobility causes these materials to exhibit high oxygen ion conductivity. The electronic conductivity is even higher and becomes metallic at high temperatures. Mizusaki et al. [81] described thermogravimetric measurements of the nonstoichiometry of La$_{1-x}$Sr$_x$FeO$_{3-\delta}$ using a model based on randomly distributed and noninteracting point defects. A similar model has also been applied to La$_{1-x}$Sr$_x$CoO$_{3-\delta}$ [82].

Petrov et al. [83] proposed a defect model associated with the defect structure of $La_{1-x}Sr_xCoO_{3-\delta}$, in which Sr ions are assumed to occupy the regular La lattice sites, constituting the predominant defect with negative effective charge. In order to maintain electrical neutrality, the substitution of Sr ions must be compensated for by the formation of equivalent positive charges, i.e., Co^{4+} ions and oxygen vacancies $[V_O^{\bullet\bullet}]$. The overall electroneutrality condition is then given by

$$[Co_{Co}^{\bullet}] + 2[V_O^{\bullet\bullet}] = [Sr_{La}'] \quad (3.11)$$

The relation between $Co_{Co}^{\bullet}$ and $V_O^{\bullet\bullet}$ may be expressed by the following defect reaction and equilibrium with increasing temperature:

$$2Co_{Co}^{\bullet} + O_O^x \rightarrow 2Co_{Co}^x + V_O^{\bullet\bullet} + \frac{1}{2}O_2(gas) \quad (3.12)$$

and

$$[V_O^{\bullet\bullet}][Co_{Co}^x]^2 = K_{Vo}[Co_{Co}^{\bullet}]^2[O_O^x]P_{O_2}^{-1/2} \quad (3.13)$$

where $Co_{Co}^{\bullet}$ and Co_{Co}^x denote Co^{4+} and Co^{3+} ions on a normal Co-site, respectively, and K_{Vo} is the equilibrium constant.

However, a detailed model for the defect structure is probably considerably more complex than that predicted by the ideal, dilute solution model. For higher-defect concentration (e.g., more than ~1%) the defect structure would involve association of defects with formation of defect complexes and clusters and formation of shear structures or microdomains with ordered defect. The thermodynamics, defect structure, and charge transfer in doped $LaCoO_3$ have been reviewed recently [84].

3.3.2 Electronic Conductivity and Thermal Expansion Coefficient

Sr-doped $LaCoO_3$ has high electronic conductivity and high ionic conductivity. Petrov et al. [83] studied the electrical conductivity of $La_{1-x}Sr_xCoO_{3-\delta}$ in air (Figure 3.9). For the sample with x = 0.0, 0.1, and 0.2 the conductivity increases with the temperature, goes through a maximum, and decreases. For samples with x ≥ 0.3 the conductivity decreases with increasing temperature. $La_{1-x}Sr_xCoO_{3-\delta}$ shows a typical p-type conductivity, which decreases with the decreasing oxygen partial pressure at high temperatures. However, the thermal expansion coefficient of Co-rich perovskites is generally too high for both YSZ and doped CeO_2 electrolytes. The average TEC of $LaCoO_3$ is about 20×10^{-6} K^{-1}, which is much higher than those of LSM and YSZ.

An alternative to the Co-rich perovskites is the Sr-doped $LaFeO_3$ which has a lower thermal expansion coefficient and a superior chemical compatibility with doped CeO_2 electrolyte. $LaFeO_3$ is expected to be more stable than Ni- and Co-based perovskites because the Fe^{3+} ion has the stable electronic configuration $3d^5$. It is, therefore, expected that compositions in the system $(La,Sr)(Co,Fe)O_3$ will have desirable properties for intermediate temperature SOFC cathode applications.

Tai et al. [79] studied the structure and electrical properties of $La_{0.8}Sr_{0.2}Co_{1-y}Fe_yO_3$ with $0.8 \leq y \leq 1$. The electronic conductivity of each composition increases with

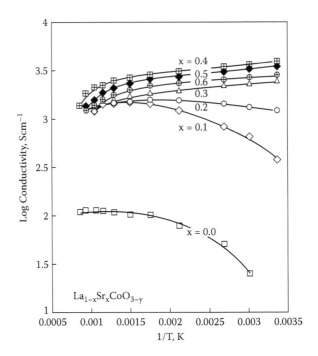

FIGURE 3.9 The electrical conductivity of $La_{1-x}Sr_xCoO_{3-\delta}$ in air as a function of Sr doping, x. (From Petrov, A.N. et al., *Solid State Ionics*, 80:189–199, 1995. With permission.)

temperature through a maximum, then decreases. This may be ascribed to the loss of lattice oxygen at elevated temperatures. The temperature of the electrical conductivity maximum shifts from approximately 200 to 920°C as the Fe content increases from $y = 0$ to $y = 0.7$. The maximum electronic conductivity of LSCF with $y = 0.2$ is 1035 Scm^{-1} at ~600°C and decreases to 875 Scm^{-1} at 1000°C. The TEC of LSCF is 20.7×10^{-6}K^{-1} for $y = 0.2$ and decreases to 15.4×10^{-6}K^{-1} when Fe content increased to 0.8. Further studies show that the electronic conductivity also shows a parabolic dependence on temperature in the $La_{1-x}Sr_xCo_{0.2}Fe_{0.8}O_3$ system [80]. The maximum electronic conductivity is 330 Scm^{-1} at ~600°C for Sr content of 0.4 and the TEC varies within the range of between 14.5 to 15.2×10^{-6}K^{-1}. The TEC behavior of LSCF was also reported by Petric et al. [85].

Kostgloudis and Ftikos [86] studied the properties of A-site deficient $La_{0.6}Sr_{0.4-z}Co_{0.2}Fe_{0.8}O_{3-\delta}$ and $La_{0.6-z}Sr_{0.4}Co_{0.2}Fe_{0.8}O_{3-\delta}$. The electrical conductivity of all compounds showed parabolic dependence on the temperature with a maximum conductivity around 600°C. The steep conductivity decrease above 600°C is attributed to the loss of oxygen. The electronic conductivity decreased with the increase in the A-site nonstoichiometry. Figure 3.10 shows an example of the electronic conductivity of $La_{0.6}Sr_{0.4-z}Co_{0.2}Fe_{0.8}O_{3-\delta}$ as a function of z in air. This indicates that the charge compensation mechanism in these A-site deficient LSCF is probably not the oxidation of B^{3+} cations to the tetravalent state, but rather the formation of oxygen vacancies. The TEC of LSCF is also affected by the A-site nonstoichiometry. The

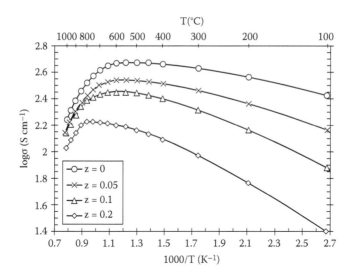

FIGURE 3.10 Electronic conductivity of $La_{0.6}Sr_{0.4-z}Co_{0.2}Fe_{0.8}O_{3-\delta}$ as a function of z in air. (From Kostgloudis, G.Ch. and Ftikos, Ch., Solid State Ionics, 126:143–151, 1999. With permission.)

TEC values at 700°C are 14.2, 14.1, and 13.8×10^{-6} K^{-1} for $La_{0.6-z}Sr_{0.4}Co_{0.2}Fe_{0.8}O_{3-\delta}$, with $z = 0.05$, and $La_{0.6}Sr_{0.4-z}Co_{0.2}Fe_{0.8}O_{3-\delta}$, with $z = 0.1$ and 0.2, respectively. The TEC of stoichiometric $La_{0.6}Sr_{0.4}Co_{0.2}Fe_{0.8}O_{3-\delta}$ is 15.3×10^{-6}K^{-1} at 700°C. In comparison, the TEC of GDC is 11.5 to 11.9×10^{-6}K^{-1} [87].

There are several key differences in the characteristics of $LaMnO_3$ and $LaCoO_3$ as the SOFC cathode materials. Firstly, $LaCoO_3$ phase is much less stable against reduction when compared to $LaMnO_3$. Secondly, the thermal expansion coefficient of $LaCoO_3$ is significantly higher than that of $LaMnO_3$. Thirdly, $LaCoO_3$ has a higher electrical conductivity than $LaMnO_3$ under similar conditions. Thus, one could tailor cathode thermal expansion and improve cathode electronic conductivity by forming a solid solution of $LaMnO_3$ and $LaCoO_3$. Aruna et al. [88] studied the electronic conductivity and thermal expansion of the solid solution of $LaMn_{1-x}Co_xO_3$. Substitution of the low concentration of Co into $LaMnO_3$ does not improve the electronic conductivity. The lowest conductivity was obtained with x = 0.5 probably due to the formation of cubic phase of the solid solution. As expected, the thermal expansion coefficient increases with increasing cobalt concentration. The solid solutions of $LaMn_{0.7}Co_{0.3}O_3$ and $LaMn_{0.4}Co_{0.6}O_3$ have TEC values of 12.5×10^{-6}K^{-1} and 16.3×10^{-6}K^{-1}, respectively. For $LaMnO_3$, the thermal expansion coefficient is 11.33×10^{-6} K^{-1} [47].

Huang et al. [89] studied the electronic conductivity and TEC of Sr- and Ni-doped $LaCoO_3$ and $LaFeO_3$ perovskites. The results show that the compositions $La_{0.8}Sr_{0.2}Co_{0.8}Ni_{0.2}O_3$ (LSCN) and $La_{0.8}Sr_{0.2}Fe_{0.8}Ni_{0.2}O_3$ (LSFN) have TEC values of 15.6×10^{-6}K^{-1} and 12.8×10^{-6}K^{-1}, respectively, better matched to that of the electrolyte than the TEC of LSC. They also show a good electronic conductivity. LSCN and LSFN systems are suggested as attractive alternatives to the conventional cathode materials for ITSOFCs.

Calcium substitution on the La site reduces the sinterability of $LaFeO_3$ and dense $La_{1-x}Ca_xFeO_3$ (LCF) can be obtained at sintering temperature of 1320°C [90]. The compositions with $x \leq 0.2$ show good thermal expansion compatibility with YSZ electrolyte and high electronic conductivity (88 Scm^{-1} at 800°C for x = 0.15). Increasing Ca substitution leads to the decomposition of the LCF and the formation of the second phase.

3.3.3 OXYGEN DIFFUSION AND SURFACE EXCHANGE COEFFICIENT

The major advantages of lanthanum cobalt ferrite–based perovskites over lanthanum manganite–based perovskites are their significantly higher oxygen diffusion and exchange properties. The oxygen self-diffusion coefficient of cobaltite-based materials is several orders of magnitude higher than that of the manganites. In comparison to the manganite-based materials (see Table 3.1), the cobaltite-based perovskites exhibit much lower activation energies. Activation energy of ~60 kJmol^{-1} was reported for pure cobaltite $La_{0.8}Sr_{0.2}CoO_{3-\delta}$ [49]. The oxygen diffusion properties in $LaFeO_3$, $LaCoO_3$, $La_{1-x}Sr_xFeO_3$, and $La_{1-x}Sr_xCoO_3$ single crystals over the temperature range of 900 to 1100°C were studied by Ishigaki et al. [91–93] using SIMS.

Teraoka et al. [94] measured the oxygen ion conductivities (σ_i) of $LaCoO_3$-based perovskites using a four-probe ionic dc technique (electron-blocking technique). The oxygen ion conductivity is 1.20 and 0.18 Scm^{-1} at 900°C for $La_{0.6}Sr_{0.4}CoO_3$ and $La_{0.6}Sr_{0.4}Co_{0.8}Fe_{0.2}O_3$. However, substitution of 20% Co with Ni and Cu enhances the ionic conductivity as compared to $La_{0.6}Sr_{0.4}CoO_3$. They reported an increase in σ_i in the order of Fe < Co < Cu < Ni. Doping of $La_{0.6}Sr_{0.4}CoO_3$ by Fe decreases σ_i. This indicates that the Fe^{3+} tends to bind O^{2-} more strongly than Co^{3+}, which seems to decrease both the mobility and the concentration of the oxygen vacancies.

On the other hand, substitution of gallium with cobalt results in a decreasing oxygen ionic conductivity and an increasing electronic conduction [95]. The oxygen ionic conductivity was reported to be 0.0034 Scm^{-1} at 882°C in air for $LaGa_{0.6}Co_{0.4}O_3$, substantially lower than that of $LaGaO_3$ and $LaCoO_3$. Doping $LaCoO_3$ with gallium leads to the formation of insulating cobalt oxides.

The conductivity and thermal stability of $Y_{0.9}Ca_{0.1}FeO_3$ was studied by Kim and Yoo [96]. The ionic conductivity was measured by ionic block technique using impedance spectroscopy and is in the range of 10^{-4} to 10^{-2} Scm^{-1} and the system becomes thermodynamically unstable below oxygen partial pressure of $10^{-13.6}$ atm at 1000°C. Table 3.2 summarizes the oxygen diffusion coefficients, surface exchange coefficients, and oxygen ionic conductivities for lanthanum cobaltite–based perovskite cathodes.

3.3.4 ELECTROCHEMICAL POLARIZATION PERFORMANCE

Horita et al. [97] studied the electrochemical polarization performance of $La_{1-x}Sr_xCoO_3$ (x = 0.2, 0.3, 0.4) cathodes on (La, Sr) (Gd, Mg)O_3, LSGM electrolyte. With an increase of Sr content in LSC, the conductivity increases above 1400 Scm^{-1} (for x = 0.3, 0.4). The temperature dependence of the conductivity shows metallic behavior, especially above x = 0.3. The polarization activity for the O_2 reduction increases with the Sr content in LSC. The cathodic polarization curves at the porous

TABLE 3.2
Oxygen Tracer Diffusion Coefficient (D^*), Oxygen Surface Exchange Coefficient (k) and Oxygen Ionic Conductivity (σ_i in air) of Doped Lanthanum Cobalt Ferrite-Based Perovskites

Composition	Temp.(°C)	$D^*(cm^2\ s^{-1})$	$K(cm\ s^{-1})$	$\sigma_i(Scm^{-2})$	Ref.
$La_{0.8}Sr_{0.2}CoO_3$	900	4×10^{-8}	2×10^{-5}		49
	800	2×10^{-8}	5×10^{-6}		49
	700	1×10^{-8}	43×10^{-6}		49
$LaFeO_3$	1000	5.28×10^{-12}	1.67×10^{-7}		91
	900	9.84×10^{-13}	3.89×10^{-8}		91
$LaCoO_3$	900	5.31×10^{-9}			92
	800	2.41×10^{-11}	3.47×10^{-7}		92
	700	9.20×10^{-13}	4.57×10^{-9}		92
$La_{0.9}Sr_{0.1}CoO_3$	800	1.99×10^{-10}	1.54×10^{-8}		93
$La_{0.9}Sr_{0.1}FeO_3$	900	2.65×10^{-9}	3.60×10^{-7}		93
$La_{0.6}Sr_{0.4}CoO_3$	900			1.20	94
$La_{0.6}Sr_{0.4}Co_{0.8}Fe_{0.2}O_3$	900			0.18	94
	900	3×10^{-7}	4×10^{-5}		49
	800	1×10^{-7}	2×10^{-5}		49
	700	2×10^{-8}	4×10^{-6}		49
$La_{0.6}Sr_{0.4}Co_{0.8}Cu_{0.2}O_3$	900			2.00	94
$La_{0.6}Sr_{0.4}Co_{0.8}Ni_{0.2}O_3$	900			3.16	94
$LaGa_{0.6}Co_{0.4}O_3$	882			0.0034	95
$Y_{0.9}Ca_{0.1}FeO_3$	1000			0.039[a]	96
	900			0.0043[a]	96

[a] Measured by electron blocking technique.

LSC/LSGM interface with different Sr content in LSC are shown in Figure 3.11 [97]. The difference of the polarization current density comes from the difference of the activity at the LSC/LSGM interface. Thus, the higher Sr content in LSC promotes the O_2 reduction and reduces the electrode polarization resistance. LSC/LSGM and LSC/CeO$_2$-YSZ systems were also studied by others [98–101].

The electrochemical performance of $La_{0.6}Sr_{0.4}Co_{0.2}Fe_{0.8}O_3$ (LSCF6428) composition was characterized by Esquirol et al. [102]. At 600°C, the conductivity of a porous LSCF coating with thickness of ~10 μm was 52.8 and 29.2 Scm^{-1} when sintered at 1000 and 850°C, respectively. This conductivity is considerably lower than the conductivity values for dense LSCF (300 to 400 Scm^{-1}). The electrode polarization resistance of LSCF sintered at 850°C was 7.5, 0.23, and 0.03 ohm cm^2 at 502, 650, and 801°C, respectively, lower than the electrode polarization resistance values at the same temperatures for the LSCF cathode sintered at 1000°C. The results show that the electrochemical activity of the LSCF electrode for the O_2 reduction at

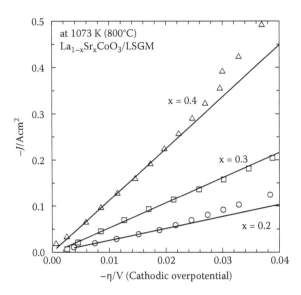

FIGURE 3.11 Cathodic polarization characteristics at the porous LSC/LSGM interface at 800°C with different Sr content in LSC. (From Horita, T. et al., *J. Electrochem. Soc.*, 148:A456–A462, 2001. With permission.)

temperatures below 600°C can be described by a typical TPB model due to the significant reduction in the ionic conductivity of the materials below 600°C.

Tu et al. [103] studied electronic conductivity and electrochemical activity of perovskite oxides $Ln_{0.4}Sr_{0.6}Co_{0.8}Fe_{0.2}O_3$ (Ln = La, Pr, Nd, Sm, Gd) for the oxygen reduction reaction in SOFCs. The electronic conductivity of $Nd_{0.4}Sr_{0.6}Co_{0.8}Fe_{0.2}O_3$ is 334 Scm^{-1} at 800°C, highest among the oxides studied. Consistent with the electrical behavior, $Nd_{0.4}Sr_{0.6}Co_{0.8}Fe_{0.2}O_3$ exhibits the best catalytic activity for oxygen reduction while $Pr_{0.4}Sr_{0.6}Co_{0.8}Fe_{0.2}O_3$ shows the highest polarization. The electrocatalytic activity for the reaction decreases with the order of dopant at the A-site as

$$Nd > Sm > La > Gd > Pr$$

This is not exactly the order as the electronic conductivity of the oxides, indicating that electrocatalytic activity of mixed ionic and electronic conducting (MIEC) oxides depends on other properties such as the oxygen exchange and ionic conductivities.

$Sm_{1-x}Sr_xCoO_3$ and $Dy_{1-x}Sr_xCoO_3$ were also investigated as potential SOFC cathode materials [104]. The perovskite phase forms in $Sm_{1-x}Sr_xCoO_3$ for x = 0.0 to 0.9, but only for x = 0.6 to 0.9 in $Dy_{1-x}Sr_xCoO_3$ system. The limited extent of the perovskite formation in $Dy_{1-x}Sr_xCoO_3$ is due to the smaller ionic radius of Dy. The electrocatalytic activity of the sputtered $Sm_{1-x}Sr_xCoO_3$ electrodes on YSZ electrolyte showed significant dependence on the Sr content with the lowest cathodic polarization at 800°C occurring at x = 0.8.

$Ba_{0.5}Sr_{0.5}Co_{0.8}Fe_{0.2}O_3$ (BSCF) shows extremely high electrochemical activity for the O_2 reduction reaction at intermediate temperatures. Shao and Haile [105] applied BSCF as cathode to a doped ceria electrolyte cell and achieved the power densities of

1010 mWcm^{-2} and 402 mWcm^{-2} at 600 and 500°C, respectively, when operated with hydrogen as the fuel and air as the cathode gas. The electrode polarization resistance is very low, 0.51 to 0.6 Ωcm^2 at 500°C. BSCF is a mixed oxygen ionic and electronic conducting oxide that is derived from SrCo$_{0.8}$Fe$_{0.2}$O$_{3-\delta}$ [106]. The thermal and electronic properties of BSCF were reported by Wei et al. [107]. The TEC of BSCF was ~20 × 10^{-6} K^{-1} and the conductivity was ~25 Scm^{-1} at 800°C. The high TEC implies the thermal compatibility may be a problem for the application of BSCF-based cathode materials in ITSOFCs.

The addition of an ionic conductive phase, such as GDC, also promotes the electrocatalytic activity of an MIEC cathode. Hwang et al. [108] studied the electrochemical activity of LSCF6428/GDC composites for the O$_2$ reduction and found that the activation energy decreased from 142 kJmol^{-1} for the pure LSCF electrode to 122 kJmol^{-1} for the LSCF/GDC composite electrodes. Thus, the promotion effect of the GDC is most effective at low-operation temperatures (Figure 3.12). This is due to the high ionic conductivity of the GDC phase at reduced temperatures.

The electrocatalytic activity of MIEC cathodes also depends strongly on the properties of the electrolyte, as shown by Liu and Wu [109]. The electrode polarization resistances, R_E, or area specific resistance (ASR) measured by the electrochemical

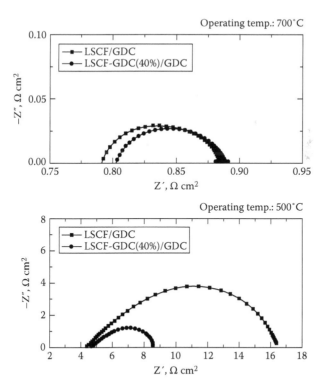

FIGURE 3.12 Impedance curves of LSCF ad LSCF/GDC composite cathodes in air measured at different temperatures. (From Hwang, H.J. et al., *J. Power Sources*, 145:243–248, 2005. With permission.)

impedance, are influenced dramatically by the electrolyte materials used. At 750°C, for example, R_E is 1.15 Ωcm^2 for a cell with a $Bi_{1.5}Y_{0.5}O_3$ (YSB), 3 Ωcm^2 for a cell with a $BaCe_{0.8}Gd_{0.2}O_3$ (BCG), 3.7 Ωcm^2 for a cell with a $Ce_{0.9}Sr_{0.1}O_2$ (SDC), and 8.0 Ωcm^2 for a cell with a YSZ electrolyte. The shapes of the impedance spectra are quite different for cells with different electrolyte materials. Steele and Bae [110] found that the ASR of an LSCF electrode was decreased by a factor of 2 to 3 times by introducing a thin dense layer (~1 μm) of LSCF adjacent to the GDC electrolyte. All these observations indicate that electrochemical activities of an MIEC electrode are significantly affected by the interfaces between the electrode and the electrolyte, not determined only by the bulk properties of the MIEC. In another words, the oxygen diffusion or transportation through the electrode–electrolyte interface can be enhanced significantly by the proper engineering of the interface.

Addition of a small amount of a noble metal, such as Pt and particularly Pd, is shown to have significant activation effect on the cathode of SOFCs. Uchida et al. [111] impregnated 0.1 $mgcm^{-2}$ Pt into a $La_{0.6}Sr_{0.4}CoO_3$ (LSC) cathode and the current output was 0.6 A/cm^2 at $\eta = 50$ mV and 800°C, 1.7 times higher than that without Pt addition. Wang and Zhong [112] reported that the addition of Pd has a significant activation effect on the electrocatalytic activity of $Sm_{0.5}Co_{0.5}CoO_3$/LSGM (SCC/LSGM) composite cathode. With an addition of 2.4 wt% Pd in the composite cathode, the electrode polarization resistance is 0.12 Ωcm^2 at 850°C, much lower than that without Pd addition. A similar activation effect has also been observed for Pd-promoted LSCF cathode [113].

3.4 OTHER PEROVSKITE OXIDES

Takeda et al. [114] suggested that replacement of lanthanum with gadolinium, on the A-site, could provide a more stable system with respect to reaction with the YSZ electrolyte. Phillipps et al. [115] studied the electronic conductivity and thermal expansion coefficient properties of $Gd_{0.8}Sr_{0.2}Co_{1-y}Mn_yO_3$ (0.1 ≤ y ≤ 1.0) and $Gd_{0.7}Ca_{0.3}Co_{1-y}Mn_yO_3$ (0.3 ≤ y≤ 1.0) system. Both systems showed highly distorted orthorhombic symmetry. In the case of $Gd_{0.7}Ca_{0.3}Co_{1-y}Mn_yO_3$ system, single-phase perovskites were not formed for y < 0.3. The $Gd_{0.8}Sr_{0.2}Co_{1-y}Mn_yO_3$ showed higher electrical conductivity as compared to that of the $Gd_{0.7}Ca_{0.3}Co_{1-y}Mn_yO_3$ system. The highest conductivity was obtained for $Gd_{0.8}Sr_{0.2}Co_{0.9}Mn_{0.1}O_3$ (271 Scm^{-1} at 800°C). However, the TEC of $Gd_{0.8}Sr_{0.2}Co_{0.9}Mn_{0.1}O_3$ was ~24 × 10^{-6} K^{-1}, which is much higher than ~10 × 10^{-6} K^{-1} of YSZ electrolyte. Increasing the Mn content could achieve the thermal expansion compatibility with YSZ, e.g., y = 0.8 for $Gd_{0.8}Sr_{0.2}Co_{1-y}Mn_yO_3$ and y = 0.6 for $Gd_{0.7}Ca_{0.3}Co_{1-y}Mn_yO_3$. The polarization performance of both materials are moderate, with the lowest overpotential measured in air at 100 $mAcm^{-2}$ and 1000°C being ~120 mV for 1.5 μm thick $Gd_{0.8}Sr_{0.2}Co_{1-y}Mn_yO_3$ and $Gd_{0.7}Ca_{0.3}Co_{1-y}Mn_yO_3$ with y = 0.4.

The Sr-doped praseodymium manganites and cobaltites have been studied by several groups as potential cathodes for ITSOFC. Kostgloudis et al. [116–118] systematically investigated the crystal structure, electrical conductivity, and thermal expansion properties of (Pr, Sr)MnO_3, (Pr, Sr)CoO_3, and (Pr, Sr)(Co, Mn)O_3 systems. All compounds have the orthorhombic perovskite $GdFeO_3$-type structure (Pbnm space group) as expected from the calculated Goldschmidt tolerance factor of the system. Among the examined oxides of the system $Pr_{1-x}Sr_xMnO_3$ ($0 \le x \le 0.5$), the composition with x = 0.5 showed best the properties required for ITSOFC; its thermal expansion coefficient was $12.2 \times 10^{-6}K^{-1}$ and electronic conductivity value was 250 Scm^{-1} at 600°C [116]. The undoped $PrCoO_3$ showed semiconducting behavior in the temperature range studied and the electronic conductivity increases with increasing temperature. On the other hand, semi-metallic behavior was observed for Sr-doped $PrCoO_3$. The electronic conductivity of $Pr_{1-x}Sr_xCoO_3$ ($0 \le x \le 0.5$) system is in the range of 1000 Scm^{-1}, but the thermal expansion coefficient is also high, 18.3 to $29.6 \times 10^{-6}K^{-1}$, depending on the composition and temperature.

$LaNiO_3$ is known to have very high electronic conductivity at room temperature [119]. However, this material is unstable above 850°C and decomposes to La_2NiO_4 and NiO, whose electronic conductivity is poor [120]. Hrovat et al. [121] studied $LaNi_{0.6}Co_{0.4}O_3$ oxides and showed that $LaNiO_3$ can be stabilized even at high temperature if some of the Ni is substituted by Co. Chiba et al. [122] investigated the system of La(Ni, M)O$_3$ (M = Al, Cr, Mn, Fe, Co, Ga) as potential cathode materials for SOFC. In these systems, $LaNi_{1-x}Fe_xO_3$ is reported to have both a high electronic conductivity and a thermal expansion coefficient which is close to that of the zirconia electrolyte. The perovskite structure becomes more stable as x in $LaNi_{1-x}Fe_xO_3$ increases. The crystal structures change from rhombohedral when x is 0.4 to orthorhombic phase when x is greater than 0.5. This system has its highest electronic conductivity of 580 Scm^{-1} at 800°C for fully dense specimen when x in $LaNi_{1-x}Fe_xO_3$ is 0.4 and its TEC is $11.4 \times 10^{-6}K^{-1}$. The electronic conductivity of $LaNi_{0.6}M_{0.4}O_3$ (M = Al, Cr, Mn, Fe, Co, Ga) in the temperature range of 600 to 1000°C decreases in order:

$$Co > Fe > Cr > Ga > Mn > Al$$

The low electronic conductivity of $LaNi_{0.6}Al_{0.4}O_3$ is probably due to the fact that $LaNi_{0.6}Al_{0.4}O_3$ is a mixture of the rhombohedral and tetragonal phases. $LaNi_{0.6}Co_{0.4}O_3$ shows the highest electronic conductivity and also high TEC of $14.3 \times 10^{-6}K^{-1}$. Hrovat et al. [121] reported a lower TEC of $11.9 \times 10^{-6}K^{-1}$ for the $LaNi_{0.6}Co_{0.4}O_3$. As will be discussed later, the perovskite La(Ni,Fe)O$_3$ also has high resistance toward Cr deposition and poisoning for SOFCs based on Fe-Cr metallic interconnect [123, 124].

The perovskite lanthanum strontium manganese chromite (LSCM), such as $La_{0.75}Sr_{0.25}Cr_{0.5}Mn_{0.5}O_3$, is a promising material as an alternative anode for SOFCs with high-electrocatalytic activity [125, 126]. LSCM can also be used as an effective cathode for SOFCs, as shown by Bastidas et al. [127]. Raj et al. [128] studied the oxygen diffusion and surface exchange on $(La_{0.75}Sr_{0.25})_{0.95}Cr_{0.5}Mn_{0.5}O_3$ and found that oxygen tracer diffusion coefficient of LSCM is 1.0×10^{-10} cm^{2}s^{-1} at 900°C, two orders of magnitude higher than that of LSM. LSCM is chemically stable with YSZ electrolyte, making it a potential candidate for the cathode-supported thin electrolyte cells. The present materials for the cathode-supported SOFCs are based primarily on the LSM/YSZ composites [129]. However, the processing temperature or cofiring temperature for the LSM-based composite cathode supports and YSZ electrolyte is limited by the fact that LSM-based materials react with YSZ electrolyte at high temperatures (1200°C or higher) to form resistive lanthanum zirconate phases at the LSM/YSZ interface. The performance of LSM-based cathode supports is also limited by the significantly reduced electrocatalytic activity of LSM resulting from its negligible oxygen ion conductivity. Jiang et al. [130] recently demonstrated a symmetric LSCM/YSZ anode and LSCM/YSZ cathode-supported YSZ thin electrolyte cell and the overall cell polarization resistance was 0.46 Ωcm^2 at 800°C, which is a reasonable value as compared to 0.2–1.0 Ωcm^2 reported for optimized Ni/YSZ anode-supported YSZ thin electrolyte cells at the same temperature [131, 132].

Lanthanum nickelate, $La_2NiO_{4+\delta}$ (LN), with the perovskite-related K_2NiF_4 structure, is expected to exhibit fast ion conduction and has attracted considerable interest as a cathode material. $Ln_2BO_{4+\delta}$ (Ln = rare-earth oxides, B = transition metal cations) structure consists of alternate rock salt rare-earth LnO and perovskite ($LnBO_3$) layers stacked in the c-axis. Many of these oxides exhibit distortion to lower symmetry at ambient temperatures. The crystal structure transforms to the original tetragonal form only at higher temperatures or when the A-site ion is partially doped with certain alkaline earth cations. In $La_2NiO_{4+\delta}$ interstitial oxide ions are mobile and its D^* and k values are 2×10^{-7} cm^2s^{-1} and 2×10^{-6} cms^{-1} at 800°C, respectively [133, 134].

Skinner [135] characterized $La_2NiO_{4+\delta}$ using *in situ* high temperature neutron powder diffraction and reported only one structural transformation from *Fmmm* orthorhombic structure to *I4/mmm* tetragonal structure above 150°C. Substitution of the relatively smaller rare-earth cations with larger Sr in $Ln_2NiO_{4+\delta}$ could increase the structural stability. Jennings and Skinner [136] also studied the thermal stability and conductivity properties of the $La_xSr_{2-x}FeO_{4+\delta}$ system. The oxygen excess is evident from the x-ray diffraction data. The electronic conductivity of the system shows dependence on the La/Sr ratio, and composition in terms of oxygen stoichiometry. At 700°C, the maximum conductivity of ~20 Scm^{-1} was obtained on the composition with x < 0.8. The conductivity is higher in pure oxygen than that in air and argon. This indicates that the conductivity of $La_xSr_{2-x}FeO_{4+\delta}$ is enhanced in more oxygen-rich atmosphere. The electrical properties of iron-substituted $La_{6.4}Sr_{1.6}Cu_8O_{20\pm\delta}$ system were also studied as a potential cathode candidate [137].

The suitability of lanthanum nickelate as an SOFC cathode has been examined by Virkar's group [138]. They showed that LN performed poorly as a single-phase cathode in an anode-supported YSZ cell. However, with an SDC/LN composite interlayer the performance of the LN cathode increased substantially and the maximum power density of the cell with a YSZ thin electrolyte (~8 μm) was ~2.2 Wcm^{-2} at 800°C, considerably higher than 0.3 to 0.4 Wcm^{-2} of similar cells with only LN or SDC interlayer. The results are significant as it shows that the composite MIEC cathodes perform much better than single-phase MIEC in the case of LN despite its mixed ionic and electronic conductivity.

3.5 INTERACTION AND REACTIVITY WITH OTHER SOFC COMPONENTS

The perovskite oxides used for SOFC cathodes can react with other fuel cell components especially with yttria-zirconia electrolyte and chromium-containing interconnect materials at high temperatures. However, the relative reactivity of the cathodes at a particular temperature and the formation of different phases in the fuel cell atmosphere critically depend on the nature of the cations presented in A- and B-site. Chemical compatibility and stability of perovskite-based materials are important due to the high operation and even higher fabrication temperatures of the SOFC components.

3.5.1 INTERACTION WITH THE ELECTROLYTE

3.5.1.1 Interaction with YSZ Electrolyte

The reactivity between LSM and the electrolyte (usually YSZ) has been extensively studied both experimentally and theoretically. Yokokawa and co-worker [139–141] examined the thermodynamic stability of various perovskite phases with respect to reaction products in the presence of zirconia, providing a framework for understanding the chemical compatibility and the thermodynamic driving force behind the formation of the phases between electrode and electrolyte. For example, their thermodynamic calculations have suggested that the formation of $La_2Zr_2O_7$ at the electrode–electrolyte interface will be high when (La+Sr) content at the A-site is > Mn^{3+} ions at B-site and when Sr/La ratio is < 0.43. Sr content > 0.43 La at the A-site leads to Sr depletion from the lattice and eventual $SrZrO_3$ formation. The substoichiometry at the A-site can help to prevent the depletion of A-site ions but Mn^{3+} ion activity is greatly enhanced when A-site deficiency is > 0.15 (i.e., {(La+Sr)/Mn} < 0.85). Under such conditions, the dissolution of manganese in YSZ electrolyte is very high. As described in a recent review by Yokokawa [142], thermodynamic calculations show that A-site deficient LSM can prevent the formation of pyrochlore at the interface. This chapter will focus on the experimental aspects of the chemical interaction and compatibility of cathode materials with YSZ electrolyte materials and its effect on the performance.

The interaction and interface phase formation between the LSM electrode and YSZ electrolyte varies with the stoichiometric composition of LSM [143, 144], the La/Sr ratio at the A-site [145] and the temperature and atmosphere of the heat treatments [146, 147]. Taimatsu et al. [148] studied the reaction of $La_{1-x}Ca_xMnO_3$ (x = 0, 0.1, 0.2) with YSZ in the temperature range of 1300 to 1425°C. The proposed mechanism suggests that the reaction proceeds by the preferential diffusion of Mn into YSZ, which depletes the LSM at the interface, so that the resulting excess La reacts with YSZ to form $La_2Zr_2O_7$. The substitution of Ca for La in $LaMnO_3$ suppresses the manganese migration, thus reducing the reactivity with YSZ. Mitterdorfer and Gauckler [144] investigated in details the mechanism of the formation of $La_2Zr_2O_7$ as a function of stoichiometry composition of LSM. Excess lanthanum oxide within the perovskite reacts immediately with YSZ to form dense $La_2Zr_2O_7$, which is characterized by the formation of cubic islands at the interface. The growth of the $La_2Zr_2O_7$ is controlled by bulk diffusion of cations.

Manganese has significant solubility in YSZ, ranging from about 11.4% at 1300°C to 5.1% at 1000°C. After cofiring YSZ and $La_{0.7}Sr_{0.3}Mn_{1.1}O_{3+x}$ at 1300°C Mn was observed to a depth of 20 μm in YSZ [149]. The manganese ions in YSZ are found to be only in divalent (about 80%) and trivalent (about 20%) states. No Mn^{4+} ions were detected in YSZ lattice suggesting that Mn^{3+} ions are reduced when manganites interact with YSZ. Waller et al. [150] determined the diffusion coefficients of manganese in single crystal and polycrystalline YSZ in the temperature range of 1100 to 1400°C. The effective diffusion coefficient of manganese in polycrystalline YSZ is far higher than that in single crystal YSZ, indicating that grain-boundary diffusion is the dominant transport process of Mn in polycrystalline YSZ. The activation energy for Mn diffusion in single crystal and polycrystalline YSZ was found to be 5.36 and 7.34 eV, respectively.

Yamamoto et al. [151] examined the influence of the Y-content in the ZrO_2 on the interface reactions and observed that cubic zirconia with a high content of yttria (12 mol%) is less active with $La_{0.9}MnO_3$ than zirconia with a low-yttria content (8 mol%). van Roosmalen and Cordfunke [152] studied the interaction between 3 mol% Y_2O_3-ZrO_2 (3YSZ) and LSM dense pellets, and found that 3YSZ was more reactive than 8 mol% Y_2O_3-ZrO_2 (i.e., yttria fully stabilized zirconia, YSZ) in the formation of pyrochlore phase with LSM. Mn ions were not detected in the zirconia phase. Jiang et al. [153] also studied the chemical interaction between LSM coating and 3YSZ electrolyte over the temperature range of 1300 to 1500°C in air. The SEM micrographs of a cross-section of an LSM/3YSZ specimen after heat treatment at high temperatures show two distinct reaction layers are formed between the LSM coating and 3YSZ electrolyte (labelled "L1" and "L2" in the figure) (Figure 3.13) [153]. EDS analysis indicates that "L1" and "L2" reaction layers are most likely owing to the formation of lanthanum zirconate pyrochlore phase and cubic zirconia solid solution at the interface of LSM/3YSZ, respectively. Similar results were also reported on the chemical interaction between (Pr, Sr)MnO$_3$ and 3YSZ electrolyte [154]. The results indicate that the formation of the fluorite-type zirconia phase is mainly due to the dissolution of Mn ions into 3YSZ, while the interaction of La ions with 3YSZ causes the formation of lanthanum zirconate pyrochlore phase.

Phase studies in the $(Zr,Y)O_2$-La_2O_3-Mn_3O_4 system show that the fluorite-type cubic zirconia phase, rather than 3YSZ phase, is in equilibrium with LSM perovskite at high temperatures [153]. Figure 3.14 is the ternary phase diagram of $(Zr,Y)O_2$-La_2O_3-Mn_3O_4 system at 1400°C based on the powder mixture experiments. In the diagram, symbols "+" are the experimental data, $(Zr,Y)O_2$ represents the 3YSZ phase and La_yMnO_3 the perovskite phase due to its nonstoichiometry on the A-site [140].

The phase relation generally agrees with the theoretical phase diagram of the ZrO_2-La_2O_3-Mn_3O_4 system at 1300°C [141]. The phase diagram basically shows that the tetragonal 3YSZ cannot be in equilibrium with LSM perovskite at high temperatures and fluorite-type cubic zirconia solid solution phase, c-$(Zr,Mn,La,Y)O_2$, can be in equilibrium with the LSM perovskite phase. Thus, from the phase stability point of view, partially stabilized zirconia, such as 3YSZ, may not be an optimum choice as the electrolyte materials in SOFC. The interaction between LSM and YSZ was also studied by others [155, 156], showing the formation of lanthanum zirconate at the interface.

The formation of $La_2Zr_2O_7$ phase at the LSM electrode/YSZ electrolyte interface is detrimental to the electrochemical activity and performance of the electrode [147, 157, 158]. Brugnoni et al. [147, 158] studied the growth coefficient and electrical resistivity of the $La_2Zr_2O_7$ layer. The formation of a $La_2Zr_2O_7$ layer was observed at $La_{0.85}Sr_{0.15}MnO_3$ electrodes treated at 1200°C. The growth of the reaction layer, i.e., $La_2Zr_2O_7$, seems to be associated with La diffusion, rather than Mn diffusion. The

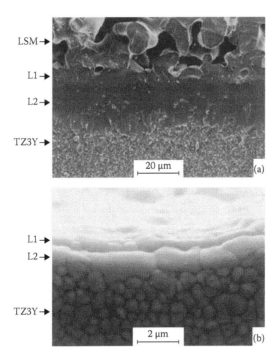

FIGURE 3.13 SEM micrographs taken from (a) a fractured cross-section of an LSM/3YSZ specimen after heat treatment at 1500°C, and (b) the polished and thermally etched cross-section of an LSM/3YSZ specimen after heat treatment at 1400°C. (From Jiang, S.P. et al., *J. Euro. Ceram. Soc.*, 23:1865–1873, 2003. With permission.)

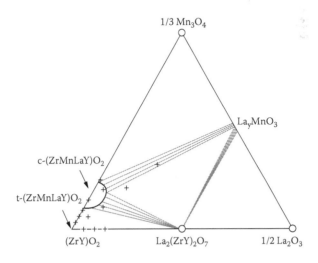

FIGURE 3.14 Ternary phase diagram of $(Zr,Y)O_2$-La_2O_3-Mn_3O_4 system at 1400°C in air. $(Zr,Y)O_2$ denotes 3 mol% Y_2O_3-ZrO_2. Symbols "+" are the experimental data. (From Jiang, S.P. et al., *J. Euro. Ceram. Soc.*, 23:1865–1873, 2003. With permission.)

growth coefficient obeys the Arrhenius relation. By extrapolation at the SOFC operation temperature (1000°C), the film growth rate $D = 8 \times 10^{-18}$ cm^{2}s^{-1} was derived. Using the Wagner equation [159], $x^2 = Dt$, where x is the average thickness as a function of the annealing time t, the growth time of 1 μm thick layer of La$_2$Zr$_2$O$_7$ at the interface would be 350,000 h. Thus, the major contribution to the La$_2$Zr$_2$O$_7$ growth at the cathode–electrolyte interface is expected during the thermal treatments of the electrodes (e.g., the sintering process). The electronic conductivity of the La$_2$Zr$_2$O$_7$ layer was measured to be 2×10^{-4} Scm^{-1} at 1000°C using impedance spectroscopy techniques on the YSZ single crystal and e-beam evaporated La$_2$O$_3$ layer [158]. Slightly lower conductivity values of $10^{-4} - 10^{-5}$ Scm^{-1} were reported for the strontium zirconates SrZrO$_3$, Sr$_4$Zr$_3$O$_{10}$, Sr$_3$Zr$_2$O$_7$, and Sr$_2$ZrO$_4$ and the lanthanum zirconates on sintered samples at 1000°C [160]. These are significantly lower than the conductivity of 0.185 Scm^{-1} at 1000°C for YSZ electrolyte [161]. The thermal expansion coefficient of La$_2$Zr$_2$O$_7$ was reported to be 9.2×10^{-6} K^{-1} also lower than 10.3×10^{-6} K^{-1} for 8YSZ electrolyte [162]. Brugnoni et al. [163] also investigated the effect of diffusion barrier layer, such as NdAlO$_3$, on the La$_2$Zr$_2$O$_7$ formation. However, introducing diffusion barrier layer could also introduce additional resistance at the electrode–electrolyte interface.

The reaction of lanthanum deficient manganite, La$_{1-x}$MnO$_3$ (x = 0–0.2), with YSZ was found to be less reactive than the stoichiometric composition. Kenjo and Nishiya [164] studied the LaMnO$_3$ and La$_{0.85}$MnO$_3$ cathodes at sintering temperatures ranging from 900 to 1300°C. In the case of stoichiometry LaMnO$_3$ electrode, the polarization increased significantly with the sintering temperature most likely due to the formation of a lanthanum zirconate layer at high-sintering temperatures, while the polarization of nonstoichiometry La$_{0.85}$MnO$_3$ decreased with the increase in the sintering temperature. Yamamoto et al. [165] found no La$_2$Zr$_2$O$_7$ at the interface between YSZ and La$_{0.8}$MnO$_3$ when the heat treatment temperature was at 1200°C. However, when the temperature was raised to 1300°C, La$_2$Zr$_2$O$_7$ was formed after 50 h of heat treatment. Mori et al. [162] studied the reaction mechanism between La$_{0.9}$MnO$_3$ and YSZ at the temperature range between 1250 and 1400°C. The induction period for the formation of La$_2$Zr$_2$O$_7$ was found to increase with the increase of the yttrium content in YSZ. The fundamental reason for the beneficial effect of A-site nonstoichiometry or Mn excess of LSM in the inhibiting of the lanthanum zirconate formation is most likely due to the fact that Mn$_3$O$_4$ does not equilibrate with lanthanum zirconate at high temperatures [153].

Mn activity is especially high when there is deficiency at the A-site in perovskites. The diffusion of Mn into YSZ causes Mn depletion of the LSM and in the case of stoichiometric LSM chemically active La$_2$O$_3$ is formed, leading to the formation of La$_2$Zr$_2$O$_7$ at the interface. This is supported by the observation of the formation of pyrochlore in 75%YSZ/25%LSM and not in 20%YSZ/80%LSM mixtures sintered at 1300°C for 2 h [166]. Excess of Mn in LSM/YSZ could delay or impede the formation of La$_2$O$_3$ species at the interface. On the other hand, there is hardly detectable La in the YSZ, indicating very low solubility of lanthanum in YSZ.

LaMO$_3$ and La$_{1-x}$Sr$_x$MO$_3$ (M = Co, Ni and Fe) perovskites are relatively unstable compared to their manganese counterparts LaMnO$_3$ and La$_{1-x}$Sr$_x$MnO$_3$. The former compounds readily react with zirconia electrolytes, leading to the formation of secondary phases at temperature as low as 1000°C in air. The Co^{3+} ions in LaCoO$_3$ are easily reduced to their divalent (e.g., Co^{2+}) state in the presence of zirconia. The activity of Co^{2+} is high and is comparatively less stable in $LaCoO_3$. Therefore, it readily comes out from the perovskite lattice and is present as CoO at the interface. Partial substitution of La with Sr in $LaCoO_3$ enhances the reaction with zirconia because the smaller Co^{4+} ions are unstable in the perovskite lattice. Similarly, the tri- and tetra-valent nickel ions are not very stable, and they are easily reduced to a divalent state. For example, La_2NiO_4 is much more stable than $LaNiO_3$.

LSCF electrodes have been reported to readily react with YSZ electrolyte [167, 168]. After annealing at 1000°C for 2h, a $La_{0.6}Sr_{0.4}Co_{0.2}Fe_{0.8}O_3$ thin film reacted with the YSZ substrate to form primarily $SrZrO_3$, while no reaction products were observed by XRD after 2 h at 800°C. However, after longer times at 800° (360 h) the interfacial resistance increased and $SrZrO_3$ was observed with X-ray diffraction (XRD). At 1200°C, the reaction between LSCF and YSZ became so rapid that LSCF film dissolved into the substrate. On the other hand, no reactions were detected between $La_{0.6}Sr_{0.4}Co_{0.2}Fe_{0.8}O_3$ thin film and $(CeO_2)_{0.8}(SmO_{1.5})_{0.2}$ electrolyte even at 1200°C. Tu et al. [103] indicated that $Ln_{0.4}Sr_{0.6}Co_{0.8}Fe_{0.2}O_3$ (Ln = La, Pr, Nd, Sm, Gd) perovskite oxides reacted with YSZ to form $SrZrO_3$ at 900°C and no reaction was observed at 800°C.

Orui et al. [169] studied the stability of La(Ni, Fe)O_3 with 10 mol% Sc_2O_3-1 mol% Al_2O_3-stabilized ZrO_2 (SASZ) and found that LNF was more reactive with SASZ electrolyte than LSM. For the cell sintered at 1100 and 1200°C, a reaction layer was clearly visible at the LNF/SASZ interface. The thickness of the reaction layer increased with the sintering temperature and the layer was identified as the oxide-containing La and Zr by the TEM/EDS analysis.

3.5.1.2   Interaction with LSGM

In addition to the YSZ electrolyte, the interaction and diffusion between SOFC cathodes and other electrolytes have also been reported. Huang et al. [170] studied the chemical reactions between cathodes, $La_{0.84}Sr_{0.16}MnO_3$ (LSM), $La_{0.5}Sr_{0.5}CoO_3$ (LSC), and the $La_{0.9}Sr_{0.1}Ga_{0.8}Mg_{0.2}O_3$ (LSGM) electrolyte. Significant interdiffusion of Co into LSGM electrolyte and Ga into LSC was found at an LSC/LSGM interface at relatively low fabrication temperatures (e.g., 1050°C for 2 h). In contrast, only a small amount of interdiffusion of Mn into LSGM and Ga into LSM was detected at the LSM/LSGM interface even though it was fired at 1470°C. Naoumidis et al. [171] investigated the chemical interaction between $La_{0.75}Sr_{0.25}Mn_{0.8}Co_{0.2}O_3$ and $Pr_{0.8}Sr_{0.2}Mn_{0.8}Co_{0.2}O_3$ cathodes and electrolyte at 1300°C. The transport of the Co from the perovskite cathode into the LSGM electrolyte is substantial and a distinct concentration of Co down to a depth of more than 100 μm was observed for the specimen sintered at 1300°C for 100 h. The Mn diffusion into the electrolyte is less pronounced. The change in the composition due to the Co and Mn diffusion into the LSGM electrolyte could also lead to the change in the electrical properties of the electrolyte. As shown by Ullmann et al. [172], the substitution of Co for Ga in the LSGM electrolyte increases the ionic conductivity and also introduces mixed conducting behavior. Considering the limited interdiffusion reactions between LSM and LSGM and similar thermal expansion coefficients, LSM could be an appropriate cathode material for LSGM-based fuel cells.

Horita et al. [173] studied the stability at the $La_{0.6}Sr_{0.4}CoO_3$ (LSC) cathode/$La_{0.8}Sr_{0.2}Ga_{0.8}Mg_{0.2}O_3$ (LSGM) electrolyte interface under SOFC operating conditions. LSC reacted with LSGM even at 1000°C for 5 h and caused shrinkage of the lattice parameters of LSGM. This is probably due to Co diffusion into LSGM. The highest electrode polarization conductivity was obtained on the LSC cathode sintered at temperatures between 1150 and 1200°C. The LSC cathode was stable at 800°C under cathodic polarization voltage of −0.3 V for ~700 h and no distinctive reaction layer was observed at the interface of LSC/LSGM. This shows that Co diffusion into LSGM does not affect the interface stability of the cell.

3.5.2 INTERACTION WITH FE-CR ALLOY METALLIC INTERCONNECT

One important issue in the development of ITSOFCs based on metallic interconnect is the interaction between the metallic interconnect and the cathode. Metallic interconnects, such as chromia-forming alloys (e.g., stainless steel), generate volatile Cr-containing species at high temperatures in oxidizing atmospheres. Without effective protective coatings, the gaseous chromium species can lead to a rapid degradation of the SOFC performance due to chemical interaction of Cr species at the Sr-doped $LaMnO_3$ (LSM) cathode side [174–176]. Figure 3.15 shows the typical polarization curves for the O_2 reduction reaction on an LSM cathode in the presence of a Fe-Cr alloy metallic interconnect at 900°C. The rapid increase in the cathodic polarization for the reaction in the presence of a Fe-Cr alloy indicates the poisoning effect of the Cr species on the LSM cathode.

Thermodynamically, chromium deposition can occur by electrochemical reduction or chemical dissociation of high valent chromium species. Thus, the focus of the debate on the deposition of chromium during the O_2 reduction on LSM based

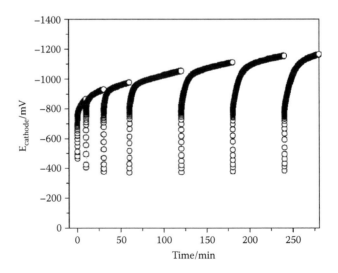

FIGURE 3.15 Polarization curves for the O_2 reduction on an LSM cathode in the presence of a Fe-Cr alloy interconnect, measured under a cathodic current passage of 200 mA cm^{-2} at 900°C.

cathode is on the nature of the deposition process. Taniguchi et al. [175] studied the interaction reaction between LSM and Inconel 600 (16% Cr, 8% Fe, and 76% Ni) at 1000°C in air. The increase in cathodic polarization was found to be related to the intensity and amount of Cr at the cathode–electrolyte interface region. Similar degradation phenomena on LSM cathodes were also reported by others [174, 177]. It was also reported that the deposition of Cr at the $La_{0.6}Sr_{0.4}Co_{0.2}Fe_{0.8}O_3$ (LSCF) cathode and SDC electrolyte is very small as compared with that at the LSM/YSZ interface [178]. Thermodynamically the negative Gibbs free energy changes for the overall fuel cell reaction and the reduction of high valence Cr species are comparable, as shown by Hilpert et al. [179] for the system at 1000°C:

$$3/4\ O_2\ (g) + 3/2\ H_2\ (g) = 3/2\ H_2O\ (g) \tag{3.14}$$

$$CrO_3\ (g) + 3/2\ H_2\ (g) = 1/2\ Cr_2O_3\ (s) + 3/2\ H_2O \tag{3.15}$$

$$CrO_2(OH)_2\ (g) + H_2 = 1/2\ Cr_2O_3\ (s) + 3/2\ H_2O\ (g) \tag{3.16}$$

The consequences of the electrochemical reduction of high valence chromium species would be the precipitation of Cr_2O_3 solid phase at the cathode–electrolyte interface boundary. These led to the hypothesis that the degradation mechanism of LSM cathode is dominated by an electrochemical reduction of high valence vapor species of chromium (CrO_3 and $Cr(OH)_2O_2$) to solid phase Cr_2O_3 in competition with the O_2 reduction reaction, followed by the chemical reaction with LSM to form $(Cr,Mn)_3O_4$ phases at the TPB, blocking the active sites [174–180]. The process is written as follows [174]:

$$Cr(OH)_2O_2(g) + CrO_3(g) + 6e^- \rightarrow Cr_2O_3(s) + H_2O(g) + 3O^{2-}{}_{(YSZ)} \tag{3.17}$$

$$3/2Cr_2O_3(s) + 3(La,Sr)MnO_3 \rightarrow 3(La,Sr)(Mn_xCr_{1-x})O_3$$
$$+ (Cr_xMn_{1-x})_3O_4 + 1/4\ O_2 \tag{3.18}$$

$$3CrO_3(g) + 3(La,Sr)MnO_3 \rightarrow 3(La,Sr)(Mn_xCr_{1-x})O_3$$
$$+ (Cr_xMn_{1-x})_3O_4 + 2.5O_2 \tag{3.19}$$

The mechanism of Cr deposition through the electrochemical reduction of high-valent Cr species in competition with O_2 reduction at the electrode–electrolyte interface is, however, challenged by Jiang et al. [181–185]. They systematically investigated the interaction reaction between LSM cathode and YSZ electrolyte in the presence of a Fe-Cr alloy interconnect and found that (a) Cr deposition on the YSZ electrolyte occurs under cathodic as well as anodic polarization; (b) the initial polarization behavior is reversible, indicating no blocking effect of the active sites for the O_2 reduction as the consequence of the deposition of the Cr_2O_3 solid phase at the early stage of the reaction; (c) no preferential deposition of Cr at the LSM cathode and YSZ electrolyte TPB area; and (d) deposition of Cr can occur on the YSZ electrolyte surface far away from the LSM electrode that is not in direct contact with the alloy.

Figure 3.16 shows typical SEM micrographs of the YSZ electrolyte surface in contact with an LSM cathode in the presence of a Fe-Cr ally at 900°C after cathodic polarization at 200 mAcm^{-2} for different periods [185]. The LSM electrode coating

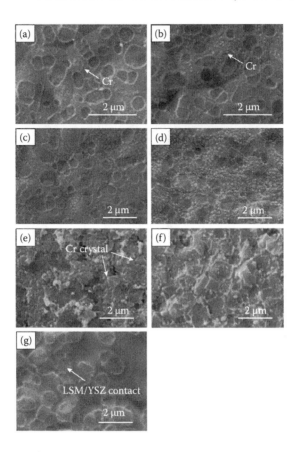

FIGURE 3.16 SEM micrographs of the YSZ electrolyte surface in contact with an LSM cathode in the presence of a Fe-Cr ally at 900°C after cathodic polarization at 200 mAcm^{-2} for (a) 5 min, (b) 15 min, (c) 30 min, (d) 4 h, (e) 20 h, and (f) 50 h. The YSZ surface in contact with LSM electrode after polarization at 200 mAcm^{-2} at 900°C for 4 h in the absence of Fe-Cr alloy is shown in (g). The LSM electrode coating was removed by HCl treatment. (From Jiang, S.P. et al., *J. Mater. Res.*, 20:747–758, 2005. With permission.)

was removed by HCl treatment. The convex rings were contact areas between LSM grains and the YSZ electrolyte surface, formed during the electrode sintering steps [186]. After cathodic current passage for 5 min, there was formation of the isolated fine Cr grains at areas between the LSM particles; see Figure 3.16 (a). As the cathodic current passage time increased to 15 min, the density of the fine grains increased (see Figure 3.16b). However, the fine Cr grains had no preferential deposition at the LSM/YSZ electrolyte interface region. Rather, the deposition of the fine Cr grains was random, simply filling the space between the LSM particles with the increase in the cathodic polarization time (see Figure 3.16 c-f).

The results show that the deposition of Cr species is not dominated by electrochemical reduction of high-valent chromium species in competition with O$_2$ reduction. The driving force for the deposition reaction was suggested to be related to the

Mn^{2+} species generated under cathodic polarization or at high temperatures. The nucleation and grain growth steps can be written as follows [182, 185]:

$$Mn^{2+} + CrO_3 \rightarrow Cr\text{-}Mn\text{-}O \text{ (nuclei)} \tag{3.20}$$

$$Cr\text{-}Mn\text{-}O \text{ (nuclei)} + CrO_3 \rightarrow Cr_2O_3 \tag{3.21}$$

$$Cr\text{-}Mn\text{-}O \text{ (nuclei)} + CrO_3 + Mn^{2+} \rightarrow (Cr, Mn)_3O_4 \tag{3.22}$$

Thus the formation of $(Cr, Mn)_3O_4$-type spinel would be facile if there is sufficient supply of Mn^{2+} ions.

The interaction between LSM coating and metallic interconnect has also been studied. Quadakkers et al. [187] studied transportation and deposition processes of Cr species on LSM, $LaCoO_3$, and Sr doped $LaCoO_3$ (LSC) films coated on alumina-forming and chromia-forming alloys at 950°C. In the case of chromia-forming alloy, Cr species were deposited over the whole width of the coating, forming spinel phases of $MnCr_2O_4$ for an LSM coating and $CoCr_2O_4$ for an LSC coating. Zhen et al. [188] recently studied the interaction between a Fe-Cr alloy interconnect and constituent oxides of LSM coating, La_2O_3, SrO and Mn_2O_3. The Cr deposition reaction between the Fe-Cr alloy and oxides varies significantly with the nature of the oxides. The interaction between the Fe-Cr alloy and La_2O_3 and Mn_2O_3 oxides primarily results in the formation of $LaCrO_3$ and $(Cr, Mn)_3O_4$ while in the case of SrO oxide, Cr_2O_3 is the main product. The interaction between LSM-based coating and metallic interconnect has also been studied by others for the potential application as the protective coating for metallic interconnect [189–193].

The Cr deposition process on MIEC cathodes, such as LSCF, is significantly different from that observed on the LSM cathode. Jiang et al. [194] studied in detail the deposition of Cr species at LSCF cathode. Figure 3.17 shows the SEM micrographs of the LSCF electrode surface under the rib of the Fe-Cr alloy interconnect after cathodic current passage of 200 $mAcm^{-2}$ for different times. Before the polarization, the LSCF electrode was already partially covered with the Cr deposits; see Figure 3.17(a). In this case, the electrode was heated to the testing temperature of 900°C and held for 1 h under open circuit. This indicates that the deposition of Cr species on the LSCF electrode has no direct relationship with the cathodic polarization. In other words, the deposition of Cr species on the LSCF electrode is not limited by the electrochemical reaction of gaseous Cr species and cannot be in direct competition with the O_2 reduction reaction. Large Cr species deposits with distinct crystal facets were formed on the LSCF electrode in the presence of Fe-Cr alloy despite the different cathodic polarization time. Nevertheless, the Cr deposition is increased with the increase of cathodic polarization time (see Figure 3.17b-d). After the polarization for 20 h, the LSCF surface is almost completely covered by the Cr deposits; see Figure 3.17 (e). On the surface of the LSCF electrode facing the channel of the alloy, the deposition of Cr species is much less. Nevertheless, with the increasing of the polarization time, Cr deposits with distinct facets were also formed on the LSCF electrode surface [194].

3.5.3 Interaction with Other SOFC Components

In addition to the YSZ electrolyte and metallic interconnect, SOFC stack also includes component materials such as seals and manifolds. Sealant materials based on glass or

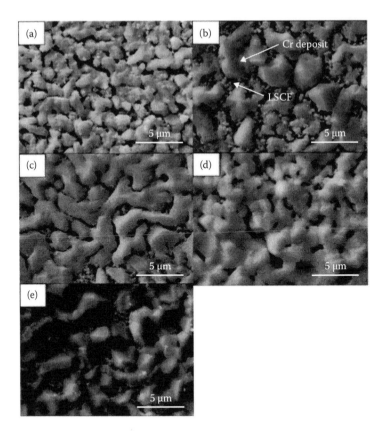

FIGURE 3.17 SEM micrographs of LSCF electrode surface under the rib of Fe-Cr alloy after cathodic current passage at 200 mAcm^{-2} and 900°C for (a) 0 min, (b) 30 min, (c) 60 min, (d) 4 h, and (e) 20 h. (From Jiang, S.P. et al., *J. Electrochem. Soc.*, 153:A127–A134, 2006. With permission.)

glass ceramic have been investigated to provide reliable seals between different SOFC components in the stack. In addition to the gas tightness, the essential requirements for a sealant material are the chemical stability in oxidizing and reducing atmospheres, low-chemical activity to the cell components with which they come in contact, high-insulating properties, matching thermal expansion coefficient, good wetting behavior and stress relaxation ability during operation. It has been shown that loss of consti-tutes such as B$_2$O$_3$ in aluminosilicate glass [195] and commercial AF 45 glass [196] is detrimental to the durability of the material. However, the information on the interac-tion between LSM cathodes and the seals and manifold materials is relatively rare. Jiang et al. [197] studied the effect of the gaseous species of the borosilicate glass system, A$_2$O-Al$_2$O$_3$-B$_2$O$_3$-SiO$_2$ (A = Na, K) on the microstructure and performance of LSM electrodes. There was a clear grain growth of the LSM electrodes exposed to the glass containing Na or K. For example, the particle size of the LSM electrode exposed to the glass containing Na was 60% larger as compared to that in the absence of glass. This shows that in addition to the essential requirements in the softening, crystallization, and chemical stability of the sealant materials in SOFC, the constitutes in sealant materials should have low volatility to minimize the adverse effect on the performance and stability of SOFC stack components.

3.6 PERFORMANCE STABILITY AND DEGRADATION

In addition to the chemical interactions described above, microstructural changes in the electrode can lead to performance degradation after long operation time. For example sintering of the porous structure can degrade electrode performance. In the case of LSM cathodes, the sintering ability is found to be related to the strontium dopant level and stoichiometric composition of $(La, Sr)_xMnO_3$. In addition, LSM with A-site deficient compositions ($x < 1$) sinters more readily than their B-site deficient counterparts ($x > 1$) [198, 199].

Jørgensen et al. [200] investigated the performance and microstructure stability of LSM/YSZ composite cathodes under a cathodic current of 300 mAcm^{-2} at 1000°C for 2000 h. The overpotential losses were doubled and there was an increase of the porosity at the composite electrode and the YSZ electrolyte interface. On the other hand, they did not observe such structural changes for the composite cathodes sintered under open circuit conditions. Wang and Jiang [201] studied the effect of polarization on the morphology of LSM materials and found that the morphology of LSM changes significantly under cathodic and anodic polarization. They suggested that the change in the morphology of LSM particles is most likely related to the lattice expansion or shrinkage under either cathodic or anodic polarization.

Therefore the sintering behavior of LSM electrodes under cathodic polarization conditions is important for the fundamental understanding of the structural and performance stability of LSM-based cathodes under SOFC operation conditions. Jiang and Wang [202] carried out a comparative study on the sintering behavior of $(La_{0.8}Sr_{0.2})_xMnO_3$ ($x = 1.0, 0.9, 0.8$) cathodes under polarization and open-circuit conditions at 1000°C for 1600 h. Figure 3.18 shows scanning electron microscope (SEM) micrographs of the surface of LSM electrodes sintered at 1000°C in air with and without a cathodic current passage of 500 mA cm^{-2} for 1600 h [202]. The particles of the LSM electrodes sintered under a constant cathodic current load were significantly smaller than those sintered under no current load (i.e., at open circuit). For example, for the $(La_{0.8}Sr_{0.2})_{0.9}MnO_3$ electrode, the particle size of the LSM grains was 0.88 ± 0.30 μm after sintering at 1000°C under a current load of 500 mAcm^{-2} for 1600 h (see Figure 3.18b). In the case of the same LSM coating sintered at open circuit under the same conditions, the LSM particles grew to 1.17 ± 0.25 μm, which is 33% larger than that sintered under current load (see Figure 3.18e). The same trend was observed for the LSM electrodes with different compositions (see Figure 3.18 a,d and c,f) . The results indicate that the sintering behavior of LSM cathodes depends on the A-site stoichiometry and on the polarization. The grain growth and sinterability of LSM electrode materials is reduced under cathodic polarization conditions as compared to those sintered under open-circuit conditions.

The interface between the cathode and electrolyte also undergoes significant changes under SOFC operating conditions. Figure 3.19 is the SEM micrograph of a $La_{0.72}Sr_{0.18}MnO_3$ cathode after polarization testing. The test was carried out in a

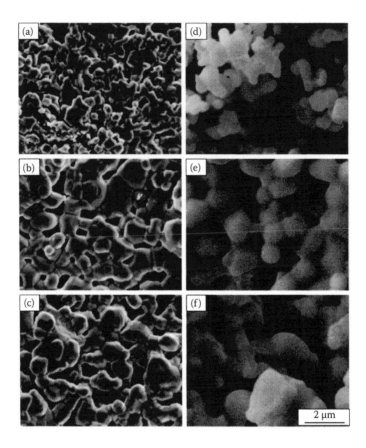

FIGURE 3.18 SEM micrographs of the electrode surface of $(La_{0.8}Sr_{0.2})_xMnO_3$ with (a) x = 1.0, (b) x = 0.9, and (c) x = 0.8 sintered under a current load of 500 mA cm^{-2} and with (d) x = 1.0, (e) x = 0.9, and (f) x = 0.8 sintered under no current load at 1000°C in air for 1600 h. (From Jiang, S.P. and Wang, W., *Solid State Ionics*, 176:1185–1191, 2005. With permission.)

50 × 50 mm cell with a Ni (50 vol%)/YSZ (50 vol%) anode and an LSM cathode under current density of 250 mAcm^{-2} in 96% H_2/4% H_2O and air at 1000°C for 2543 h. Two features can be seen from the cross-section of the electrode after the long-term polarization test. Isolated dark particles with high Mn content appeared in the electrode. This indicates the formation of the Mn-rich phase, probably Mn_3O_4, inside the LSM electrode after the cathodic polarization at 500 mAcm^{-2} for 2543 h. The formation of a Mn-rich region in the LSM electrode is most likely due to the significant diffusion and migration of manganese species under the cathodic polarization conditions. The second feature is the appearance of voids at the LSM electrode and YSZ electrolyte. However, the overall performance degradation of the cell was 70 mV over 2500 h and the contribution from the LSM cathode side should be small [202]. The formation of voids at the LSM/YSZ interface region was also observed by Kuznecov et al. [203].

FIGURE 3.19 SEM micrograph of the cross-section of an LSM electrode after the fuel cell testing on a 50 × 50 mm cell with Ni (50 vol%)/YSZ (50 vol%) anode under a current density of 250 mAcm^{-2} in 96% H$_2$/4% H$_2$O and air at 1000°C for 2543 h. (From Jiang, S.P. and Wang, W., *Solid State Ionics*, 176:1185–1191, 2005. With permission.)

Sholklapper et al. [204] studied the structure and performance stability of a nanoparticle-infiltrated LSM in porous scandia-stabilized zirconia (SSZ) electrode at 650°C. The infiltrated LSM particles were ~100 nm in diameter and some orientation alignment of the LSM nanoparticles within SSZ was observed after the cathodic polarization at 150 mAcm^{-2} for 500 h. However, there is no voltage degradation under the conditions studied.

Despite the importance of the stability or shelf-life of the electrode powder, ink, and coating in the stability and reproducibility of the cell performance, the shelf-life studies of electrode materials are rare. Bergsmark et al. [205] studied the stability of LSM powder with a small lanthanum excess (A-site over-stoichiometry) and found that the pellets prepared from the powder disintegrated after exposure to air, due to the formation of La(OH)$_3$. Stability of LSM powder decreased with increasing (La + Sr)/Mn ratio. Jiang et al. [206] investigated in detail the shelf-life of Sr doped LaMnO$_3$ (LSM) materials of varying stoichiometric compositions under storage conditions of low and high humidity at ambient temperature for ~350 days. The adhesion, chemical stability, and stability of the electrode performance of LSM materials are critically dependent on the A-site stoichiometry (i.e., (La + Sr)/Mn ratio) of the LSM materials and much less dependent on storage conditions (humidity level and storage time) over the period investigated. The results demonstrate that LSM materials with A-site substoichiometry ((La + Sr)/Mn ≤ ~0.9) are effective in achieving good adherence, high chemical stability, and high stability of electrode performance.

Mai et al. [207] investigated the performance stability of anode-supported cells with La$_{0.58}$Sr$_{0.4}$Co$_{0.2}$Fe$_{0.8}$O$_{3-\delta}$ cathodes over a period of 1000 h under different operation conditions. Figure 3.20 shows the cell voltage drop at 0.3 Acm^{-2} calculated from the *I-V* curves [207]. The results show that the operating temperature has the strongest effect on degradation and cells operated at 800°C show a degradation rate which

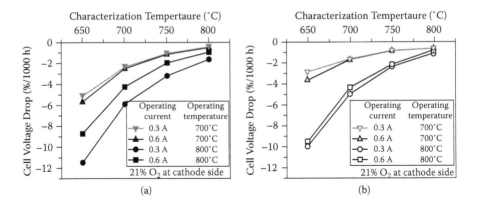

FIGURE 3.20 Cell voltage drop calculated from *I-V* curves at 0.3 Acm^{-2} before and after tested for 1000 h. Operating temperature: temperature during the long-term test; characterization temperature: temperature at which the *I-V* curves before and after the long-term tests were measured. (From Mai, A. et al., *Solid State Ionics*, 177:1965–1968, 2006. With permission.)

is about twice as high as for the cells operated at 700°C. The degradation rate is slightly higher for cells operated with air than that operated with 5% oxygen on the cathode side. They observed no clear influence of the current density on the degradation rate of the cells with LSCF cathodes.

Simner et al. [208] also examined the stability of anode-supported YSZ cells with $(La_{0.60}Sr_{0.4})_{0.98}Co_{0.2}Fe_{0.8}O_{3-\delta}$ cathodes with a SDC interlayer and reported a performance degradation rate of 30% for the cell tested at 750°C and 0.7 V for 500 h. The postmortem examination of the cell indicates no significant changes in microstructure of the cathode. However, the observed enrichment of Sr at the LSCF/SDC interface and LSCF/current collector interface was suggested to be responsible for the high degradation. It appears that LSCF electrodes show a much higher degradation rate than LSM materials.

3.7 SUMMARY AND CONCLUSIONS

Despite the significant advances in materials for SOFCs, lanthanum strontium manganite (LSM) and lanthanum strontium cobalt ferrite (LSCF) based perovskites are still the materials of choice for cathodes of SOFCs for YSZ and doped ceria electrolytes, respectively. LSM materials show the high thermal stability and compatibility with YSZ electrolyte at SOFC operating conditions and most important among the cathode materials, LSM has an excellent microstructural stability and its long-term performance stability has been well established. The electrocatalytic activity of LSM-based materials can be substantially enhanced by the nano-structured approach, e.g., by infiltration and impregnation of catalytic active and MIEC nano-particles to the stable LSM porous structure. LSCF materials show acceptable performance for SOFC operating at intermediate temperatures; however, the long-term stability of the LSCF-based materials with doped ceria is largely unknown, and the mechanism of the performance degradation of the LSCF-based cathodes is poorly understood. In the case of ITSOFCs using a Fe-Cr alloy metallic interconnect, the Cr poisoning and deposition are serious concerns for the both LSM- and LSCF-based perovskite cathodes, though the use of composite cathodes can significantly reduce the effect of the Cr poisoning [209].

To develop an alternative MIEC cathode not only the *ex situ* properties, e.g., σ, TEC, D^*, and k, but also the electrocatalytic activity, structural and chemical stability, and Cr-tolerance must be considered. Beyond testing in small SOFC button cells, the viability of new cathode materials must ultimately be proven in large-scale stack cells under practical current and temperature gradients. The issues involved in the development of cathode materials for large-scale stacks are significantly more complex than those in the small button cells briefly reviewed in this chapter. However, this does provide serious challenges as well as opportunities for materials scientists and engineers in the development of commercially viable ITSOFCs.

REFERENCES

1. Adler SB. *Chem Rev* 2004; 104: 4791–4843.
2. Fergus JW. *J Power Sources* 2005; 147: 46–57.
3. De Jonghe LC, Jacobson CP, Visco SJ. *Annu Rev Mater Res* 2003; 33: 169–182.
4. Ivers-Tiffée E, Weber A, Herbstritt D. *J Eur Ceram Soc* 2001; 21: 1805–1811.
5. Koyama M, Wen CJ, Yamada K. *J Electrochem Soc* 2000; 147: 87–91.
6. Tofield BC, Scott WR. *J. Solid State Chem* 1974; 10: 183.
7. Li Z, Behruzi M, Fuerst L, Stöver D. In: Singhal SC, Iwahara H. SOFC-III, The Electrochem Soc, Pennington, NJ, 1993; 171.
8. Zhang ZT, Lin OY, Tang ZL. In: Dokiya M, Tagawa H, Singhal SC. Editor, SOFC-IV, The Electrochem Soc, Pennington, NJ, 1995; 502.
9. Zheng F, Pederson LR. *J Electrochem Soc* 1999; 146: 2810–2816.
10. Jiang SP, Love JG, Zhang JP, Hoang M, Ramprakash Y, Hughes AE, Badwal SPS. *Solid State Ionics* 1999; 121: 1–10.
11. Mizusaki J, Tagawa H, Naraya K, Sasamoto T. *Solid State Ionics* 1991; 49: 111–118.
12. Mizusaki J. *Solid State Ionics* 1992; 52: 79–91.
13. Mizusaki J, Mori N, Takai H, Yonemura Y, Minamiue H, Tagawa H, et al. *Solid State Ionics* 2000; 129: 163–177.
14. Mizusaki J, Yonemura Y, Kamata H, Ohyama K, Mori N, Takai H, et al. *Solid State Ionics* 2000; 132: 167–180.
15. Miyoshi S, Hong J-O, Yashiro K, Kaimai A, Nigara Y, Kawamura K, Kawada T, Mizusaki J. *Solid State Ionics* 2002; 154: 257–263.
16. van Roosmalen JAM, Cordfunke EHP, Helmholdt RB, Zandbergen HW. *J. Solid Sate Chem* 1994; 110: 100–105.
17. Mitchell JF, Argyriou DN, Potter CD, Hinks DG, Jorgense JD, Bader SD. *Phys Rev B* 1996; 54: 6172–6183.
18. Nowotny J, Rekas M. *J Am Ceram Soc* 1998; 81: 67–80.
19. van Roosmalen JAM, Cordfunke EHP. *J Solid State Chem* 1994; 110: 106–108.
20. van Roosmalen JAM, Cordfunke EHP. *J Solid State Chem* 1994; 110: 109–112.

21. van Roosmalen JAM, Cordfunke EHP. *J Solid State Chem* 1994; 110: 113–117.
22. Nakamura K, Xu M, Klaeser M, Linker G. *J Solid State Chem* 2001; 156: 143–153.
23. Nakamura K, Ogawa K. *J Solid State Chem* 2002; 163: 65–76.
24. Yokokawa H, Horita T. in: Singhal SC, Kendall K, editors. High Temperature Solid Oxide Fuel Cells – Fundamentals, Design, and Applications, Elsevier 2003; 119–147.
25. Alonso JA, Martínez-Lope MJ, Casais MT. *Euro J Solid State Inorg Chem* 1996; 33: 331–341.
26. Alonso JA, Martínez-Lope MJ, Casais MT, MacManus-Driscoll JL, de Silva PSIPN, Cohen LF, Fernández-Díaz MT. *J Mater Chem* 1997; 7: 2139–2144.
27. Miyoshi S, Hong J-O, Yashiro K, Kaimai A, Nigara Y, Kawamura K, Kawada T, Mizusaki J. *Solid State Ionics* 2003; 161: 209–217.
28. Yasumoto K, Mori N, Mizusaki J, Tagawa H, Dokiya M. *J Electrochem Soc* 2001; 148: A105–A111.
29. Yasumoto K, Shiono M, Tagawa H, Dokiya M, Hirano K, Mizusaki J. *J Electrochem Soc* 2002; 149: A531–A536.
30. Lee HY, Cho WS, Oh SM, Wiemhöfer H-D, Göpel W. *J Electrochem Soc* 1995; 142: 2659–2664.
31. Hammouche A, Siebert E, Hammou A, Kleitz M. *J Electrochem Soc* 1991; 138: 1212–1216.
32. Jiang Y, Wang S, Zhang Y, Yan J, Li W. *J Electrochem Soc* 1998; 145: 373–378.
33. Jiang Y, Wang S, Zhang Y, Yan J, Li W. *Solid State Ionics* 1998; 110: 111–119.
34. Chen XJ, Chan SH, Khor KA. *Electrochem Solid-State Lett* 2004; 7: A144–A147.
35. Wang W, Jiang SP. *Solid State Ionics* 2006; 177: 1361–1369.
36. Jiang SP, Love JG. *Solid State Ionics* 2001; 138: 183–190.
37. Egdell RG, Harrison MR, Hill MD, Porte L, Wall G. *J Phys C: Soli State Phys* 1984; 17: 2889–2900.
38. Kemp JP, Beal DJ, Cox PA. *J Solid State Chem* 1990; 86: 50–58.
39. Howlett JF, Flavell WR, Thomas AG, Hollingworth J, Warren S, Hashim Z, et al. Faraday Discuss 1996; 105: 337–354.
40. Jiang SP. *J Solid State Electrochem* 2007; 11: 93–102.
41. Kuo JH, Anderson HU, Sparlin DM. *J Solid State Chem* 1990; 87: 55–63.
42. Yokokawa H, Sakai N, Kawada T, Dokiya M. *Solid State Ionics* 1990; 40–41: 398–401.
43. Kamata H, Yonemura Y, Mizusaki J, Tagawa H, Naraya K, Sasamoto T. *J Phys Chem Solids* 1995; 56: 943–950.
44. Yang C-CT, Wei W-CJ, Roosen A. *Mater Chem Phys* 2003; 81: 134–142.
45. Yamamoto O, Takeda Y, Imanishi N, Sakaki Y. In: Singhal SC, Iwahara H, editors. SOFC-III. The Electrochem. Soc., Pennington, NJ: 1993: 205.
46. Mori M, Yamamoto T, Itoh H, Inaba H, Tagawa H. In: Singhal SC, Dokiya M, editors. SOFC-VI. Electrochem. Soc., Pennington, NJ, 1999: 347.
47. Aruna ST, Muthuraman M, Patil KC. *J Mater Chem* 1997; 7: 2499–2503.
48. Nasrallah MM, Carter JD, Anderson HU, Koc R. In: Grosz F, Zegers P, Singhal SC, Yamamoto O, editor. SOFC-II. Luxemburg, Commission of the European Communities, 1991; 637.
49. Carter S, Seluk A, Chater RJ, Kaida J, Kilner JA, Steele BCH. *Solid State Ionics* 1992; 53–56: 597–605.
50. De Souza RA, Kilner JA, Walker JF. *Mater Lett* 2000; 43:43–52.
51. Horita T, Tsunoda T, Yamaji K, Sakai N, Kato T, Yokokawa H. *Solid State Ionics* 2002; 152–153: 439–446.
52. Yasuda I, Ogasawa K, Hishinuma M, Kawada T, Dokiya M. *Solid State Ionics* 1996; 86–88: 1197–1201.
53. Ji Y, Kilner JA, Carolan MF. *Solid State Ionics* 2005; 176: 937–943.

54. Jiang SP. *Solid State Ionics* 2002; 146: 1–22.
55. Fleig J. *Annu Rev Mater Res* 2003; 33: 361–382.
56. McIntosh S, Adler SB, Vohs JM, Gorte RJ. *Electrochem Solid-State Lett* 2004; 7: A111–A114.
57. Kim J-D, Kim G-D, Moon J-W, Park Y-I, Lee W-H, Kobayashi K, Nagai M, Kim C-E. *Solid State Ionics* 2001; 143: 379–389.
58. Leng YJ, Chan SH, Khor KA, Jiang SP. *J Appl Electrochem* 2004; 34: 409–415.
59. Youngblood GE, Rupaal AS, Pederson LR, Bates JL. In: Singhal SC, Iwahara H, editors. SOFC-III. The Electrochemical Society, Pennington, NJ 1993; 585.
60. Gharbage B, Pagnier T, Hammou A. *J Electrochem Soc* 1994; 141: 2118–2121.
61. van Heuveln FH, Bouwmeester HJM. *J Electrochem Soc* 1997; 144: 134–140.
62. Jiang SP, Zhang JP, Foger K. *J Electrochem Soc* 2000; 147: 3195–3205.
63. Jiang SP, Love JG, Ramprakash Y. *J Power Sources* 2002; 110: 201–208.
64. Ishihara T, Kudo T, Matsuda H, Takita Y. *J Am Ceram Soc* 1994; 77: 1682–1684.
65. Ishihara T, Kudo T, Matsuda H, Takita Y. *J Electrochem Soc* 1995; 142: 1519–1524.
66. Huang X, Liu J, Lu Z, Liu W, Pei L, He T, et al. *Solid State Ionics* 2000; 130: 195–201.
67. Sasaki Y, Takeda Y, Kato A, Imanishi N, Yamamoto O, Hattori M, et al. *Solid State Ionics* 1999; 118: 187–194.
68. Murray EP, Tsai T, Barnett SA. *Solid State Ionics* 1998; 110: 235–243.
69. Murray EP, Barnett SA. *Solid State Ionics* 2001; 143: 265–273.
70. Xia C, Zhang Y, Liu M. *Electrochem Solid-State Lett* 2003; 6: A290–A292.
71. Xia CR, Rauch W, Wellborn W, Liu M. *Electrochem Solid-State Lett* 2002; 5: A217–A220.
72. Hart NT, Brandon NP, Day MJ, Shemilt JE. *J Mater Sci* 2001; 36: 1077–1085.
73. Liu Y, Zha S, Liu M. *Adv Mater* 2004; 16: 256.
74. Liu Y, Compson C, Liu M. *J. Power Sources* 2004; 138: 194–198.
75. Jiang SP. *Mater Sci Eng* A 2006; 418: 199–210.
76. Jiang SP, Wang W. *Solid State Ionics* 2005; 176: 1351–1357.
77. Jiang SP, Wang W. *J Electrochem Soc* 2005; 152: A1398–A1408.
78. Tsai T, Barnett SA. *Solid State Ionics* 1997; 98: 191–196.
79. Tai LW, Nasrallah MM, Anderson HU, Sparlin DM, Sehlin SR. *Solid State Ionics* 1995; 76: 259–271.
80. Tai LW, Nasrallah MM, Anderson HU, Sparlin DM, Sehlin SR. *Solid State Ionics* 1995; 76: 273–283.
81. Mizusaki J, Yoshihiro M, Yamauchi S, Fueki K. *J Solid State Chem* 1987; 67:1–8.
82. Lankhorst MHR, Bouwmeester HJM. *J Electrochem Soc* 1997; 144: 1268–1273.
83. Petrov AN, Kononchuk OF, Andreev AV, Cherepanov VA, Kofstad P. *Solid State Ionics* 1995; 80: 189–199.
84. Petrov AN, Cherepanov VA, Zuev AYu. *J Solid State Electrochem* 2006; 10: 517–537.
85. Petric A, Huang P, Tietz F. *Solid State Ionics* 2000; 135: 719–725.
86. Kostogloudis GCh, Ftikos Ch. *Solid State Ionics* 1999; 126: 143–151.
87. Kharton VV, Figueiredo FM, Navarro L, Naumovich EN, Kovalevsky AV, Yaremchenko AA, et al. *J Mater Sci* 2001; 36: 1105–1117.
88. Aruna ST, Muthuraman M, Patil KC. *Mater Res Bull* 2000; 35: 289–296.
89. Huang K, Lee HY, Goodenough JB. *J Electrochem Soc* 1998; 145: 3220–3227.
90. Hung M-H, Rao MVM, Tsai D-S. *Mater Chem Phys* 2007; 101: 297–302.
91. Ishigaki T, Yamauchi S, Mizusaki J, Fueki K. *J Solid State Chem* 1984; 55: 50–53.
92. Ishigaki T, Yamauchi S, Mizusaki J, Fueki K. *J Solid State Chem* 1984; 54: 100–107.
93. Ishigaki T, Yamauchi S, Kishio K, Mizusaki J, Fueki K. *J Solid State Chem* 1988; 73: 179–187.

94. Teraoka Y, Nobunaga T, Okamoto K, Miura N, Yamazoe N. *Solid State Ionics* 1991; 48: 207–212.
95. Kharton VV, Viskup AP, Naumovich EN, Lapchuk NM. *Solid State Ionics* 1997; 104: 67–78.
96. Kim CS, Yoo HI. *J Electrochem Soc* 1996; 143: 2863–2870.
97. Horita T, Yamaji K, Sakai N, Yokokawa H, Weber A, Ivers-Tiffee E. *J Electrochem Soc* 2001; 148: A456–A462.
98. Huang K, Tichy RS, Goodenough JB. *J Am Ceram Soc* 1998; 81: 2576–2580.
99. Maric R, Ohara S, Fukui T, Yoshida H, Nishimura M, Inagaki T, Miura K. *J Electrochem Soc* 1999; 146: 2006–2010.
100. Uchida H, Arisaka S, Watanabe M. *Electrochem Solid-State Lett* 1999; 2: 428–430.
101. Gödickemeier M, Sasaki K, Gauckler LJ. *J Electrochem Soc* 1997; 144: 1635–1646.
102. Esquirol A, Brandon NP, Kilner JA, Mogensen M. *J Electrochem Soc* 2004; 151: A1847–A1855.
103. Tu HY, Takeda Y, Imanishi N, Yamamoto O. *Solid State Ionics* 1999; 117: 277–281.
104. Tu HY, Takeda Y, Imanishi N, Yamamoto O. *Solid State Ionics* 1997; 100: 283–288.
105. Shao Z, Haile SM. *Nature* 2004; 431: 170–173.
106. Teraoka Y, Zhang HM, Furukawa S, Yamazoe N. *Chem Lett* 1985; 11: 1743.
107. Wei B, Lu Z, Li S, Liu Y, Liu K, Su W. *Electrochem Solid-State Lett* 2005; 8: A428–A431.
108. Hwang HJ, Moon J-W, Lee S, Lee EA. *J Power Sources* 2005; 145: 243–248.
109. Liu M, Wu Z. *Solid State Ionics* 1998; 107: 105–110.
110. Steele BCH, Bae J-M. *Solid State Ionics* 1998; 106: 255–261.
111. Uchida H, Arisaka S, Watanabe M. *Solid State Ionics* 2000; 135: 347–351.
112. Wang S, Zhong H. *J Power Sources* 2007; 165: 58–64.
113. Sahibzada M, Benson SJ, Rudkin RA, Kilner JA. *Solid State Ionics* 1998; 113–115: 285–290.
114. Takeda Y, Ueno H, Imanishi N, Yamamoto O, Sammes N, Phillipps MB. *Solid State Ionics* 1996; 86–88: 1187–1190.
115. Phillipps MB, Sammes NM, Yamamoto O. *Solid State Ionics* 1999; 123: 131–138.
116. Kostogloudis GCh, Vasilakos N, Ftikos Ch. *J Euro Ceram Soc* 1997; 17: 1513–1521.
117. Kostogloudis GCh, Vasilakos N, Ftikos Ch. *Solid State Ionics* 1998; 106: 207–218.
118. Kostogloudis GCh, Fertis P, Ftikos Ch. *J Euro Ceram Soc* 1998; 18: 2209–2215.
119. Rajeev KP, Raychaudhuri AK. *Phys Rev* 1992; 46: 1309–1320.
120. Drennan J, Travares CP, Steele BCH. *Mater Res Bull* 1982; 17: 621.
121. Hrovat M, Katsarakis N, Reichmann K, Bernik S, Kuscer D, Holc J. *Solid State Ionics* 1996; 83: 99–105.
122. Chiba R, Yoshimura F, Sakurai Y. *Solid State Ionics* 1999; 124: 281–288.
123. Zhen YD, Tok AIY, Jiang SP, Boey FYC. *J Power Sources* 2007; 170: 61–66.
124. Komatsu T, Arai H, Chiba R, Nozawa K, Arakawa M, Sato K. *Electrochem Solid-State Lett* 2006; 9: A9–A12.
125. Tao S, Irvine JTS. *Nature Materials* 2003; 2: 320–323.
126. Jiang SP, Chen XJ, Chan SH, Kwok JT. *J Electrochem Soc* 2006; 153: A850–A856.
127. Bastidas DM, Tao S, Irvine JTS. *J Mater Chem* 2006; 16: 1603–1605.
128. Raj ES, Kilner JA, Irvine JTS. *Solid State Ionics* 2006; 177: 1747–1752.
129. Yamahara K, Jacobson CP, Visco SJ, De Jonghe LC. *Solid State Ionics* 2005; 176: 451–456.
130. Jiang SP, Zhang L, Zhang Y. *J Mater Chem* 2007; 17: 2627–2635.

131. Leng YJ, Chan SH, Khor KA, Jiang SP. *Intern J Hydrogen Energy* 2004; 29: 1025–1033.
132. Reitz TL, Xiao H. *J Power Sources* 2006; 161: 437–443.
133. Skinner SJ, Kilner JA. *Solid State Ionics* 2000; 135: 719–712.
134. Vashook VV, Yushkevich II, Kokhanovsky LV, Makhnach LV, Tolochko SP, Kononyuk IF, *et al. Solid State Ionic* 1999; 119: 23–30.
135. Skinner SJ. *Solid State Sci* 2003; 5: 419–426.
136. Jennings AJ, Skinner SJ. *Solid State Ionics* 2002; 152–153: 663–667.
137. Skinner SJ, Munnings C. *Mater Lett* 2002; 57: 594–597.
138. Laberty C, Zhao F, Swider-Lyons KE, Virkar AV. *Electrochem Solid-State Lett* 2007; 10: B170–B174.
139. Yokokawa H, Sakai N, Kawada T, Dokiya M. *J Electrochem Soc* 1991; 138: 2719–2727.
140. Yokokawa H, Horita T, Sakai N, Dokiya M, Kawada T. *Solid State Ionics* 1996; 86–88: 1161–1165.
141. Yokokawa H, Sakai N, Kawada T, Dokiya M. In: Badwal SPS, Bannister MJ, Hannink RHJ. Editors. Science and Technology of Zirconia V, Technomic Publishing Company, Inc., Lancaster, Pennsylvania, 1993: 752.
142. Yokokawa H. *Annu Rev Mater Res* 2003; 33: 581–610.
143. Tsai T, Barnett SA. *Solid State Ionics* 1997; 93: 207–217.
144. Mitterdorfer A, Gauckler LJ. *Solid State Ionics* 1998; 111: 185–218.
145. Stochniol G, Syskakis E, Naoumidis A. *J Am Ceram Soc* 1995; 78: 929–932.
146. Lau SK, Singhal SC. In: Corrosion 85, The National Association of Corrosion Engineers, Houston, TX, 1985; 345/1–9.
147. Brugnoni C, Ducati U, Scagliotti M. *Solid State Ionics* 1995; 76: 177–182.
148. Taimatsu H, Wada K, Kaneko H, Yamamura H. *J Am Ceram Soc* 1992; 75: 401–405.
149. Kawada T, Sakai N, Yokokawa H, Dokiya M, Anzai I. *Solid State Ionics* 1992; 50: 189–196.
150. Waller D, Sirman JD, Kilner JA. In: Stimming U, Singhal SC, Tagawa H, Lehnert W, editors. SOFC-V. Electrochemical Soc. Inc., Pennington, NJ, 1997: 1140.
151. Yamamoto O, Shen GQ, Takeda Y, Imanishi N, Sakaki Y. *Electrochem Soc Proc* 1991; 91–6: 158.
152. van Roosmalen JAM, Cordfunke EHP. *Solid State Ionics* 1992; 52: 303–312.
153. Jiang SP, Zhang J-P, Foger K. *J Euro Ceram Soc* 2003; 23: 1865–1873.
154. Zhang J-P, Jiang SP, Love JG, Foger K, Badwal SPS. *J Mater Chem* 1998; 8: 2787–2797.
155. Khandkar A, Elangovan S, Liu M. *Solid State Ionics* 1992; 52: 57–68.
156. Mitsuyasu H, Eguchi K, Arai H. *Solid State Ionics* 1997; 100: 11–15.
157. Labrincha JA, Frade JR, Marques FMB. *J Mater Sci* 1993; 28: 3809–3815.
158. Brugnoni C, Scagliotti M. *Solid State Ionics* 1994; 73: 265.
159. Wagner C. *Z Physik Chem B* 1933; 21: 25.
160. Poulsen FW, van der Puil N. *Solid State Ionics* 1992; 53–56: 777–783.
161. Ciacchi FT, Crane KM, Badwal SPS. *Solid State Ionics* 1994; 73: 49–61.
162. Mori M, Abe T, Itoh H, Yamamoto O, Shen GO, Takeda Y, Imanishi N. *Solid State Ionics* 1999; 123: 113–119.
163. Brugnoni C, Ducati U, Chemelli C, Scagliotti M, Chiodelli G. *Solid State Ionics* 1995; 76: 183–188.
164. Kenjo T, Nishiya M. *Solid State Ionics* 1992; 57: 295–302.
165. Yamamoto O, Takeda Y, Kanno R, Kojima T. In Singhal SC. editor, SOFC-I, PV89-11, The Electrochem Soc, Pennington, NJ, 1989; 242.

166. Clausen C, Bagger C, Bilde-Sørensen JB, Horesewell A. *Solid State Ionics* 1994; 70/71: 59–64.
167. Chen CC, Nasrallah MM, Anderson HU. In: Singhal SC, Iwahara H, editors. SOFC-III. The Electrochemical Society, Pennington, NJ, 1993; 252.
168. Chen CC, Nasrallah MM, Anderson HU. In: Singhal SC, Iwahara H, editors. SOFC-III. The Electrochemical Society, Pennington, NJ 1993; 598–612.
169. Orui H, Watanabe K, Chiba R, Arakawa M. *J Electrochem Soc* 2004; 151: A1412–A1417.
170. Huang K, Feng M, Goodenough JB, Schmerling M. *J Electrochem Soc* 1996; 143: 3630–3636.
171. Naoumidis A, Ahmad-Khanlou A, Samardzija Z, Kolar D. Fresenius *J Anal Chem* 1999; 366: 277–281.
172. Ullmann H, Trofimenko N, Naoumidis A, Stöver D. *J Euro Ceram Soc* 1999; 19: 791–796.
173. Horita T, Yamaji K, Sakai N, Yokokawa H, Weber A, Ivers-Tiffee E. *Solid State Ionics* 2000; 138: 143–152.
174. Badwal SPS, Deller R, Foger K, Ramprakash Y, Zhang JP. *Solid State Ionics* 1997; 99: 297–310.
175. Taniguchi S, Kadowaki M, Kawamura H, Yasuo T, Akiyama Y, Miyake Y, Saitoh T. *J Power Sources* 1995; 55: 73–79.
176. Paulson SC, Birss VI. *J Electrochem Soc* 2004; 151: A1961–A1968.
177. Matsuzaki Y, Yasuda I. *Solid State Ionics* 2000; 132: 271–278.
178. Matsuzaki Y, Yasuda I. *J Electrochem Soc* 2001; 148: A126–A131.
179. Hilpert K, Das D, Miller M, Peck DH, Weiβ R. *J Electrochem Soc* 1996; 143: 3642–3647.
180. Konysheva E, Penkalla H, Wessel E, Mertens J, Seeling U, Singheiser L, Hilpert K. *J Electrochem Soc* 2006; 153: A765–A773.
181. Jiang SP, Zhang JP, Apateanu L, Foger K. *Electrochem Comm* 1999; 1: 394–397.
182. Jiang SP, Zhang JP, Apateanu L, Foger K. *J Electrochem Soc* 2000; 147: 4013–4022.
183. Jiang SP, Zhang JP, Foger K. *J Electrochem Soc* 2001; 148: C447–C455.
184. Jiang SP, Zhang JP, Zheng XG. *J Euro Ceram Soc* 2002; 22: 361–373.
185. Jiang SP, Zhang S, Zhen YD. *J Mater Res* 2005; 20: 747–758.
186. Jiang SP, Wang W. *Electrochem Solid-State Lett* 2005; 8: A115–A118.
187. Quadakkers WJ, Greiner H, Hansel M, Pattanaik A, Khanna AS, Mallener W. *Solid State Ionics* 1996; 91: 55–67.
188. Zhen YD, Jiang SP, Zhang S, Tan V. *J Euro Ceram Soc* 2006; 26: 3253–3264.
189. Kim J-H, Song R-H, Hyun S-H. *Solid State Ionics* 2004; 174: 185–191.
190. Fujita K, Ogasawara K, Matsuzaki Y, Sakurai T. *J Power Sources* 2004; 131: 261–269.
191. Huang K, Hou PY, Goodenough JB. *Solid State Ionics* 2000; 129: 237–250.
192. Chen X, Hou PY, Jacobson CP, Visco SJ, De Jonghe LC. *Solid State Ionics* 2005; 176: 425–433.
193. Li JQ, Xiao P. *J Eur Ceram Soc* 2001; 21: 659–668.
194. Jiang SP, Zhang S, Zhen YD. *J Electrochem Soc* 2006; 153: A127–A134.
195. Lahl N, Singheiser L, Hilpert K. In: Singhal SC, Dokiya M, editors. SOFC-VI, The Electrochem Soc, Pennington, NJ, 1999; 1057.
196. Günther C, Hofer G, Kleinlein W. In Stimming U, Singhal SC, Tagawa H, Lehnert W. Editors, SOFC-V, The Electrochem Soc, Pennington, NJ, 1997; 746.
197. Jiang SP, Christiansen L, Hughan B, Foger K. *J Mater Sci Lett* 2001; 20: 695–697.
198. Meixner DL, Cutler RA. *Solid State Ionics* 2002; 146: 273–284.
199. Van Roosmalen JAM, Cordfunke EHP, Huijsmans JPP, *Solid State Ionics* 1993; 66: 285–293.

200. Jørgensen MJ, Holtappels P, Appel CC. *J Appl Electrochem* 2000; 30: 411–418.
201. Wang W, Jiang SP. *J Solid State Electrochem* 2004; 8: 914–922.
202. Jiang SP, Wang W. *Solid State Ionics* 2005; 176: 1185–1191.
203. Kuznecov M, Otschik P, Obenaus P, Eichler K, Schaffrath W. *Solid State Ionics* 2003; 157: 371–378.
204. Sholklapper TZ, Radmilovic V, Jacobson CP, Visco AJ, De Jonghe LC. *Electrochem Solid-State Lett* 2007; 10: B74–B76.
205. Bergsmark E, Furuseth S, Dyrlie O, Norby T, Kofstad P. In: Grosz F, Zegers P, Singhal SC, Yamamoto O. editors, SOFC-II, Commission of the European Communities, Luxembourg, 1991; 473.
206. Jiang SP, Zhang JP, Ramprakash Y, Milosevic D, Wilshier K. *J Mater Sci* 2000; 35: 2735–2741.
207. Mai A, Becker M, Assenmacher W, Tietz, F, Hathiramani D, Ivers-Tiffée E, *et al.* *Solid State Ionics* 2006; 177: 1965–1968.
208. Simner SP, Anderson MD, Engelhard MH, Stevenson JW. *Electrochem Solid-State Lett* 2006; 9: A478–A481.
209. Jiang SP, Zhen YD, Zhang S. *J Electrochem Soc* 2006; 153: A1511–1517.

4 Interconnects

Zhenguo (Gary) Yang and Jeffrey W. Fergus

CONTENTS

4.1 INTRODUCTION

4.1.1 INTERCONNECT REQUIREMENTS

The identification and fabrication of suitable interconnect materials is a major challenge in the development of solid oxide fuel cells (SOFCs) [1, 2]. The primary function of the interconnect is to carry electrical current from the electrochemical cell to the external circuit, so the inteconnect material must have good electrical conductivity. In addition, the interconnect must be chemically and mechanically stable for operation at high temperatures for long periods of time. This is especially challenging because the interconnect is exposed to both oxidizing conditions at the cathode and reducing conditions at the anode. Thus, interconnect materials should not undergo significant dimensional changes with changes in either temperature or oxygen partial pressure. One such change is thermal expansion, which must be matched with the expansion of other fuel cell components to avoid the generation of thermal stresses during heating or cooling. Similarly, some materials undergo expansion or contraction with changes in oxygen partial pressure, which can also lead to stresses in the fuel cell. Thus, although the interconnect is not always a load-bearing component, moderate mechanical strength of the interconnect material is required to avoid fracture when stresses are generated during operation. Even higher strength is required in some planar designs where the interconnect provides structural support. In such designs, the interconnect also separates the fuel from air, so oxygen and hydrogen permeability should be low. The interconnect material must also be chemically stable with other fuel cell components. This includes not only stability during fuel cell operation, but also during fabrication, which typically involves higher temperatures. Finally, ease of fabrication and cost are important so that application of the fuel cell is economically feasible.

4.1.2 MATERIALS USED FOR INTERCONNECTS

Both ceramic and metallic materials have been used as interconnects for SOFCs. Ceramic interconnects [3, 4] are made from semiconducting oxides, which have good stability in air and are generally compatible with other ceramic fuel cell components. The conductivity of semiconducting oxides increases with increasing temperature, which is well-suited for the high operating temperature of SOFCs. However, the decrease in conductivity with decreasing temperature creates a challenge in the development of intermediate temperature SOFCs. Most metals are inherently unstable in air, so oxidation resistance is a major challenge for metallic interconnects [5, 6]. Although metals are excellent electrical conductors, the oxide scales formed during high-temperature exposure have much lower conductivity, and thus determine the overall resistance. Unfortunately, many doping additions that reduce the electrical resistance of the scale by increasing the concentration or mobility of charge carriers also increase the transport rates leading to oxide scale growth, and thus can degrade the oxidation resistance. However, because of their low cost and good fabricability, metallic interconnects are more widely used than ceramic interconnects, which are more expensive and difficult to fabricate.

4.2 CERAMIC INTERCONNECTS

4.2.1 STABILITY

Ceramic interconnects are oxides and are thus stable in the oxidizing conditions at the cathode side, but some conducting oxides can be reduced at low oxygen partial pressures. However, pure $LaCrO_3$ is stable to oxygen partial pressures as low as 10^{-21} atmosphere at 1000°C [7] and is thus stable in typical fuel environments. The solubilities of dopant ions, which are added to increase the conductivity, decrease with increasing oxygen partial pressure [8–12]. Of the two common dopants, calcium and strontium, calcium has the higher solubility at high-oxygen partial pressures, whereas strontium has the higher solubility at low-oxygen partial pressures

[10–13]. Enthalphy measurements indicate that strontium stabilizes the higher oxidation states on the chromium site [14], which is important for good conductivity at high oxygen partial pressures. To fill all oxygen vacancies, Cr^{4+} must compensate for the divalent alkaline earth on the lanthanum site. The stabilizaton of the higher valence chromium is further supported by the observation of Cr^{6+} in strontium doped lanthanum chromite [15]. The change in valence of the chromium ion results in a significant change in the ionic radius—0.55 to 0.615 Å for Cr^{4+} to Cr^{3+} [16]. This increase in ionic radius leads to expansion of the lattice during reduction in the oxygen partial pressure and can lead to the generation of stress and mechanical instabilities [17]. This and other factors affecting thermal expansion will be discussed in a later section.

4.2.1.1   Volatilization

Another stability issue for SOFC components is volatilization during cell operation. Most SOFC interconnect materials contain chromium, which can be oxidized to Cr^{6+}-containing species (CrO_3 or $CrO_2(OH)_2$). These species diffuse to the cathode and are reduced, so that Cr_2O_3 deposits on the cathode surface and degrades fuel cell performance. The partial pressures of these vapor species are proportional to the square root of the activity of chromia. The activity of chromia in lanthanum chromite is low, so that the vapor pressure are orders of magnitude lower than those for the same species in equilibrium with chromia [18,19]. Thus, cathode poisoning is a more serious problem with metallic interconnects, where the chromia activity of the scale is one, or close to one, so a more detailed discussion of cathode poisoning is included in the section on metallic interconnects.

4.2.1.2   Chemical Compatibility

The interconnect material is in contact with both electrodes at elevated temperatures, so chemical compatibility with other fuel cell components is important. Although, direct reaction of lanthanum chromite based materials with other components is typically not a major problem [2], reaction between calcium-doped lanthanum chromite and YSZ has been observed [20–24], but can be minimized by application of an interlayer to prevent calcium migration [25]. Strontium doping, rather than calcium doping, tends to improve the resistance to reaction [26], but reaction can occur with strontium doping, especially if $SrCrO_4$ forms on the interconnect [27].

4.2.2   TRANSPORT PROPERTIES

Lanthanum chromite is a p-type conductor so divalent ions, which act as electron acceptors on the trivalent (La^{3+} or Cr^{3+}) sites, are used to increase the conductivity. As discussed above, the most common dopants are calcium and strontium on the lanthanum site. Although there is considerable scatter in the conductivities reported by different researchers due to differences in microstrucure and morpohology, the increase in conductivity with calcium doping is typically higher than that with strontium doping [4]. The increase in conductivity at 700°C in air with calcium additions is shown in Figure 4.1 [1, 2, 28–44]. One of the advantages of the perovskite structure is that it

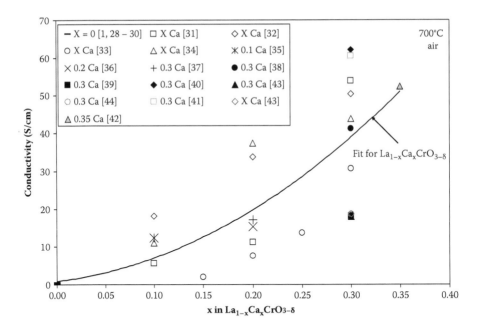

FIGURE 4.1 Conductivity of calcium-doped lanthanum chromite in air at 700°C [1, 2, 28–44].

contains two differently sized ions, so large cations, such as calcium and strontium can replace lanthanum, while smaller cations, such as nickel, copper, and aluminum, can replace chromium (see Figure 4.2) [43–52]. In most cases, divalent dopants are used as electronic acceptors to increase the conductivity by increasing the hole concentration (e.g., copper and nickel in Figure 4.2). Other divalent dopants that have been shown to increase the conductivity of lanthanum chromite include magnesium [34] and cobalt [53]. However, Figure 4.2 shows that increasing the concentration of trivalent aluminum also increases the conductivity. Since doping trivalent aluminum on a trivalent chromium site would not increase the hole concentration, this increase is presumably due to an increase in carrier mobility, rather than carrier concentration.

Lanthanum chromite is the most common base for SOFC interconnects, but chromites of other lanthanide elements have also been used [43, 45, 46, 48, 54, 55]. Although the conductivity of calcium-doped gadolinium chromite for low calcium contents is in the upper range of conductivities for lanthanum chromite, other non-lanthanum chromites typically have lower conductivities. However, the use of other lanthanides provides benefits in controlling the phase transformation temperature and in potential cost savings [48].

The addition of doped cerium oxide to lanthanum chromite has recently been shown to produce large increases in conductivity [44, 49–52]. The conductivity increases to a maximum at 3 to 5% ceria addition, depending on the dopant used in the ceria (Sm, Gd, or Y), and the highest conductivity in air is for 5% $Ce_{0.8}Sm_{0.2}O_{1.9}$. The oxygen permeation rate increases with ceria addition, indicating that at least some of the increase in conductivity is due to oxygen ion conductivity. At low oxygen partial pressures, the

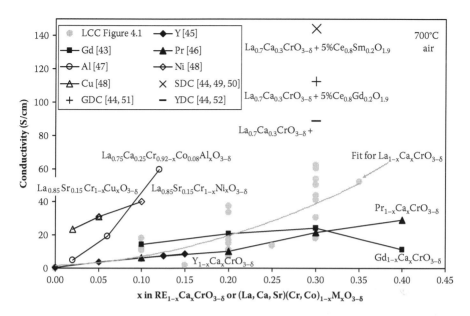

FIGURE 4.2 Conductivity of chromite compounds in air at 700°C [43–52].

maximum in conductivity occurs at a lower additive concentration, which is presumably due to the reduction of Ce^{4+} to Ce^{3+} creating additional charge carriers.

Since the interconnect is exposed to both the air and the fuel, the conductivity at low oxygen partial pressures is important for fuel cell performance. The conduction in doped lanthanum chromite occurs by a small polaron hopping mechanism, in which the negative charge resulting from Sr^{2+} replacing La^{3+} is compensated for by the chromium being oxidized to Cr^{4+} [56–60]. With decreasing oxygen partial pressure, the loss of oxygen results in the formation of oxygen vacancies, which are compensated for by the reduction of Cr^{4+} to Cr^{3+}. The resulting decrease in charge carriers leads to a decrease in conductivity. The effect of doping on the conductivity at low oxygen partial pressures is different from that at high oxygen partial pressures. Figure 4.3 shows that, not only does the conductivity decrease with decreasing oxygen partial pressure (e.g., compare $La_{0.7}Ca_{0.3}CrO_3$ in air, Ar-5%H_2 and Ar-20%H_2), but the effect of dopant additions is different [41, 43, 47]. While increasing the calcium or aluminum content results in an increase in the conductivity in air, the conductivity in a fuel environment decreases with increases in both dopants. Titanium, which does not improve the conductivity high oxygen partial pressures, can improve the conductivity at low oxygen partial pressures due to the reduction of Ti^{4+} to Ti^{3+} [61–63]. To improve the conductivity at low oxygen partial pressures, donor dopants, such as niobium, can be used [64]. The combination of a *p*-type conductor on the air side and *n*-type conductor on the fuel side has recently been reported [65]. Although, the relative thicknesses of the two layers must be carefully controlled to achieve the appropriate oxygen partial pressure in each of the two materials, such an approach could produce an interconnect with a low electrical resistance.

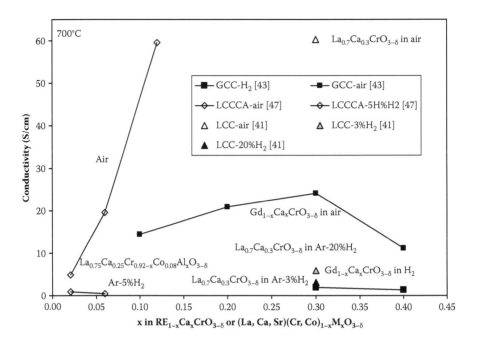

FIGURE 4.3 Conductivity of chromite compounds in reducing atmospheres at 700°C [41, 43, 47].

4.2.3 PHYSICAL PROPERTIES

4.2.3.1 Dimensional Changes

Heating to and cooling from the high operating temperature can create thermal stresses, so the coefficient of thermal expansion (CTE) of the interconnect should match those of the other fuel cell components. As shown in Figure 4.4, doping lanthanum with either calcium or strontium increases the CTE, which is desirable, since the CTE of yttria-stabilized zirconia (YSZ) (broken horizontal line) is higher than that of undoped lanthanum chromite [1, 2, 34, 35, 37, 40, 48, 54, 66–68]. Although there are significant variations between different reports, the CTE of strontium-doped lanthanum chromite is typically larger than that of calcium-doped chromite, so less strontium is required to increase the CTE to match that of YSZ. Other dopants, such as nickel, can also be used to control the CTE [48, 67, 69]. The ceria additions to calcium-doped lanthanum chromite, which as shown above lead to large increases in conductivity, also increase the CTE, so the dopant level may need to be limited to maintain an acceptable level of thermal stress [44]. Fortunately, among the ceria materials used, the smallest increase in CTE is for $Ce_{0.8}Sm_{0.2}O_{1.9}$, which also provides the largest increase in conductivity.

As mentioned above, when the oxygen partial pressure is reduced, the lattice expands as Cr^{4+} is oxidized to Cr^{3+}. If this expansion is too large, just as with thermal expansion, stresses can be generated and lead to cracking. Increasing the dopant

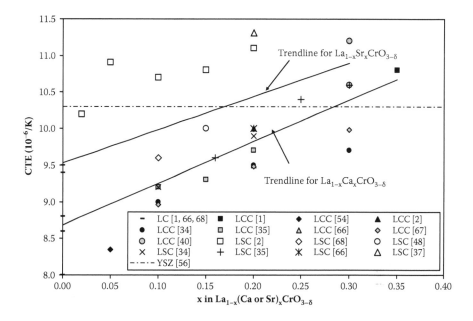

FIGURE 4.4 Coefficients of thermal expansion of calcium- and strontium-doped lanthanum chromite [1, 2, 34, 35, 37, 40, 48, 54, 66–68].

concentration increases the oxygen nonstoichiometry, which increases the amount of expansion during reduction in oxygen partial pressure [70]. Dopants on the chromium site, such as nickel [17, 48, 71], titanium [48, 68, 71, 72], and vanadium [73] can be used to decrease the amount of expansion associated with the reduction of the chromium ion. Yttrium chromite has lower conductivity than lanthanum chromite, but expands less during reduction [35].

4.2.3.2 Mechanical Properties

The stresses associated with changes in temperatures and oxygen partial pressure cannot be completely eliminated, so the mechanical properties of the interconnect are important for durability. Comparison of the high temperature strength of lanthanum chromite with various dopants [4] shows that the highest strength is for doping with magnesium [74], followed by strontium [34, 66, 75, 76] and then calcium [76, 77], which is the opposite trend of that observed for conductivity. Strontium-doped lanthanum chromite also maintains strength at low oxygen partial pressures better than calcium-doped lanthanum chromite [1, 66]. One of the reasons for the better mechanical properties of srontium-doped lanthanum chromite is the presence of more grain-boundary phases [66]. There are reports in which calcium-doped lanthanum chromite is stronger than strontium-doped lanthanum chromite, but this has been attributed to its superior sintering behavior [78]. The mechanical properties also depend on oxygen partial pressures as the strength, toughness, and elastic modulus increase with decreasing oxygen partial pressure, which is attributed to the increase in oxygen vacancy concentration [79]. Calcium-doped yttrium chromite is stronger than calcium-doped lanthanum chromite, which, especially when combined with the smaller expansion during reduction mentioned above, results in good durability [80].

4.2.4 PROCESSING

The fabrication of an interconnect requires synthesis of the desired phase, forming the desired shape and sintering to achieve adequate strength and conductivity. Synthesis of the desired phase affects the subsequent processes because, for example, reduction in particle size tends to reduce the required sintering temperature. To attain small sinterable powder, polymer precursor methods, such as the Pechini method, are typically used to produce powders of ceramic interconnect materials [81–83]. In the Pechini method, nitrates of the desired cations are mixed with citric acid and ethylene glycol, but other organic compounds, such as urea [84, 85] and glycine [86–88], are also used. Nitrates can also be simply dissolved in an aqueous solution and when processed in an autoclave produce small non-agglomerated particles [89].

Interconnects are formed into the desired shape using ceramic processing techniques. For example, bipolar plates with gas channels can be formed by tape casting a mixture of the ceramic powder with a solvent, such as trichloroethylene (TCE)-ethanol [90]. Coating techniques, such as plasma spray [91] or laser ablation [92] can also be used to apply interconnect materials to the other fuel cell components.

The final stage of the fabrication process is to sinter the interconnect material to adequate density, so a major challenge in the use of the implementation of ceramic interconnects in SOFCs is the development of suitable and cost-effective fabrication processing. Ceramics typically require high-temperature sintering to achieve adequate density. In the case of lanthanum chromite the formation of chromia during processing inhibits sintering of the powders [93], so chromium-deficient compositions have been used to reduce the amount of chromia formation and improve sintering [94]. For strontium-doped lanthanum chromite, the beneficial effect of chromium deficiency on sinterability has been attributed to the formation of a strontium-containing liquid [95]. While strontium is beneficial for sintering, calcium is generally more effective than strontium in improving sinterability [78, 96]. At high calcium contents, a liquid phase can form during sintering and enhance densification [97]. Other dopants can also improve the sintering behavior. For example, vanadium [48, 72, 98] and nickel [48, 72, 76, 96] have been shown to improve sinterability. Doping on the anion site can also be beneficial, since small amounts of CaF_2 have been shown to improve the sintering behavior of lanthanum chromite [99].

Doping affects material properties, so other approaches to improving sinterability, without changing the composition, have been developed. For example, microwave-assisted sintering can be used to improve densification [100]. Spark plasma sintering can also be used to rapidly produce dense materials [101]. Combination of synthesis and densification through self-propogating high-temperature synthesis, which can rapidly produce dense materials, has been shown to be thermodynamically feasible for the production of lanthanum chromite, but may not be practical for integration with other SOFC fabrication processes [102].

The difficulty and high cost of the fabrication of ceramic interconnect materials is their primary disadvantage and has led to recent emphasis on metallic interconnects, which will be discussed in the next section.

4.3 METALLIC INTERCONNECT MATERIALS

4.3.1 CANDIDATE ALLOYS

Considering the materials requirements for interconnect applications, the metallic materials of interests potentially include Ni, Fe, and Cr base oxidation resistant alloys, as well as precious metals. With the possible exception of silver, precious metals are too costly for this particular application. Given its lower cost compared to other precious metals, limited use of silver may be acceptable, but there are concerns over its low melting point (960°C) and high volatility at SOFC operating temperatures [103–106] as well as intrinsic instability under air/hydrogen dual exposure conditions [107, 108]. Therefore, most attention has been given to the transition metal-based oxidation resistant alloys which may include Ni(-Fe)-Cr base heat resistant alloys, Cr base alloys, and Fe-Cr base alloys—austenitic stainless steels and ferritic stainless steels (FSSs) [1, 5, 6]. These different groups of alloys are schematically presented in Figure 4.5 and their structure and properties relevant to the interconnect application are listed in Table 4.1. All of these alloys contain "active" constituents, mainly Cr and Al, which are preferentially oxidized at the alloy surface and form a protective scale to minimize further environmental attack during high-temperature exposure [109–111]. Since alumina (Al_2O_3) is electrically insulating [112], alloys which form a semi-conductive chromia scale (with a conductivity of 10^{-2} S·cm^{-1} at 800°C in air [112–116] are the preferred candidates. It has to be noted, however, if an insulating scale can be excluded from the electrical current path, alumina formers, which usually demonstrate higher oxidation resistance than chromia-forming alloys, can be considered for interconnect applications. Among the different groups of chromia-forming alloys, Cr base alloys and FSSs have a

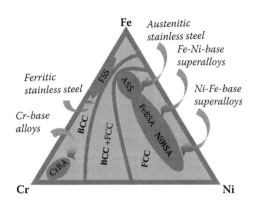

FIGURE 4.5 Schematic of alloy design for SOFC applications.

TABLE 4.1

Comparison of Key Properties of Different Alloy Groups for SOFC Applications

Alloy	Matrix Structure	Coefficient of Thermal Expansion $\times 10^{-6} \cdot K^{-1}$ (RT-800°C)	Oxidation Resistance	Mechanical Strength	Manufactur-ability	Cost
Cr base alloys	BCC	11-12.5	Good	High	Difficult	Very expensive
Ferritic stainless steels	BCC	11.5-14	Good	Low	Fairly readily	Less expensive
Austenitic stainless steels	FCC	18-20	Good	Fairly high	Readily	Less expensive
Fe-Ni-Cr base alloys	FCC	15-20	Good	High	Readily	Fairly expensive
Ni(-Fe)-Cr base alloys	FCC	14-19	Good	High	Readily	Expensive

body-centered cubic (BCC) crystal structure and a CTE in the range of 11.0 to 12.5 × 10⁻⁶K⁻¹ [6] which matches well with that of ceramic cells in the range of 10.5 to 12.5 × 10⁻⁶K⁻¹, depending on whether the cell is electrolyte- or anode-supported [2]. For this reason, the BCC base alloys have been the most widely investigated category of alloys for interconnect applications. Cr base alloys were the early favorites as a replacement for doped lanthanum chromites as the interconnect material in high temperatures (900 to 1000°C) SOFCs. The leading candidate in this category is Plansee Ducralloy, an oxide dispersion-strengthened composition that contains 94% Cr, 5% Fe, and 1% Y_2O_3 (as $Cr_5FeY_2O_3$) [117, 118]. In addition to appropriate CTE (11.8 × 10⁻⁶ K⁻¹ over 20 to 1000°C) [119, 120], Ducralloy demonstrates excellent oxidation resistance and high-temperature mechanical strength. However, it is difficult and costly to fabricate, particularly in comparison with Fe-Cr and Ni(-Fe)-Cr base high-temperature oxidation resistant alloys. In contrast, Fe-Cr base FSSs not only demonstrate good oxidation resistance and the ability to match the CTE of cell components [6, 120], but also are among the most cost-effective and fabricable alloys. With the reduction of SOFC operating temperatures, FSSs were proposed and tested as interconnect materials in stacks in the late 1990s [120–123]. In recent years, they have become the most widely studied group of oxidation resistant alloys for interconnect applications.

Other categories of chromia forming alloys—including Ni(-Fe)-Cr base and Fe(-Ni)-Cr base alloys (e.g., austenitic stainless steels)—have a face-centered cubic (FCC) substrate structure. In comparison to the FSS, the FCC base alloys, in particular the Ni(-Fe)-Cr base alloys, are generally much stronger and potentially more oxidation resistant in the SOFC interconnect operating environment [6, 123–129]. However, the FCC Ni(-Fe)-Cr base alloys with sufficient Cr for an appropriate oxidation resistance often exhibit a high CTE, typically in the range of 15.0 to 20.0 × $10^{-6}K^{-1}$ (except Invar alloys [133]) from room temperature to 800°C, and are much more expensive than the FSSs. Due to the CTE mismatch, significant power loss or degradation in performance has been observed during thermal cycling test of stacks using Ni(-Fe)-Cr base alloy interconnects [130–132].

Nevertheless, Ni(-Fe)-Cr base alloys may find application as interconnect materials through the use of innovative SOFC stack and seal designs and novel interconnect structures. For example, a cladding approach has been applied to fabricate a stable composite interconnect structure consisting of FCC Ni-Cr base alloy claddings on a BCC FSS substrate [134, 135]. The clad structure appeared to be stable over 1000 hours at 800°C in air and exhibited a linear CTE close to that of the FSS, but needs further long-term stability evaluation before its commercial use.

Traditional alloy design emphasizes surface and structural stability, but not the electrical conductivity of the scale formed during oxidation. In SOFC interconnect applications, the oxidation scale is part of the electrical circuit, so its conductivity is important. Thus, alloying practices used in the past may not be fully compatible with high-scale electrical conductivity. For example, Si, often a residual element in alloy substrates, leads to formation of a silica sublayer between scale and metal substrate. Immiscible with chromia and electrically insulating [112], the silica sublayer would increase electrical resistance, in particular if the subscale is continuous.

With an emphasis on scale electrical conductivity (surface stability as well), a number of new alloys have been recently developed specifically for SOFC interconnect applications. The one that has received wide attention is Crofer 22 APU, an FSS developed by Quadakkers et al. [136, 137] at Julich and commercialized by Thyssen Krupp of Germany. Crofer 22 APU, which contains about 0.5% Mn, forms a unique scale, as shown in Figure 4.6, comprised of a $(Mn,Cr)_3O_4$ spinel top layer and a chromia sublayer [137–139]. The electrical conductivity of $(Mn,Cr)_3O_4$ has been reported

FIGURE 4.6 Transmission electron microscopy (TEM) bright field image of the scale grown on a Crofer 22 APU coupon during an isothermal oxidation at 800°C in air after 300 h.

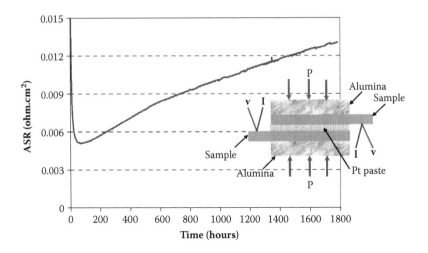

FIGURE 4.7 ASR of oxide scale on Crofer 22 APU as a function of time at 800°C in air. Samples were preoxidized in air for 100 h at 800°C [139].

to be up to 2 orders of magnitude higher than that of chromia [6], so its formation in the scale formed on Crofer 22 APU leads to a higher scale conductivity [136–139]. The measured scale area specific resistance (ASR) of Crofer 22 APU was lower than that of traditional FSSa including E-brite (26–27% Cr), despite the fact that the scale grown on E-brite was thinner. The advantage of the newly developed alloys is especially obvious when compared with normal grade compositions such as AISI 446 (16–17% Cr), which exhibit a relatively high ASR due to the formation of a thin, continuous insulating silica layer beneath the chromia scale. Nevertheless, the ASR of the scale on Crofer 22 APU still increases with time, as shown in Figure 4.7, due to the continuous growth of the scale (particularly the chromia subscale). Extrapolation from the test would suggest an unacceptable ASR over thousands of hours stack lifetime (assuming no scale spallation occurring). In addition, the newly developed alloy may need further improvements in reduction of chromium volatility and compatibility with adjacent components such as sealing materials, in spite of substantial progress in both these areas.

Overall, the newly developed alloys, such as Crofer 22 APU and ZMG 232 (developed by Hitachi Metal, Ltd.) [142, 143], are promising candidate alloys for SOFC interconnect applications, but their stability during long-term operation at 650 to 800°C remains questionable.

4.3.2 SURFACE STABILITY: OXIDATION AND CORROSION

The oxidation and corrosion behavior of metals and alloys has been widely investigated in a range of environments for a myriad of applications. Recently, oxidation resistant alloys have been studied particularly for SOFC interconnect applications.

Early works were typically carried out using single atmosphere exposure conditions, either air (or moist air) representing the cathode side environment [124–129, 139, 142, 144–162] or a reducing atmosphere simulating the anode side environment [124, 125, 127–129, 144, 145]. Lately, studies have been also performed to determine the oxidation/corrosion behavior of metal and alloys under dual-atmosphere exposure conditions that closely simulate the interconnect exposure conditions during SOFC operation [154–159]. The alloys studied include both Fe-Cr base FSSs and Ni or Ni-Cr base heat-resistant alloys, as well as Cr or Cr base alloys.

When exposed in air or cathode-side environment, active elements, e.g., Cr, in alloy substrates are preferentially oxidized forming an oxide scale on the alloy surface to protect it from further environment attack. Though the scale growth on an alloy substrate is affected by factors such as scale vaporization and grain-boundary diffusion [163–165], it is often approximated by a parabolic relationship with time t:

$$(\Delta w)^2 = k_p \cdot t \qquad (4.1)$$

where Δw is the weight gain, and k_p is the parabolic rate constant. Table 4.2 lists the parabolic oxidation rate of different alloys [5]. Among the compositions of interest, Ni(-Fe)-Cr base alloys usually demonstrate higher oxidation resistance than the Fe-Cr base FSSs due to the relatively noble nature of nickel. In general, alloys with higher Cr% possess higher oxidation resistance and, therefore, exhibit a lower scale growth rate. To remain capable of self-healing and prevent breakaway oxidation [166–168], alloy substrates are required to maintain a Cr reservoir with Cr% above a critical value that marks the transition from external to internal oxidation. In particular, for an Fe-Cr substrate, the critical Cr% is 17 to 20%, depending on temperature, minor alloy additions, impurities (e.g., S, P, C), etc. It has been found that reactive elements such as rare-earth metals in a trace amount (a few hundredths or tenths of a percent) can significantly increase alloy oxidation resistance by improving scale adherence and suppressing scale growth via Cr outward diffusion [169–172], so most if not all newly developed alloys for the interconnect application contain reactive element additions, e.g., La in Crofer 22 APU.

In comparison to the oxidizing, cathode side environment, the reducing, anode side environment is more complex particularly when a hydrocarbon fuel is used. The presence of high water vapor partial pressure, carbon activity, and residual components such as sulfur, makes metallic interconnects susceptible to varied forms of corrosion. Though the oxygen partial pressure (10^{-18} to 10^{-12} bar) at the anode side is much lower than that at the cathode side, formation of oxides such as chromia and $(Mn,Cr)_3O_4$ is nevertheless still thermodynamically feasible. In hydrogen or moist hydrogen, FSSs exhibit a scale growth rate that is comparable to that in air [139, 142, 147, 151, 152]. The scale grown in the hydrogen fuel is usually composed of the same major phases as found in air, although their morphology and minor components can be different. For example, x-ray diffraction analysis on a scale grown on Crofer 22 APU in moist hydrogen and that in moist air indicated that both scales were composed of Cr_2O_3 and $(Mn,Cr)_3O_4$ phases [139, 155]. The spinel phase tended to grow with a large aspect

TABLE 4.2
Parabolic Rate Constant of Selected Heat Resistant Alloys [5]

Alloy	Temperature (°C)	Rate Constant ($\times 10^{-13}$ g$^2\cdot$cm$^{-4}\cdot$s^{-1})	Notes
G-30	800	0.416	
	700	0.041	
Nicrofer 6025	800	0.022	Ni-base, alumina former
	700	0.005	
Haynes 214	800	0.031	
	700	0.002	
Haynes 230	800	0.361	Ni-base, chromia former
	700	0.05	
Rene 41	800	1.967	
	700	0.126	
Haynes 188	800	0.712	Co-Ni-base, chromia former
	700	0.327	
Haynes 556	800	0.01	
	700	0.058	
Haynes 120	800	0.352	Fe-Ni-base, chromia former
	700	0.012	
Pyromet	800	0.793	
	700	0.062	
E-brite	800	0.353	
	700	0.015	
446	800	1.332	Ferritic, chromia former
	700	0.088	
AL 453	800	1.984	
	700	0.128	
Fecralloy	800	0.035	
	700	0.004	Ferritic, alumina former
Alpha-4	800	0.088	
	700	0.015	

ratio in moist hydrogen, in comparison to well-defined, equiaxed crystallites in air. For Ni-Cr base alloys, NiO formation is suppressed in the scales since oxidation of nickel is thermodynamically unfavorable in the low oxygen partial pressure fuels. Overall it appears Ni-Cr base alloys exhibit higher oxidation resistance in moist hydrogen than in air, and even those with a relatively low Cr% appear to be promising for interconnect applications in terms of oxidation resistance [127, 148]. Similarly, differences in oxidation resistance between Fe-Cr base and Ni(-Fe)-Cr base alloys are also observed in high water vapor environments. For Fe-Cr base alloys, exposure to high water vapor environments potentially leads to breakaway oxidation through formation Fe or Fe-rich oxides [173–176]. The breakaway oxidation was sensitive to the water vapor content in the atmosphere and %Cr in the Fe-Cr alloys: the higher the water vapor content or the lower the Cr%, the earlier the breakaway oxidation took place. In contrast, Ni(-Fe)-Cr base alloys exhibited enhanced scale adherence and suppressed formation of NiO in the scale after exposure to high vapor environments [177, 178].

When a hydrocarbon fuel is fed, either directly or after reformation, into the anode chamber, metallic interconnects are exposed to an environment with a carbon activity and therefore potentially could suffer carbon-induced corrosion. In carbon-bearing gas environments, alloys including Fe(-Ni)-Cr and Ni-Fe-Cr base alloys are susceptible to metal dusting at temperatures in the 400 to 850°C range, which leads to alloy degradation by disintegration of the metal matrix into tiny particles [179–181]. Metal dusting likely occurs in a hydrocarbon fuel with a carbon activity ≥ 1 [149–151] and the resistance to metal dusting depends on alloy composition. For example, with electro-polished surfaces, chromia-forming austenitic alloys, including Alloy 800, Inconel 601, 690, 693, and Alloy 602CA, suffered a rapid metal dusting, which did not occur until after about 50 one-hour cycles for the ferritic steel Fe-27Cr-0.0001Y. In comparison to the chromia-forming alloys, alumina-forming alloys showed superior performance, with no apparent degradation due to metal dusting after 1200 cycles. In addition, Ni base alloys, which developed scales with less spinel content, performed better than the Fe base alloys, which formed scales with higher spinel content. When exposed to a carbonaceous environment with a carbon activity less than unity, oxidation resistant alloys are much less likely to suffer metal dusting or carbon-induced corrosion, as confirmed by recent studies, typically in short terms. Long-term stability of metals or alloys in carbonaceous fuels under SOFC operating conditions and effects of local chemistry variance on the anode side during SOFC operation are still not clear.

During SOFC operation, interconnects are performing in neither an air-only nor fuel-only environment, but under a dual exposure as they are simultaneously exposed to air on the cathode side and fuel on the anode side and thus experience a hydrogen partial pressure or chemical potential gradient from the fuel side to the air side. The oxidation/corrosion of metals and alloys, including Fe-Cr base stainless steels, Ni-Cr base alloys, under hydrogen/air dual exposures at high temperatures can be very different from that under single exposure conditions [127, 155–158, 160, 182]. In particular, the composition and microstructure of the scale grown on the air side differs significantly from the behavior when exposed to air on both sides, while the oxidation/corrosion behavior on the hydrogen fuel side is typically comparable to that when exposed to the hydrogen fuel on both sides.

Fe-Cr base FSSs, in particular those with a relatively low Cr%, were reported to be susceptible to breakaway oxidation due to hematite (Fe_2O_3) phase nodular growth in the scale grown on the air side of the air/hydrogen sample. For example, AISI 430, with 17% Cr, formed hematite nodules (see Figure 4.8) on the air side of the air/hydrogen sample during isothermal heating at 800°C after 300 hours, potentially resulting in localized attack. In comparison, there was no hematite phase formation on the air/air sample, on the hydrogen side of the air/hydrogen sample, nor on the hydrogen/hydrogen sample. The oxidation behavior on the hydrogen side of the air/hydrogen sample was similar to that on the hydrogen/hydrogen sample. The potential detrimental effects of the dual-atmosphere exposures appeared to be dependent on

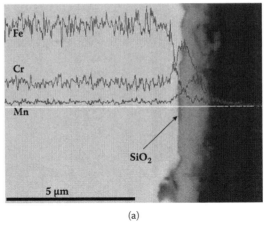

(a)

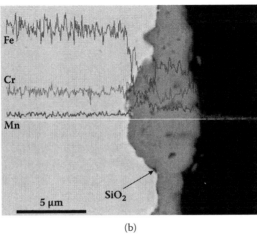

(b)

FIGURE 4.8 Scanning electron microscopy (SEM) cross-sections of AISI 430 coupons after 300 h of oxidation at 800°C in air under different exposure conditions: (a) both sides exposed to air and (b) on the air side of the air/($H_2 + 3\%H_2O$) exposures [154].

the alloy composition, in particular Cr% in the Fe-Cr substrate. For Crofer 22 APU, with about 23% Cr, no hematite phase formation or nodule growth was observed under the same test conditions as for AISI 430. Instead, the spinel top layer of the scale on the air side of the hydrogen/air sample was enriched in iron and grew into a different morphology from that on the air/air sample under the same conditions. This anomalous oxidation thus appears to be a result of combined effects from both the hydrogen flux from the fuel side to the air side and increased water vapor partial pressure on the air side. E-brite, with 27% Cr, appeared to be more resistant to formation of hematite nodules in the scale grown on the air side of the air/hydrogen sample, though the surface microstructure of the scale was different from the air only sample. The anomalous oxidation/corrosion behavior of oxidation-resistant alloys observed under the dual-exposure conditions appears to be similar to that found in a high partial pressure water vapor single environment [183].

In addition to the Fe-Cr base stainless steels, dual exposure may also affect the oxidation behavior of Ni-Cr base alloys [127, 184]. For Ni-Cr based alloys, the dual exposures also resulted in different oxidation/corrosion behavior from that in a single atmosphere exposure. But unlike the ferritic chromia-forming alloys, nickel and Ni-Cr base alloys formed a uniform, well-adherent scale on the air side of the air/hydrogen sample. Overall, it appears that the anomalous oxidation/corrosion on the air side of the air/fuel sample is sensitive to surrounding environments, alloy composition, preexisting conditions, and other factors. Further systematic work is required to gain more insight into the oxidation/corrosion behavior under the dual environment and its effects on metallic interconnect long-term stability.

4.3.3 CHROMIA SCALE VOLATILITY AND CELL POISONING

The electrical conductivity requirement for interconnect applications necessitates the use of chromia-forming (or Cr-rich spinel) oxidation-resistant alloys. One drawback of the chromia-forming alloys for this particular application, however, is the Cr volatility of the chromia or Cr-rich scale. As indicated by many studies [185–189], during high-temperature exposure Cr_2O_3 (s) reacts with O_2 via the following reaction

$$Cr_2O_3(s) + \frac{3}{2}O_2(g) = 2CrO_3(g) \tag{4.2}$$

forming CrO_3 (g) that becomes the most abundant vapor species from the chromia scale in dry air. The presence of water vapor in the air significantly increases the overall vapor partial pressure of Cr vapor species through formation of $CrO_2(OH)_2$ (g), formed via the following reaction:

$$\frac{1}{2}Cr_2O_3(s) + \frac{1}{2}O_2(g) + H_2O(g) = CrO_2(OH)_2(g) \tag{4.3}$$

At high temperatures, the partial pressure of $CrO_2(OH)_2$ (g) can be one or two orders of magnitude higher than that of CrO_3, which is the second highest among Cr vapor species. For the chromia-forming alloys, the Cr volatility depends on scale composition and microstructure, which depends on the alloy composition. For example, alloys with Mn as a residual component or an additive, such as Crofer 22 APU, can form a scale with a $(Mn,Cr)_3O_4$ spinel top layer that helps minimize Cr volatility. Formation of the spinel top layer on Crofer 22 APU leads to a chromium volatility that is a factor of 2 to 3 lower than that of pure chromia forming alloys, such as $Cr5FeY_2O_3$ and E-brite [190–192]. However the spinel forming alloys, such as Crofer 22 APU, still release Cr due to the fact that the Cr-containing spinel still forms

Cr^{+6}-containing vapor species, albeit at a lower rate, and the fact that the spinel does not fully cover the alloy surface during the early stages of oxidation [139].

Chromia vapor species can migrate into cathodes and deposit in the electrode and at the electrode–electrolyte interface, thereby causing a rapid degradation in SOFC performance. Early studies on Inconel 600 (Ni16Cr8Fe) and Ducralloy Cr5Fe1Y_2O_3 using (La,Sr)MnO_3 (LSM) cathodes on Y_2O_3-ZrO_2 (YSZ) electrolyte concluded that electrochemical reduction of chromium vapor species at the triple-phase boundaries (TPBs) blocked active sites and inhibited oxygen reduction at the cathode [187, 193, 194]. Recently, works were carried out on Fe-Cr base stainless steels, on LSM cathodes, and found that Cr species preferentially deposited at the LSM/YSZ interface region, forming deposit rings or bands on the YSZ electrolyte surface close to the edge of the LSM electrode [195–198]. It was proposed that the driving force for deposition of Cr species at the LSM cathode is the generation of Mn^{2+} species under cathodic polarization or at high temperatures and subsequent reaction between Mn^{2+} and the Cr species, forming Cr-Mn-O nuclei and then $(Cr,Mn)_3O_4$. The formation of chromium manganese spinel was also observed by other works using an LSM cathode on YSZ electrolyte and stainless steels as chromium sources [199, 200]. The chromium poisoning has been attributed to (a) blocking of the electrochemically active sites by electrochemical reduction of chromium, giving rise to the formation of a solid chromium oxide, and (b) the decomposition of the cathode by formation of a chromium containing mixed oxide driven by thermodynamics without any influence of electrical potentials. Degradation with the most common cathode material, LSM, is due primarily to mechanism. In addition to LSM cathodes, recent publications reported and discussed the chromium poisoning effects on (La,Sr)(Co,Fe)O_3 (LSCF) [200–204], and (La,Sr)FeO_3 (LSF) [204] cathodes with various electrolytes, using chromia forming alloys as a chromium source.

In addition to electrode and electrolyte materials, the extent of Cr poisoning on the cell is dependent on the type of alloy used for the interconnect. Chromia formers led to Cr poisoning and rapid performance degradation of cells having LSCF cathodes and YSZ electrolyte with (Sm,Ce)O_2 interlayer, as shown in Figure 4.9. In contrast, while no degradation in cell performance was observed in the presence of alumina formers such as Haynes 214™ (Ni-16Cr-4.5Al-3Fe) and Kanthal alloy (Fe-22Cr-5.8Al). It appeared that, to some extent, preoxidation helped mitigate the Cr poisoning and cell degradation. Though marginal improvement in cell performance in the presence of Crofer 22 APU was reported by others [200, 204], the improvement, however, was not sufficient to limit the cell degradation from Cr poisoning to an acceptable level.

4.3.4 CHEMICAL COMPATIBILITY

In addition to the aforementioned interactions with the surrounding gas environments, metallic interconnects also interact with adjacent components at their interfaces, potentially causing degradation of metallic interconnects and affecting the stability of the interfaces. One typical example is the rigid glass-ceramic seals, in particular those made from barium-calcium-aluminosilicate (BCAS) base glasses [205–209]. FSS interconnect candidates have been shown to react extensively with

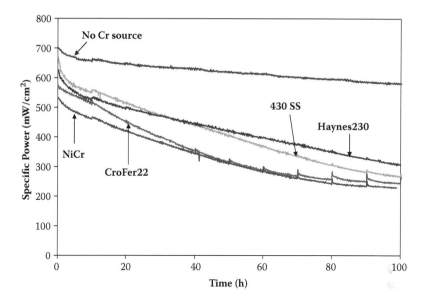

FIGURE 4.9 Electrochemical performance of the LSCF cells tested with various raw chromia-forming metal meshes. Performance of the LSCF cell tested with no mesh was also plotted for comparison [203].

the BCAS sealing glass-ceramic resulting in an interface that was more prone to defects [210–212]. For traditional chromia-forming stainless steels, the extent and nature of their interactions with the glass-ceramic depends on the exposure conditions and proximity of the interface of sealing glass and FSS to the ambient air. At or near the edges, where oxygen from the air is accessible, the chromia scale grown on the steel and its vapor species can react with BaO in the glass-ceramic, forming $BaCrO_4$, presumably via the following reactions:

$$2Cr_2O_3 \, (s) + 4BaO(s) + 3O_2(g) = 4BaCrO_4(s) \tag{4.4}$$

$$CrO_2(OH)_2(g) + BaO(s) = BaCrO_4(s) + H_2O(g) \tag{4.5}$$

Due to the large thermal expansion mismatch between barium chromate and the sealing glass or FSS (e.g., AISI 446) [213], the extensive formation of barium chromate has been shown to result in crack initiation and growth between the sealing glass and alloy coupons.

In the interior seal regions, where access of oxygen from the air was blocked, chromium or chromia dissolves into the BCAS sealing glass to form chromium-rich solid solutions. The stainless steel can also react with residual species in the sealing glass-ceramic to generate porosity in the glass-ceramic along the interface in the interior regions. Recent work further investigated the compatibility of FSSs and sealing glasses under air/hydrogen dual exposures [214, 215]. The corrosion at the interface of the sealing glass and the chromia-forming steel appears quite different from that when exposed to hydrogen or air only. At the air side, iron oxide nodules are present on the FSS near or at the triple phase boundary (TPB) of air/glass/metal, causing short-circuiting of the glass-ceramic seals. In contrast, no iron oxide is formed at the interface between the glass-ceramics and the ferritic steel on the hydrogen side. However, it is not clear how the dual exposures led to the iron oxide formation and the subsequent seal degradation. Besides the BCAS sealing glass, PbO-containing glass-ceramics have been investigated for their compatibility with FSSs, and extensive internal and external oxidation of FSSs was observed [216, 217].

Alloy composition also affects the interactions at the seal/interconnect interfaces. Crofer 22 APU was found to exhibit improved chemical compatibility and bonding with the BCAS based glass-ceramics due to the growth of its unique scale on the alloy during high-temperature exposure [212]. Under prolonged heating, however, the alloy still reacted with the sealing glass-ceramic, leading to the formation of barium chromate at the edge areas of the joints, solid solution phases, and porosity in the interior regions. The detrimental effects due to formation of $BaCrO_4$ were substantially mitigated by the use of an alumina-forming alloy, but porosity was still observable along the glass-ceramic/metal interface [211]. Thus, engineering the alloy surface, e.g., by aluminizing, has been adopted to improve the sealing effectiveness [218].

Besides the glass seal interfaces, interactions have also been reported at the interfaces of the metallic interconnect with electrical contact layers, which are inserted between the cathode and the interconnect to minimize interfacial electrical resistance and facilitate stack assembly. For example, perovskites that are typically used for cathodes and considered as potential contact materials have been reported to react with interconnect alloys. Reaction between manganites- and chromia-forming alloys lead to formation of a manganese-containing spinel interlayer that appears to help minimize the contact ASR [219, 220]. Sr in the perovskite conductive oxides can react with the chromia scale on alloys to form $SrCrO_4$ [219, 221].

4.4 PROTECTIVE COATINGS FOR METALLIC INTERCONNECTS

Newly developed alloys have improved properties in many aspects over traditional compositions for interconnect applications. The remaining issues that were discussed in the previous sections, however, require further materials modification and optimization for satisfactory durability and lifetime performance. One approach that has proven to be effective is surface modification of metallic interconnects by application of a protection layer to improve surface and electrical stability, to modify compatibility with adjacent components, and also to mitigate or prevent Cr volatility. It is particularly important on the cathode side due to the oxidizing environment and the susceptibility of SOFC cathodes to chromium poisoning.

Functionally, as shown in Figure 4.10, the protection layer is intended first to serve as a mass barrier to both chromium cation outward and oxygen inward transport

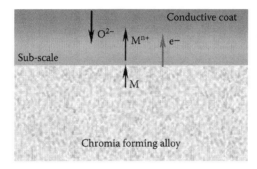

FIGURE 4.10 Schematic of mass transport in a conductive oxide coating on a chromia-forming alloy.

(via solid state diffusion if the barrier contains no open porosity). The difference in chromium chemical potential across the protection layer drives chromium cations (e.g., Cr^{3+}) to potentially diffuse into and through the protection layer. If this occurs, chromium can volatilize from the protection layer surface. Therefore, the material for the protection layer ought to exhibit little or no Cr solubility and possess very low chromium cation diffusivity. Alternatively, the coating can react with the alloy to form a reaction product layer that can function as a Cr barrier. Opposite to the chromium outward diffusion, there is potential oxygen inward diffusion driven by the oxygen chemical potential gradient across the protection layer. This oxygen flux leads to selective oxidation of the substrate alloy and therefore subsequent growth of a chromia or chromia-rich scale between the protection layer and the bulk alloy. Extensive growth of the oxide interlayer may increase the electrical resistance and likelihood of spallation, especially during thermal cycling, and degrade the mechanical integrity of the interconnect. Thus, the protection layer should have low oxygen ion conductivity, σ_o, which is related to the chemical diffusion coefficient of oxygen, D_o, via the Nernst-Einstein equation,

$$k_B T \sigma_o = 4e^2 D_o C_o \qquad (4.6)$$

where e is the electron charge, C_o is the oxygen concentration in the protection layer, k_B is the Boltzman constant, and T is the absolute temperature. In addition to the mass transport properties, thermo-mechanical and chemical stability of the protection layer are essential to maintain its structural integrity during SOFC operation. Accordingly, candidate materials for the protection layer are also required to have a good thermal expansion match to the substrate alloy and be thermo-chemically stable and compatible to adjacent stack components in SOFC stacks.

The aforementioned requirements on surface stability are typical for all exposed areas of the metallic interconnect, as well as other metallic components in an SOFC stack; e.g., some designs use metallic frames to support the ceramic cell. In addition, the protection layer for the interconnect or in particular the "active" areas that interface with electrodes and are in the path of electric current must be electrically conductive. This conductivity requirement distinguishes the interconnect protection layer from many traditional surface modifications, as well as nonactive areas of interconnects, where only surface stability is emphasized. While the electrical conductivity is usually dominated by their electronic conductivity, conductive oxides for protection layer applications often demonstrate a non-negligible oxygen ion conductivity, as well, that leads to the scale growth beneath the protection layer. With this in mind, a high electrical conductivity is always desirable for the protection layers, along with low chromium cation and oxygen anion diffusivity.

Reported examples of protection layers include overlay coatings of the same conductive perovskite compositions as are often used as cathode and interconnect materials in SOFCs. Overall, it appears that the chromites, which exhibit a lower oxygen ion conductivity than many other perovskite compositions, such as cobaltites, provided better protection to the metal substrates by more effectively inhibiting the scale growth beneath the perovskite protection layer [222–224]. Also, the chromites offer a better CTE match to alloy substrates than cobaltites if a BCC substrate alloy such as Ducralloy or an FSS is used, thus providing better thermomechanical stability [222, 225]. One potential concern is the fact that the chromites will release Cr via vaporization, albeit at a relatively low rate [187], which may still lead to an unacceptable degradation in cell performance. In comparison, non-Cr containing perovskites—such as cobaltites, with higher electrical conductivity and higher oxygen ion conductivity than the chromites—offer more effective improvement in surface conductivity. However, the higher ionic conductivity leads to a higher growth rate of the scale beneath the protection layer, thus negatively affecting the surface stability of the coated metallic interconnect. Furthermore there is a concern regarding potential Cr solid state transport through the non-chromium perovskite layers during high-temperature exposure, which can eventually lead to the presence of Cr at the surface of the protection layer and subsequent migration into cells and cell poisoning [226].

In addition to the perovskites, spinel protective layers have also been investigated. As an extension of the beneficial effect of the $(Mn,Cr)_3O_4$ spinel top layer, which grows on Crofer 22 APU, that spinel composition has been applied or thermally grown as a coating on oxidation-resistant alloys to improve their surface and electrical stability [227]. However, to prevent release of Cr via surface vaporization, non-Cr spinels may be more favorable candidates for protection layer applications. Recently, $(Mn,Co)_3O_4$ spinel layers have been applied on FSS interconnects, as shown in Figure 4.11. In particular, $Mn_{1.5}Co_{1.5}O_4$ fabricated with slurry-based approaches has demonstrated promising performance [228–237]. The Mn-Co spinel protection layer appears to be an effective mass barrier to both chromium outward and oxygen inward transport, as indicated by long-term evaluations. Recent work [238] indicated that, after about one year of isothermal heating in air at 800°C, the scale beneath the $Mn_{1.5}Co_{1.5}O_4$ protection layer on Crofer 22 APU grew to only ~3.0 μm thick, compared to a 13 to 15 μm thick scale on bare Crofer 22 APU. Importantly, neither spallation nor Cr penetration across the protection layer was observed. The stability of the spinel

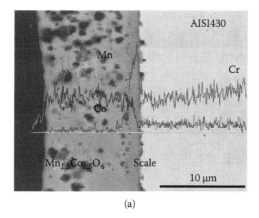

(a)

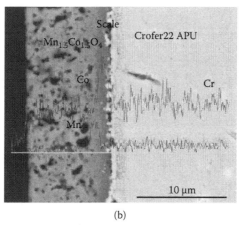

(b)

FIGURE 4.11 SEM images of cross-sections of $Mn_{1.5}Co_{1.5}O_4$ protection layers on (a) AISI430 and (b) Crofer 22 APU after heat treatment at 800°C for 24 h in 2.75%H_2 + Ar and subsequent oxidation at 800°C in air for 100 h [219, 230].

protection layer and its effectiveness as a chromium mass barrier were not affected by thermal cycling. With an electrical conductivity (see Figure 4.12) that is 3 to 4 and 1 to 2 orders of magnitude higher than that of chromia and $MnCr_2O_4$, respectively, [239–241]. Another potential mixed spinel, $(Mn,Cu)_3O_4$, has been reported to have an even higher conductivity of 200 S·cm⁻¹ [242]. Because of this, high-conductivity spinel protection layers drastically improved interfacial contact resistance. Electrochemical evaluation showed that the $Mn_{1.5}Co_{1.5}O_4$ protection layers of FSSs were effective in blocking Cr migration, which resulted in long-term stability of the $(La,Sr)MnO_3$ cathode [243]. Overall, it appears that the $(Mn,Co)_3O_4$ spinels fabricated via optimized approaches are promising coating materials to improve the surface stability of FSS interconnects, minimize contact resistance, and seal off chromium in the metal substrates.

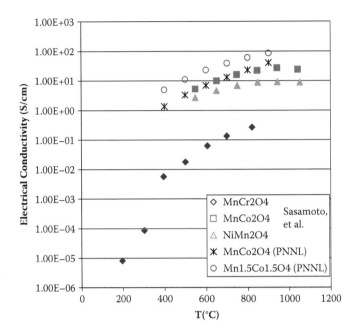

FIGURE 4.12 Electrical conductivity of transition metal oxide spinels.

4.5 SUMMARY AND CONCLUSIONS

As a critical component of SOFC stacks, interconnects perform in a very challenging environment at high temperatures for up to thousands of hours. For a stable performance, the materials for interconnects must satisfy a number of stringent requirements, while maintaining low material and manufacturing costs. Depending on operating temperature, both chromites and high-temperature oxidation resistant alloys can be promising candidates. In the high-temperature range of 900 to 1000°C, the doped chromites with a perovskite structure appear to be suitable, considering their excellent high-temperature stability, satisfactory conductivity, and a good CTE matching. There are, however, challenges for sintering to high density, reduction at the anode side and thus induced size instability, as well as a relative high cost associated with the raw materials and fabrication. For SOFCs operating in a range of 650 to 800°C, cost-effective high-temperature oxidation resistant alloys, such as FSSs, are more favorable. Traditional alloys, however, do not satisfy all of the requirements, so new alloys, which demonstrate improved properties and suitability for the interconnect applications, have been developed but challenges remain in terms of Cr-volatility and long-term surface and electrical stability. As an alternative approach to bulk alloy development, surface modification of metallic interconnects via application of protection layers has proven to be viable for improvement of their stability and mitigation of adverse interactions with cell and stack components. In particular, protection layers fabricated from non-Cr containing conductive oxides, such as $(Mn, Co)_3O_4$ spinels, on the FSS are among the most promising candidates.

Further progress in terms of materials understanding and optimization will be necessary to achieve satisfactory cost-effective, long-term interconnect performance.

REFERENCES

1. Zhu WZ and Deevi SC. Development of interconnect materials for solid oxide fuel cells. *Mater. Sci. Eng. A* 2002:A348;227–243.
2. Minh NQ and Takahashi T. *Science and Technology of Ceramic Fuel Cells.* Amsterdam: Elsevier, 1995;165–198.
3. Sakai N, Yokokawa H, Horita T, and Yamaji K. Lanthanum chromite-based interconnects as key materials for SOFC stack development. *Int. J. Appl. Ceram. Technol.* 2004:1;23–30.
4. Fergus JW. Lanthanum chromite based materials for solid oxide fuel cell interconnects. *Solid State Ionics* 2004:171;1–15.
5. Yang Z, Weil KS, Paxton DM, and Stevenson JW. Selection and evaluation of heat-resistant alloys for SOFC interconnect applications. *J. Electrochem. Soc.* 2003:150;A1188-A1201.
6. Fergus JW. Metallic interconnects for solid oxide fuel cells. *Mater. Sci. Eng. A* 2005:A397;271–283.
7. Nakamura T, Petzow G, and Gauckler LJ. Stability of the perovskite phase $LaBO_3$ (B = V, Cr, Mn, Fe, Co, Ni) in reducing atmosphere. I. Experimental results. *Mater. Res. Bull.* 1979:14;649–659.
8. Onuma S, Miyoshi S, Yashiro K, Kaimai A, Kawamura K, Nigara Y, Kawada T, Mizisaki J, Sakai N, and Yokokawa H. Phase stability of $La_{1-x}Ca_xCrO_{3-\delta}$ in oxidizing atmosphere. *J. Sol. St. Chem.* 2003:170;68–74.
9. Miyoshi S, Onuma S, Kaimai A, Matsumoto H, Yashiro K, Kawada T, Mizusaki J, and Yokokawa H. Chemical stability of $La_{1-x}Sr_xCrO_3$ in oxidizing atmospheres. *J. Solid State Chem.* 2004:177; 4112–4118.
10. Sfeir J. $LaCrO_3$-based anodes: Stability considerations. *J. Power Sources* 2003: 118; 276–285.
11. Peck DH, Miller M, and Hilpert K. Phase diagram study in the $CaO-Cr_2O_3-La_2O_3$ system in air and under low oxygen partial pressure. *Solid State Ionics* 1999:123;47–57.
12. Peck DH, Miller M, and Hilpert K. Phase diagram study in the $SrO-Cr_2O_3-La_2O_3$ system in air and under low oxygen partial pressure. *Solid State Ionics* 1999:123;59–65.
13. Ianculescu A, Braileanu A, Pasuk I, and Zaharescu M. Phase formation study of alkaline earth-doped lanthanum chromites. *J. Therm. Anal. Calorimetry* 2001:66;501–507.
14. Cheng J and Navrostky A. Energetics of $La_{1-x}A_xCrO_{3-\delta}$ perovskites (A = Ca or Sr). *J. Solid State Chem.* 2005:178;234–244.
15. Liu X, Su W, Lu Z, Liu J, Pei L, Liu W, and He L. Mixed valence state and electrical conductivity of $La_{1-x}Sr_xCrO_3$. *J. Alloys Comp.* 2000:305;21–23.
16. Shannon RD. Revised effective ionic radii and systematic studies of interatomic distances in halides and chalcogenides. *Acta Cryst. A* 1976:32;751–767.
17. Armstrong TJ, Stevenson JW, Pederson LR, and Raney PE. Dimensional instability of doped lanthanum chromite. *J. Electrochem. Soc.* 1996:143;2919–2925.
18. Peck D-H, Miller M, and Hilpert K. Vaporization and thermodynamics of $La_{1-x}Ca_xCrO_{3-\delta}$ investigated by Knudsen effusion mass spectrometry. *Solid State Ionics* 2001:143;391–400.
19. Peck D-H, Miller M, and Hilpert K. Vaporization and thermodynamics of $La_{1-x}Sr_xCrO_{3-\delta}$ investigated by Knudsen effusion mass spectrometry. *Solid State Ionics* 2001:143;401–412.

20. Smith DS, Sayer M, and Odier P. The formation and characterization of a ceramic-ceramic interface between stabilized zirconia and lanthanum chromite. *J. Physique–Colloque* 1986:C1;150–157.
21. Carter JD, Appel CC, and Mogensen M. Reactions at the calcium doped lanthanum chromite-yttria stabilized zirconia interface. *J. Sol. St. Chem.* 1996:122;407–415.
22. Yamamoto T, Itoh H, Mori M, Mori N, Watanabe T, Imanishi N, Takeda Y, and Yamamoto O. Chemical stability between NiO/8YSZ cermet and alkaline-earth metal substituted lanthanum chromite. *J. Power Sources* 1996:61;219–222.
23. Nishiyama H, Aizawa M, Sakai N, Yokokawa H, Kawada T, and Dokiya M. Property of (La,Ca)CrO$_3$ for interconnect in solid oxide fuel cell (part 2). Durability. *J. Ceram. Soc. Japan* 2001:109;527–534.
24. Carter JD, Appel CC and Mogensen M. Reactions at the calcium doped lanthanum chromite—yttria stabilized zirconia interface. *J. Sol. St. Chem.* 1996:122;407–415.
25. Yasumoto K, Itoh H, and Yamamoto T. Anode supported interconnect for electrolyte membrane SOFC. *Electrochem. Soc. Proc.* 2003:2003(07);832–840.
26. Mori M, Itoh H, Mori N, Abe T, Yamamoto O, Takeda Y, and Imanishi N. Reaction between alkaline earth metal doped lanthanum chromite and yttria stabilized zirconia: In Badwal SPS, Bannister MJ, and Hannink RHJ. *Science and Technology of Zirconia V.* Lancaster, PA: Technomic Publishing Co., 1993:776–785.
27. Hatchwell CE, Sammes NM, Tompsett GA, and Brown IWM. Chemical compatibility of chromium-based interconnect related materials with doped cerium oxide electrolyte. *J. Eur. Ceram. Soc.* 1999:19;1697–1703.
28. Webb JB, Sayer M, and Mansingh A. Polaronic conduction in lanthanum strontium chromite. *Can. J. Phys.* 1977:55;1725–1731.
29. Karim DP and Aldred AT. Localized level hopping transport in La(Sr)CrO$_3$. *Phys. Rev. B* 1979:20;2255–2263.
30. Meadowcroft DB. Some properties of strontium-doped lanthanum chromite. *Brit. J. Appl. Phys.* 1969: D2;1225–1233.
31. Sakai N, Kawada T, Yokokawa H, Dokiya M, and Iwata T. Sinterability and electrical conductivity of calcium-doped lanthanum chromites. *J. Mater. Sci.* 1990:25;4531–4534.
32. Yasuda I and Hikita T. Electrical conductivity and defect structure of calcium-doped lanthanum chromites. *J. Electrochem. Soc.* 1993:140;1699–1704.
33. Armstrong TR, Stevenson JW, Hasinska K, and McCready DE. Synthesis and properties of mixed lanthanide chromide perovskites. *J. Electrochem. Soc.* 1998:147;4282–4289.
34. Mori M, Yamamoto T, Itoh H, and Watanabe T. Compatibility of alkaline earth metal (Mg,Ca,Sr)-doped lanthanum chromites as separators in planar-type high-temperature solid oxide fuel cells. *J. Mater. Sci.* 1997:32;2423–2431.
35. Schafer W and Schmidberger R. Ca and Sr doped LaCrO$_3$: Preparation, properties, and high temperature applications. In: Vincenzini P, editor. *High Tech Ceramics.* Amsterdam, the Netherlands: Elsevier Science Publishers, 1987:1737–1742.
36. Stakkestad G, Faaland S, and Sigvartsen T. Investigation of electrical conductivity and Seebeck coefficient of Ca- and Sr-doped LaCrO$_3$. *Phase Transitions* 1996:58;159–173.
37. Kim J-H, Peck D-H, Song R-H, Lee G-Y, Shin D-R, Hyun S-H, Wackerl J, and Hilpert K. Synthesis and sintering properties of (La$_{0.8}$Ca$_{0.2-x}$Sr$_x$)CrO$_3$ perovskite materials for SOFC interconnect. *J. Electroceramics* 2006:17;729–733.
38. Chakraborty A, Basu RN, and Maiti HS. Low temperature sintering of La(Ca)CrO$_3$ prepared by an autoignition process. *Mater. Lett.* 2000:45;162–166.
39. Tanasescu S, Orasanu A, Berger D, Jitaru I, and Shoonman J. Electrical conductivity and thermodynamic properties of some alkaline earth-doped lanthanum chromites. *Int. J. Thermophysics* 2005:26;543–557.

40. Ghosh S, Sharma AD, Basu RN, and Maiti HS. Synthesis of $La_{0.7}Ca_{0.3}CrO_3$ SOFC interconnect using a chromium source. *Electrochem. Solid-State Lett.* 2006:9;A516–A519.

41. Kumar A, Devi PS, and Maiti HS. A novel approach to develop dense lanthanum calcium chromate sintered ceramics with very high conductivity. *Mater. Chem.* 2004:16;5562–5563.

42. Homma K, Nakamura F, Ohba N, Mitsui A, and Hashimoto T. Improvement of sintering property of $LaCrO_3$ system by simultaneous substitution of Ca and Sr. *J. Ceram. Soc. Japan* 2007:115;81–84.

43. Zhong H-H, Zhou X-L, Liu X-Q, and Meng G-Y. Synthesis and electrical conductivity of perovskite $Gd_{1-x}Ca_xCrO_3$ ($0 \leq x \leq 0.3$) by auto-ignition process. *Solid State Ionics* 2005:176;1057–1061.

44. Zhou X, Deng F, Zhu M, Meng G, and Liu X. High performance composite interconnect $La_{0.7}Ca_{0.3}CrO_3$/20 mol% $ReO_{1.5}$ doped CeO_2 (Re = Sm, Gd, Y) for solid oxide fuel cells. *J. Power Sources* 2007:164;293–299.

45. Weber WJ, Griffin CW, and Bates JL. Effects of cation substitution on electrical and thermal transport properties of $YCrO_3$ and $LaCrO_3$. *J. Am. Ceram. Soc.* 1987:70;265–270.

46. Liu X, Su W, and Lu Z. Study on synthesis of $Pr_{1-x}Ca_xCrO_3$ and their electrical properties. *Mater. Chem. Phys.* 2003:82;327–330.

47. Zhong Z. Stoichiometric lanthanum chromite based ceramic interconnects with low sintering temperature. *Solid State Ionics* 2006:177;757–764.

48. Simner SP, Hardy JS, and Stevenson JW. Sintering and properties of mixed lanthanide chromites. *J. Electrochem. Soc.* 2001:148;A351–A360.

49. Zhou X-L, Ma J-J, Deng F-J, Meng G-Y, and Liu X-Q. A high performance interconnecting ceramic for solid oxide fuel cells (SOFCs). *Solid State Ionics* 2006:177;3461–3466.

50. Zhou X, Deng F, Zhu M, Meng G, and Liu X. Novel composite interconnecting ceramics $La_{0.7}Ca_{0.3}CrO_{3-\delta}$/ $Ce_{0.2}Sm_{0.8}O_{1.9}$ for solid oxide fuel cells. *Mater. Res. Bull.* 2007:42; 1582–1588.

51. Zhou X, Ma J, Deng F, Meng G, and Liu X. Preparation and properties of ceramic interconnecting materials, $La_{0.7}Ca_{0.3}CrO_{3-\delta}$ doped with GDC for IT-SOFCs. *J. Power Sources* 2006:162;279–285.

52. Zhou X, Zhu M, Deng F, Meng G, and Liu X. Electrical properties, sintering and thermal expansion behavior of composite interconnecting materials $La_{0.7}Ca_{0.3}CrO_{3-\delta}$/ $Y_{0.2}Ce_{0.8}O_{1.9}$ for SOFCs. *Acta Mater.* 2007:55;2113–2118.

53. Koc R and Anderson HU. Electrical conductivity and Seebeck coefficient of (La,Ca) $(Cr,Co)O_3$. *J. Mater. Sci.* 1992:27;5477–5482.

54. Kharton VV, Yaremchenko AA, and Naumovich EN. Research on the electrochemistry of oxygen ion conductors in the former Soviet Union. II. Perovskite-related oxides. *J. Sol. St. Electrochem.* 1999:3;303–326.

55. Liu Z, Dong D, Huang Z, Lü Z, Sui Y, Wang X, Miao J, Shen ZX, and Su W. A novel interconnect material for SOFCs. *Electrochem. Solid-State Lett.* 2005:8;A250–A252.

56. Hilpert K, Steinbrech RW, Boroomand F, Wessel E, Meschke F, Zuev A, Teller O, Nickel N, and Singheiser L. Defect formation and mechanical stability of perovskites based on $LaCrO_3$ for solid oxide fuel cells (SOFC). *J. Eur. Ceram. Soc.* 2003:23;3009–3020.

57. Zuev A, Singheiser L, and Hilpert K. Defect structure and isothermal expansion of A-site and B-site substituted lanthanum chromites. *Solid State Ionics* 2002:147;1–11.

58. Anderson HU, Nasrallah MM, Flandermeyer BK, and Agrawal AK. High-temperature redox behavior of doped $SrTiO_3$ and $LaCrO_3$. *J. Sol. St. Chem.* 1985:56;325–334.

59. Anderson HU, Kuo JH, and Sparlin DM. Review of defect chemistry of $LaMnO_3$ and $LaCrO_3$. *Electrochem. Soc. Proc.* 1989:89(11);111–128.

60. Yasuda I and Hishinuma M. Electrical conductivity and chemical diffusion coefficient of Sr-doped lanthanum chromites. *Solid State Ionics* 1995:80;141–150.

61. Vashook V, Müller R, Zosel J, Ahlborn K, Ullmann H, and Guth U. Catalytic and electrical properties of SOFC anode material based on lanthan-chromite-titanate. *Mat.-wiss. U. Werkstofftech.* 2002:33;335–338.

62. González-Cuenca M, Zipprich W, Boukamp BA, Pudmich G, and Tietz F. Impedance studies on chromite-titanate porous electrodes under reducing conditions. *Fuel Cells* 2001:1;256–264.

63. Bansal KP, Kumari S, Das BK, and Jain BC. Electrical conduction in titania-doped lanthanum chromite ceramics. *J. Mater. Sci.* 1981:16;1994–1998.

64. Yu CJ, Anderson HU, and Sparlin DM. High-temperature defect structure of Nb-doped $LaCrO_3$. *J. Sol. St. Chem.* 1989:78;242–249.

65. Huang W and Gopalan S. Bi-layer structures as solid oxide fuel cell interconnections. *J. Power Sources* 2006:154;180–183.

66. Paulik SW, Baskaran S, and Armstrong TR. Mechanical properties of calcium- and strontium-substituted lanthanum chromite. *J. Mater. Sci.* 1997:33;2397–2404.

67. Yasuda I and Hishinuma M. Lattice expansion of acceptor-doped lanthanum chromites under high-temperature reducing atmospheres. *Electrochemistry (Tokyo)* 2000:68;526–530.

68. Mori M and Hiei Y. Thermal expansion behavior or titanium-doped La(Sr)CrO₃ solid oxide fuel cell interconnects. *J. Am. Ceram. Soc.* 2001:84;2573–2578.

69. Höfer HE and Kock WF. Crystal chemistry and thermal behavior in the La(Cr,Ni)O₃ perovskite system. *J. Electrochem. Soc.* 1993:140;2889–2894.

70. Boroomand F, Wessel E, Bausinger H, and Hilpert K. Correlation between defect chemistry and expansion during reduction of doped LaCrO₃ interconnects for SOFCs. *Solid State Ionics* 2000:129;251–258.

71. Mori K, Miyamoto H, Takenobu K, and Matsudaira T. Controlling the chromite expansion in reducing atmosphere at high temperature. In: A.J. McEvoy, editor. *European Solid Oxide Fuel Cell Forum Proceedings* Vol. 2. Lucerne, Switzerland: The European Fuel Cell Forum, 2000;875–880.

72. Mori M. Enhancing effect on densification and thermal expansion compatibility for $La_{0.8}Sr_{0.2}Cr_{0.9}Ti_{0.1}O_3$-based SOFC interconnect with C-site doping. *J. Electrochem. Soc.* 2002:149;A797–A803.

73. Larsen PH, Hendriksen PV, and Mogensen M. Dimensional stability and defect chemistry of doped lanthanum chromites. *J. Thermal Analysis* 1997:49;1263–1275.

74. Montross CS, Yokokawa H, Dokiya M, and Bekessy L. Mechanical properties of magnesia-doped lanthanum chromite versus temperature. *J. Am. Ceram. Soc.* 1995:78;1869–1872.

75. Sammes NM and Ratnaraj R. High temperature mechanical properties of $La_{1-x}Sr_xCr_{1-y}Co_yO_3$ for SOFC interconnect. *J. Mater. Sci.* 1995:30;4523–4526.

76. Yasuda I, Ogiwara T, and Yakabe H. Accumulation of detailed property data for LaCrO₃-based interconnector materials. *Electrochem. Soc. Proc.* 2001:2001(16); 783–792.

77. Montross CS, Yokokawa H, and Dokiya M. Toughening in lanthanum chromite due to metastable phase. *Scripta Mater.* 1996:34;913–917.

78. Sammes NM, Ratnaraj R, and Fee MG. The effect of sintering on the mechanical properties of SOFC ceramic interconnect materials. *J. Mater. Sci.* 1994:29; 4319–4324.

79. Meschke F, Singheiser L, and Steinbrech RW. Mechanical properties of doped LaCrO₃ interconnects after exposure to SOFC-relevant conditions. In: McEvoy AJ, editor. *European Solid Oxide Fuel Cell Forum Proceedings* Vol. 2. Lucerne, Switzerland: The European Fuel Cell Forum, 2000:865–873.

80. Paulik SW, Baskaran S, and Armstrong TR. Mechanical properties of calcium-subsituted yttrium chromite. *J. Mater. Lett.* 1999:18;819–822.

81. Mori M and Sammes NM. Sintering and thermal expansion characterization of Al-doped and Co-doped lanthanum strontium chromites synthesized by the Pechini method. *Solid State Ionics* 2002:146;301–312.

82. Tai TW and Lessing PA. Modified resin-intermediate of perovskite powders. Part II. Processing for fine, nonagglomerated strontium-doped lanthanum chromite powders. *J. Mater. Res.* 1992:7;511–519.

83. Ding X, Liu Y, Gao L, and Guo L. Synthesis and characterization of doped $LaCrO_3$ perovskite prepared by EDTA-citrate complexing method. *J. Alloys Compounds* 2007: 458; 346–350.

84. Berger D, Jitaru I, Stanica N, Perego R, and Schoonman J. Complex precursors for doped lanthanum chromite synthesis. *J. Mater. Synth. Proc.* 2001:9;137–142.

85. Marinho ÉP, Souza AG, de Melo DS, Santos IMG, Melo DMA, and da Silva WJ. Lanthanum chromites partially substituted by calcium strontium and barium synthesized by urea combustion—Thermogravimetric study. *J. Thermal Analysis Calorimetry* 2007:87;801–804.

86. Yang YJ, Wen T-L, Tu H, Wang D-Q, and Yang J. Characteristics of lanthanum strontium chromite prepared by glycine nitrate process. *Solid State Ionics* 2000: 135;475–479.

87. Wen T-L, Wang D, Chen M, Tu H, Lu Z, Zhang Z, Nie H, and Huang W. Material research for planar SOFC stack. *Solid State Ionics* 2002:148;513–519.

88. Deshpande K, Mukasyan A, and Varma A. Aqueous combustion synthesis of strontium-doped lanthanum chromite ceramics. *J. Am. Ceram. Soc.* 2003:86;1149–1154.

89. Ovenstone J, Chan KC, and Ponton B. Hydrothermal processing and characterisation of doped lanthanum chromite for use in SOFCs. *J. Mater. Sci.* 2002:37;3315–3322.

90. Tai L-W and Lessing PA. Tape casting and sintering of strontium-doped lanthanum chromite for a planar solid oxide fuel cell bipolar plate. *J. Am. Ceram. Soc.* 1991:74;155–160.

91. George RA and Bessette NF. Reducing the manufacturing cost of tubular SOFC technology. *J. Power Sources* 1998:71;131–137.

92. Suzuki M, Sasaki H, and Kajimura A. Oxide ion conductivity of doped lanthanum chromite thin film interconnects. *Solid State Ionics* 1997:96;83–88.

93. Yokokawa H, Sakai N, Kawada T, and Dokiya M. Chemical thermodynamic considerations in sintering of $LaCrO_3$-based perovskites. *J. Electrochem. Soc.* 1991:138;1018–1027.

94. Yokokawa H. Materials thermodynamics: A bridge from fundamentals to applications. *Electrochem. Soc. Proc.* 2002:2002(26);1–15.

95. Paulik SW, Hardy J, Stevenson JW, and Armstrong TR. Sintering of non-stoichiometric strontium doped lanthanum chromite, *J. Mater. Lett.* 2000:19;863–865.

96. Ding X, Liu Y, Gao L, and Guo L. Effects of cation substitution on thermal expansion and electrical properties of lanthanum chromites. *J. Alloys Compounds* 2006:425;318–322.

97. Chick LA, Kiu J, Stevenson JW, Armstrong TA, McCready DE, Maupin GD, Coffey GW, and Coyle CA. Phase transitions and transient liquid-phase sintering in calcium-substituted lanthanum chromite. *J. Am. Ceram. Soc.* 1997:80;2109–2120.

98. Simner SP, Hardy JS, Stevenson JW, and Armstrong TR. Sintering of lanthanum chromite using strontium vanadate. *Solid State Ionics* 2000:128;53–63.

99. Ding X and Guo L. Effect of CaF_2 on the sintering and thermal expansion of $La_{0.85}Sr_{0.15}Cr_{0.95}O_3$. *J. Mater. Sci.* 2004:41;6185–6188.

100. Subasri R, Mathews T, Swaminathan K, and Sreedharan OM. Microwave assisted synthesis of $La_{1-x}Sr_xCrO_3$ (x=0.05, 0.15 and 0.30) and their thermodynamic characterization by fluoride emf method. *J. Alloys Compounds* 2003:354;193–197.

101. Takeuchi T, Takeda Y, Funahashi R, Aihara T, Tabuchi M, Kageyama H, Nomura K, Tanimot K, and Miyazaki Y. Spark-plasma sintering of interconnector material $(La_{0.9}Sr_{0.1})CrO_3$. *Electrochem. Soc. Proc.* 2001:2001(16);865–871.

102. Shiryaev AA, Nersesyan MD, Ming Q, and Luss D. Thermodynamic feasibility of SHS of SOFC materials. *J. Mater. Synth. Proc.* 1999:7;83–90.

103. Wei TC and Phillips J. *Adv. in Catalysis* 1996:41;359–421.

104. Bao X, Lehmpfuhl G, Weinberg G, Schlogl R, and Ertl G. *J. Chem. Soc. Faraday Trans.* 1992:88;865–872.

105. Hondros ED, Moore AJW. *Acta Metall.* 1960:8;647–653.

106. Fromm E and Gebhardt E. *Gase und Kohlenstoff in Metallen*, 1976, Springer-Verlag, Berlin/Heidelberg.

107. Klueh RL and Mullins WW. *Trans. Met. Society* 1968:242;237–244.

108. Singh, P, Yang Z, Viswanathan V, and Stevenson JW. *J. Mater. Eng. Perform.* 2004:13;287–294.

109. Sims CT, Stoloff NS, and Hagel WC. *Superalloys II*, 1987, John Wiley & Sons, New York.

110. Davis JR. *ASM Specialty Handbook: Stainless Steels*, 1994, ASM International®, Materials Park, OH.

111. Wasielewski GE and Robb RA. *High Temperature Oxidation in The Superalloys*, Sims CS, Hagel W, (eds.) p. 287, 1972, John Wiley & Sons, Inc., New York.

112. Kofstad P. *Nonstoichiometry, Diffusion, and Electrical Conductivity in Binary Metal Oxides*, 1983, Robert E. Krieger Publishing Company, Florida.

113. Kofstad P and Bredesen R. *Solid State Ionics* 1992:52;69–75.

114. Holt A and Kofstad P. *Solid State Ionics* 1994:69;137–143.

115. Holt A and Kofstad P. *Solid State Ionics* 1994:69;127–136.

116. Kofstad P. In *Proc. 2nd European SOFC Forum*, Thorstensen B. (ed.) pp. 479–490, 1996, The European Fuel Cell Forum, Lucerne, Switzerland.

117. Quadakkers WJ, Greiner H, and Kock W. In *Proc. 1st European SOFC Forum*, Bossel U, (ed.) pp. 525-544, 1994, The European Fuel Cell Forum, Lucerne, Switzerland.

118. Kock W, Martinz H-P, Greiner H, and Janousek M. In *SOFC IV-Electrochem. Soc. Proc. PV 95-1*, Dokiya M, Yamamoto O, Tagawa H, Singhal SC, (eds.) pp. 841–849, 1995, The Electrochemical Society, Pennington, NJ.

119. Greiner H, Grogler T, Kock W, and Singer RF. In *SOFC IV-Electrochem. Soc. Proc. PV 95-1*, Dokiya M, Yamamoto O, Tagawa H, Singhal SC, (eds.) pp. 879–888, 1995, The Electrochemical Society, Pennington, NJ.

120. Buchkremer HP, Diekmann U, de Haart LGJ, Kabs H, Stimming U, and Stover D. In *SOFC V- Electrochem. Soc. Proc. PV 95-1*, Stimming U, Singhal SC, Tagawa H, Lehnert W, (eds.) pp. 160–170, 1997, The Electrochemical Proceedings Series, Pennington, NJ.

121. Malkow T, Crone UVD, Laptev AM, Koppitz T, Breuer U, and Quadakkers WJ. In *SOFC V- Electrochem. Soc. Proc. PV 95-1*, Stimming U, Singhal SC, Tagawa H, Lehnert W, (eds.) pp. 1244–1252, 1997, The Electrochemical Proceedings Series, Pennington, NJ.

122. Taniguchi S, Kadowaki M, Yasuo T, Akiyama Y, Miyake Y, and Nishio K. *Denki Kagaku* 1997:65;574–579.

123. Buchkremer HP, Diekmann U, de Haart LGJ, Kabs H, Stover D, and Vinke IC. In *Proc. 3rd European SOFC Forum*, P. Stevens, (ed.) Vol. 1, pp. 143–150, 1998, The European Fuel Cell Forum, Lucerne, Switzerland.

124. Linderoth S, Hendriksen PV, Mogensen M, and Langvad, N. *J. Mater. Sci.* 1996:31;5077–5082.

125. England DM and Virkar AV, *J. Electrochem. Soc.* 1999:146;3196–3202.

126. England DM and Virkar AV, *J. Electrochem. Soc.* 2001:148;A330–338.

127. Yang Z, Xia G-G, Singh P, and Stevenson JW. *J. Power Sources* 2006:160;1104–1110.

128. Geng SJ, Zhu JH, and Lu ZG. *Solid State Ionics* 2006:177;559–568.

129. Geng SJ, Zhu JH, and Lu ZG. *Script. Mater.* 2006:55;239–242.

130. Miyake Y, Yasuo T, Akiyama Y, Tangiguchi S, Kadowaki M, Kawamura H, and Saitoh T. In *SOFC IV-Electrochem. Soc. Proc. PV 1995-1*, Dokiya M, Yamamoto O, Tagawa H, Singhal SC, (eds.) pp.100–109, 1995, The Electrochemical Society, Pennington, NJ.

131. Iwata T, Kadokawa N, and Takenoiri S. In *SOFC IV-Electrochem. Soc. Proc. PV 95-1*, (eds.), Dokiya M, Yamamoto O, Tagawa H, and Singhal SC, pp.110–119, 1995, The Electrochemical Society, Pennington, NJ.

132. Shiomitsu T, Kadowaki T, Ogawa T, and Maruyama T. In *SOFC IV-Electrochem. Soc. Proc. PV 95-1*, Dokiya M, Yamamoto O, Tagawa H, Singhal SC, (eds.) pp. 850–857, 1995, The Electrochemical Society, Pennington, NJ.

133. Mankins WL and Lamb S. *ASM Handbook*, Vol. 2, pp. 428–445, 1990, ASM International.

134. Chen L, Yang Z, Jha B, Xia G, and Stevenson JW. *J. Power Sources* 2005:152;40–45.

135. Chen L, Jha B, Yang Z, Xia G-G, Stevenson JW, and Singh P. *J. Mater. Eng. Perform.* 2006:15;399–403.

136. Quadakkers WJ, Shemet V, and Singheiser L. U.S. Patent No. 2003059335, 2003.

137. Quadakkers WJ, Malkow T, Piron-Abellan J, Flesch U, Shemet V, and Singheiser L. In *Proc. 4th European SOFC Forum*, Vol. 2, p. 827-836, 2000, The European Fuel Cell Forum, Lucerne, Switzerland.

138. Buchkremer HP, Diekmann U, de Haart LGJ, Kabs H, Stover D, and Vinke IC. In *Proc. 3rd European SOFC Forum*, Stevens P, (ed.) Vol. 1, pp. 143–150, 1998, The European Fuel Cell Forum, Lucerne, Switzerland.

139. Yang Z, Hardy JS, Walker MS, Xia G, Simner SP, and Stevenson JW. *J. Electrochem. Soc.* 2004:151;A1825–A1831.

140. Fava FF, Barraille I, Lichanot A, Larrieu C, and Dovesi R. *J. Phys. Condense Mater.* 1997:9;10715–10724.

141. Lu Z, Zhu J, Payzant EA, and Paranthaman MP. *J. Am. Ceram. Soc.* 2005:88;1050–1053.

142. Horita T, Xiong Y, Yamaji K, and Sakai N. *J. Electrochem. Soc.* 2003:150;A243–248.

143. Uehara T, Ohno T, and Toji A. in *Proc. 5th European SOFC Forum*, Huijsmans J, (ed.) pp. 281-288, 2002, The European Fuel Cell Forum, Lucerne, Switzerland.

144. Quadakkers WJ, Greiner H, Kock W, Buchkremer HP, Hilpert K, and Stover D. In *Proc. 2nd European SOFC Forum*, Thorstensen B, (ed.) pp. 297–305, 1996, The European Fuel Cell Forum, Lucerne, Switzerland.

145. Huang K, Hou PY, and Goodenough, JB. *Solid State Ionics* 2000:129;237–250.

146. Meulenberg WA, Uhlenbruck S, Wessel E, Buchkremer HP, and Stover D. *J. Mater. Sci.* 2003:38;507–513.

147. Brylewski T, Nanko M, Maruyama T, and Przybylski K. *Solid State Ionics* 2001:143;131–150.

148. Geng SJ, Zhu JH, and Lu ZG. *Electrochem. Solid-State Lett.* 2006:9;A211–214.

149. Zeng Z and Natesan K. *Solid State Ionics* 2004:167;9–16.

150. Toh CH, Munroe PR, Young DJ, and Foger K, *Mater. High Temp.* 2003:20;129–139.

151. Horita T, Xiong Y, Kishimoto H, Yamaji K, Sakai N, Brito ME, and Yokokawa H. *J. Electrochem. Soc.* 2005:152;A2193–2198.

152. Horita T, Xiong Y, Kishimoto H, Yamaji K, Sakai N, and Yokokawa H. *Surf. Interface Anal.* 2004:36;973–976.

153. Jian L, Huezo J, and Ivey DG. *J. Power Sources* 2003:123;151–162.

154. Yang Z, Walker MS, Singh P, and Stevenson JW. *Electrochem. Solid State Lett.* 2003:6;B35–37.

155. Yang Z, Walker MS, Singh P, Stevenson JW, and Norby T. *J. Electrochem. Soc.* 2004:151;B669–678.
156. Singh P, Yang Z, Viswanathan V, and Stevenson JW. *J. Mater. Perform. Eng.* 2004:13;287–294.
157. Yang Z, Xia G-G, Singh P, and Stevenson JW. *Solid State Ionics* 2005:176;1495–1503.
158. Ziomek-Moroz M, Cramer SD, Holcomb GR, Covino, Jr BS, Bullard SJ, and Singh P. In *Corrosion 2005*, paper No. 10, NACE International, Houston, TX.
159. Holcomb GR, Ziomek-Moroz M, Cramer SD, Covino, Jr. BS, and Bullard SJ. *J. Mater. Eng. Perform.* 2006:15;404–409.
160. Kurokawa H, Kawamura K, and Maruyama T. *Solid State Ionics* 2004:168;13–21.
161. Quadakkers WJ, Hansel M, and Rieck T. *Mater. & Corro.* 1999:49;252–257.
162. Larring Y, Hangsrud R, and Norby T. *J. Electrochem. Soc.* 2003:150;B374–379.
163. Kofstad P. *High Temperature Corrosion*, 2nd ed., 1988, Elsevier Applied Science, London.
164. Bongartz K, Quadakkers WJ, Pfeifer JP, and Becker JS. *Surf. Sci.* 1993:292; 196–208.
165. Rapp RA. *Metall. Trans. A* 1984:15A;765–782.
166. Huczkowski P, Shemet V, Piron-Abellan J, Singheiser L, Quadakkers WJ, and Christiansen N. *Mater. Corrosion* 2004:55;825–830.
167. Huczkowski P, Ertl S, Piron-Abellan J, Christiansen N, Höflerb T, Shemet V, Singheiser L, and Quadakkers WJ. *Mater. High Temp.* 2005:22;253–262.
168. Huczkowski P, Christiansen N, Shemet V, Piron-Abellan J, Singheiser L, and Quadakkers WJ, *J. Fuel Cell Sci. Tech.* 2004:1;30–34.
169. Hou P and Stringer J. *Oxid. Met.* 1992:38;323–345
170. Golightly F, Stott H, and Wood G. *Oxid. Met.* 1976:10;163–187.
171. Pint B. *Oxid. Met.* 1996:45;1–37.
172. Pieraggi B and Rapp RA. *J. Electrochem. Soc.* 1993:140;2844–2850.
173. Kofstad P. *Oxid. Met.* 1995:44;3–27.
174. Douglass DL, Kofstad P, Rahmel A, and Wood GC. *Oxid. Met.* 1996:45;529–620.
175. Kvernes I, Oliveira M, and Kofstad P. *Corrosion Sci.* 1977:17;237–252.
176. Shen J, Zhou L, and Li T. *Oxid. Met.* 1997:38;347–356.
177. Pint BA. *J. Eng. Gas Turbine and Power* 2006:128;370–376.
178. Fujii CT and Meussner RA. *Corro. Iron & Steel* 1964:111;1215–1221.
179. Lefrancois PA and Hoyt WB. *Corrosion* 1963:19;360–368.
180. Grabke HJ, Muller-Lorenz EM, Eltester B, and Lucas M. *Mater. High Temp.* 2000:17;339–345.
181. Toh CH, Munroe PR, and Young DJ. *Mater. High Temp.* 2003:20;527–534.
182. Nakagawa K, Matsunaga Y, and Yanagisawa T. *Mater. High Temp.* 2003:20;67–73.
183. Wood GC, Wright IG, Hodgkiess T, and Whittle DP. *Werkst Korros* 1970:21;900–910.
184. Yang Z, Xia G-G, Singh P, and Stevenson JW. *J. Electrochem. Soc.* 2006:153;A1873–1879.
185. Graham HG and Davis HH. *J. Am. Ceram. Soc.* 1971:54;89–93.
186. Das D, Miller M, Nickel H, and Hilpert K. In *Proc.1st European SOFC Forum*, Bossel U, (ed.) pp. 703–715, 1994, The European Fuel Cell Forum, Lucerne, Switzerland.
187. Hilpert K, Das D, Miller M, Peck DP, and Weiß R. *J. Electrochem. Soc.* 1996:143;3642–3647.
188. Jacobson N, Myers D, Opila E, and Copland E. *J. Phys. Chem. Sol.* 2005:66;471–478.
189. Gindorf C, Singheiser L, and Hilpert K. *J. Phys. Chem. Sol.* 2005:66;384–387.
190. Konycheva E, Penkalla H, Wessel E, Seeling U, Singheiser L, and Hilpert K. In *SOFC IX-Electrochem. Soc. Proc. PV-2005-07*, Singhal SC, Mizusaki J, (eds.) pp. 1874–1184, 2005, The Electrochemical Society, Pennington, NJ.

191. Hojda R and Paul L. In *Proc. MS&T2005: Materials For Hydrogen Economy*, Petrovic JJ, Anderson IE, Adams TM, Sandrock G, Legzdins CF, Stevenson JW, Yang Z. (eds.) pp. 155–163, 2005, The Minerals, Metals & Materials Society, Warrendale, PA.

192. Yang Z, Xia G-G, Maupin GD, and Stevenson JW. *Surf. Coat. Tech.* 2006: 201; 4476–4483.

193. Taniguchi S, Kadowaki M, Kawamura H, Yasuo T, Akiyama Y, Miyake Y, and Saitoh T. *J. Power Sources* 1995:55;73–79.

194. Badwal SPS, Deller R, Foger K, Ramprakash Y, and Zhang JP, Solid *State Ionics* 1997:99;297–310.

195. Jiang SP, Zhang JP, and Foger K. *J. Electrochem. Soc. 2000:*147;3195–3205.

196. Jiang SP, Zhang JP, and Foger K. *J. Electrochem. Soc.* 2001:148; C447–455.

197. Jiang SP, Zhang S, and Zhen YD. *J. Mater. Res.* 2005:20;747–758.

198. Jiang SP, Zhang JP, and Zheng XG. *J. Eur. Ceram. Soc.* 2002:22;361–373.

199. Paulson SC and Birss VI. *J. Electrochem. Soc.* 2004:151;A1961–1968.

200. Konysheva E, Penkalla H, Wessel E, Mertens J, Seeling U, Singheiser L, and Hilpert K. *J. Electrochem. Soc.* 2006:153;A765–773.

201. Matsuzaki Y and Yasuda I. *Solid State Ionics* 2000:132;271–278.

202. Jiang SP, Zhang S, and Zhen YD. *J. Electrcochem. Soc.* 2006:153;A127–A134.

203. Kim JY, Sprenkle VL, Canfield NL, Meinhardt KD, and Chick LA. *J. Electrochem. Soc.* 2006:153;A880–A886.

204. Simner SP, Anderson MD, Xia G-G, Yang Z, Pederson LR, and Stevenson JW. *J. Electrochem. Soc.* 2006:152;A740–A745.

205. Eichler K, Solow G, Otschik P, and Schafferath W. *J. Eur. Ceram. Soc.* 1999:19; 1101-1104.

206. Meinhardt KD, Vienna JD, Armstrong TR, and Peterson LR. U.S. patent No. 6430966, 2001.

207. Sohn SB, Choi SY, Kim GH, Song HS, and Kim GD. *J. Non-Cryst. Solids* 2002:297;103–112.

208. Schwickert T, Geasee P, Janke A, Diekmann U, and Conradt R. In *Proc. Int. Brazing and Soldering Conf.*, p. 116, 2000, Albuquerque, New Mexico.

209. Sohn S-B, Choi S-Y, Kim G-H, Song H-S, and Kim G-D. *J. Amer. Ceram. Soc.* 2004:87;254–260.

210. Yang Z, Meinhardt KD, and Stevenson JW. *J. Electrochem. Soc.* 2003:150; A1095–A1101.

211. Yang Z, Stevenson JW, and Meinhardt KD. *Solid State Ionics* 2003:160;213–225.

212. Yang Z, Xia G-G, Meihardt KD, Weil KS, and Stevenson JW. *J. Mater. Eng. Perform.* 2004:13;327–334.

213. Pistorius CWFT and Pistorius MC. *Z. Krist* 1962:117;259–272.

214. Haanapel VAC, Shemet V, Vinke IC, and Quadakkers WJ. *J. Power Sources* 2005:141;102–107.

215. Haanapel VAC, Shemet V, Vinke IC, Gross SM, Koppitz Th, Menzler NH, Zahid M, and Quadakkers WJ. *J. Mater. Sci.* 2005:40;1583–1592.

216. Haanapel VAC, Shemet V, Gross SM, Koppitz Th, Menzler NH, Zahid M, and Quadakkers WJ. *J. Power Sources* 2005:150;86–100.

217. Batfalsky P, Haanapel VAC, Malzbender J, Menzler NH, Shemet V, Vinke IC, and Steinbrech RW. *J. Power Sources* 2005:155;128–137.

218. Yang Z, Coyle CA, Baskaran S, and Chick LA. U.S. Patent No. 6843406, 2005.

219. Yang Z, Xia G-G, Singh P, and Stevenson JW. *J. Power Sources* 2006:155;246–252.

220. Quadakkers WJ, Greiner H, Hansel M, Pattanaik A, Khanna AS, and Mallener W. *Solid State Ionics* 1996:91;55–67.

221. Maruyama T, Inoue T, and Nagata K. In *SOFC VII-Electrochem. Soc. Proc. PV-2001-16*, Singhal SC and Dokiya M, (eds.), pp. 889–894, 2001, The Electrochemical Society, Pennington, NJ.

222. Ullmann H and Trofimenko N. *Solid State Ionics* 1999:119;1–8.

223. Elangovan S, Balagopal S, Timper M, Bay I, Larsen D, and Hartvigsen J. *J. Mater. Eng. Perform.* 2004:13;265–273.

224. Linderoth S. Surf. Coating Tech. 1996:80;185.

225. Ullmann H, Trofimenko N, Tietz F, Stover D, and Ahmad-Khanlou A. *Solid State Ionics* 2000:138;79–90.

226. Yang Z, Xia G-G, Maupin GD, and Stevenson JW. *J. Electrochem. Soc.* 2006:153;A1852–A1858.

227. Qu W, Jian L, Hill JM, Ivey DG. J. Power Sources 2006:153;114–124.

228. Yang Z, Xia G, and Stevenson JW. *Electrochem. Solid State Lett.* 2005:8;A168–A170.

229. Yang Z, Xia G-G, Simner SP, and Stevenson JW. *J. Electrochem. Soc.* 2006:152;A1896–A1901.

230. Yang Z, Xia G-G, Li X-H, and Stevenson JW. *Int. J. Hydrogen Energy* 2006:32; 3648–3654.

231. Larring Y and Norby T. *J. Electrochem. Soc.* 2000:147;3251–3256.

232. Chen X, Hou PY, Jacobson CP, Visco SJ, and De Jonghe LC. *Solid State Ionics* 2005:176;425–433.

233. Zahid M, Tietz F, Sebold D, and Buchkremer HP. In *Proc. 6th European SOFC Forum*, Mogensen M, (ed.) pp. 820–827, 2004, The European Solid Oxide Fuel Cell Forum, Lucerne, Switzerland.

234. Burriel M, Garcia, Santiso J, Hansson AN, Linderoth S, and Figueras A. *Thin Solid Films* 2005:473;98–103.

235. Gorokhovsky VI, Gannon PE, Deibert MC, Smith RJ, Kayani A, Kopczyk M, VanVorous D, Yang Z, Stevenson JW, Visco S, Jacobson C, Kurokawa H, and Sofiee SW. *J. Electrochem. Soc.* 2005:153;A1886–1893.

236. Basu R, Knott N, and Petric A. In *SOFC IX-Electrochem. Soc. Proc. PV2005-07*, Singhal SC, Mizusaki J, (eds.) pp.1859-1865, 2005, The Electrochemical Society, Pennington, NJ.

237. Wei P, Zhitomirsky I, and Petric A. In *SOFC IX-Electrochem. Soc. Proc. PV2005-07*, Singhal SC and Mizusaki J, (eds.) pp.1851–1858, 2005, The Electrochemical Society, Pennington, NJ.

238. Yang Z, Xia G-G, Maupin G, Simner S, Li X, Stevenson J, and Singh P. In *Fuel Cell Seminar*, paper No. 253, 2006, Courtesy Associate, Washington.

239. Yokoyama T, Abe Y, Meguro T, Komeya K, Kondo K, Kaneko S, Sasamoto T. Japan J. Appl. Phys. 1996:35;5775–5780.

240. Yang Z, Li X-H, Maupin GD, Singh P, Simner SP, Stevenson JW, Xia G-G, and Zhao X-D. *Ceram. Sci. Eng. Proc.* 2006:27;231–240.

241. Ling H and Petric A. In *SOFC IX-Electrochem. Soc. Proc. PV2005-07*, Singhal SC, Mizusaki J, (eds.) pp. 1866–1873, 2005, The Electrochemical Society, Pennington, NJ.

242. Martin BE and Petric A. Electrical properties of copper-manganese spinal solutions and their cation valence and cation distribution. *J. Phys. Chem. Solids* 2007: 68; 2262–2270.

243. Simner SP, Anderson MD, Xia G-G, Yang Z, and Stevenson JW. *Ceram. Eng. Sci. Proc.* 2005:26;83–90.

5 Sealants

P.A. Lessing, J. Hartvigsen, and S. Elangovan

CONTENTS

5.1 INTRODUCTION

Sealing of high-temperature solid oxide fuel cell (SOFC) stacks is a critical issue for maintaining the electrical performance of the fuel cell over a long period of time. Thermal stresses due to repeated heating and cooling cycles can degrade or fracture seals. Long-time material interactions and corrosion can also degrade seals. These are ongoing problems for developers of SOFCs that are used to generate electricity from hydrogen or hydrocarbon fuels. However, sealing is an even greater concern for solid oxide stacks used in a high-temperature steam electrolysis mode or solid oxide electrolysis cells (SOECs) to generate hydrogen. This steam electrolysis method essentially involves inputting electrical power into an SOFC in a reverse polarity mode [1, 2]. Because hydrogen is the primary valuable product for electrolysis and, due to hydrogen's small size, the gas molecules can quickly leak out through various high-temperature seals.

Planar configuration SOFCs require several types of seals. In many instances, an edge seal will be between a zirconia electrolyte and a high-temperature metal frame [3]. Seals can also be made to metal interconnect plates. Depending on the design, this seal will be exposed to operating temperatures of about 750 to 850°C. In fuel cells, this seal will always be exposed to an oxidizing atmosphere on the cathode side (e.g., air) and a wet fuel gas containing various ratios of hydrocarbons (e.g., CH_4), H_2, CO, CO_2, and H_2O on the anode side. For steam electrolysis, the cathode will be exposed to a mixture of H_2O in the form of steam and H_2 where the ratio will vary depending on the position within the cell. The inlet will be high in H_2O. Some H_2 must always be maintained in the cathode compartment in order to prevent oxidation of the Ni to NiO (nonconductive). Therefore, electrode atmospheres are different for a high-temperature electrolysis cell than for a fuel cell. Comprehensive review articles have recently been published that cover sealant materials used for SOFCs [4] and SOECs [5].

For fuel cells, a small amount of leakage does not cause a large drop in the stack efficiency and is generally tolerated. DC electricity is the valued product, and any unconverted fuel, such as H_2 and CO, that is leaked out of the anode chamber is combusted and contributes heat to the cells. Leakage hydrodynamics is covered later in this chapter.

Sealing is much less of a problem for tubular-type SOFCs. However, most tubular designs do use some high-temperature seals to join single tubes, or flattened tubes, into bundles for attachment to a gas manifold, which is usually the fuel gas. Many tubular designs are said to be "seal-less"; however, this only means that the gas seals can be located out of the high-temperature areas and can be executed using fairly conventional metal seals. There are several very different tubular designs that are being developed. Their designed operating temperatures vary from about 700 to 900°C, depending on the materials of construction.

Figure 5.1 shows schematic drawings (generic) of a simple planar stack of cells. The upper drawing is an exploded view where the light-gray colored strips are the edge seals. The lower drawing of Figure 5.1 shows a manifold seal in a stack that has a cross-flow configuration for the fuel and air gas streams. The external manifolds are illustrative since many designs have internal manifolds. Figure 5.2 is a schematic of seals typically found in a stack with metallic internal gas manifolds and a metallic bipolar plate. Common seals include: (a) cell to metal frame (S1) which could include sealing of the edges of the cells and sealing to a particular cell layer, (b) metal frame to metal interconnect (S2), (c) frame/interconnect pair to electrically insulating spacer (S3), (d) stack to base manifold plate (S4), and (e) cell electrode edge or electrolyte to interconnect edge.

5.2 TYPES OF HIGH-TEMPERATURE SEALS

5.2.1 GLASS SEALS

Historically, one of two techniques has typically been used to seal a planar SOFC stack: glass joining or compressive sealing. Glass was originally used because it is simple to make and apply. The first requirement for a rigid seal is that the seal's

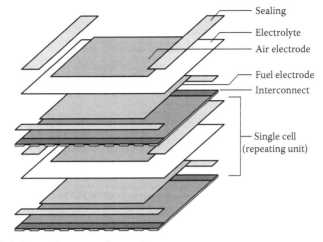

Exploded view, planar cross-flow stack concept.

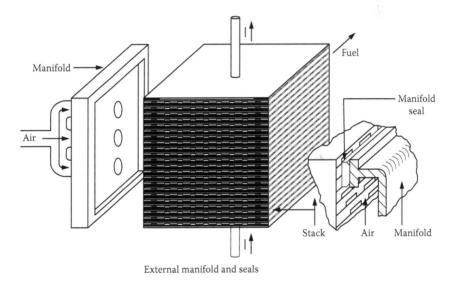

External manifold and seals

FIGURE 5.1 Schematics of edge sealing of planar cells (above) and external gas manifold seals (below) used for a simple cross-flow SOFC stack design.

thermal expansion should closely match the thermal expansion of the cell components that it is joining. High stresses can result from thermal gradients or thermal expansion mismatches between the glass and cell materials during heating and cooling. Fully stabilized zirconia has a relatively high coefficient of thermal expansion (CTE) for a ceramic (10–$11 \times 10^{-6}K^{-1}$), but this CTE is lower than most common metals. After melting and cooling, glass seals are brittle and nonyielding, making them very susceptible to cracking because of tensile stresses. If the sealing glass does not crystallize,

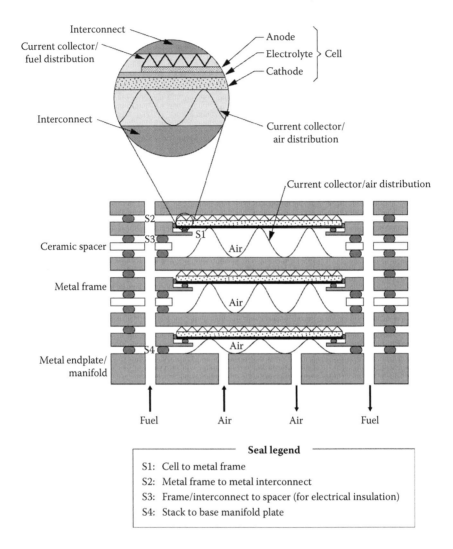

FIGURE 5.2 Schematic of seals typically found in a planar design SOFC stack with metallic interconnect and metallic internal gas manifold channels (possibly for counter flow pattern of fuel and air gases).

then cracks that may form because of thermal cycling can "self-heal" if the glass seals are periodically heated to temperatures near the glass melting point.

Glass is relatively low in strength when compared to polycrystalline ceramics. The amorphous or vitreous structure of some glasses with high CTEs often tended to crystallize when held at high temperatures for a long time, and the CTE would change. This has led to the use of special compositions to form purposefully crystallized glasses, also known as glass-ceramics. Glass-ceramics are much stronger than glass and the designed, resultant crystalline structure will have a known CTE.

5.2.2 Glass-Ceramic Seals

Glass-ceramics are special compositions of glasses that are amorphous when melted but are specifically designed to partially or fully crystallize when held in a high-temperature range in conjunction with a carefully designed cooling profile that is below the melting/solidification temperature range. Many developers fabricate their own glass-ceramic compositions based on silica as the glass former.

Many barium aluminosilicate–based compositions will eventually react with the chromium oxide or aluminum oxide scales on the metal interconnect or metal edge rails to form barium chromate or a celsian phase at the interface [6]. This can cause a mechanical weakness that is easily delaminated. Also, compositions that contain boron can react over time with water (steam) to produce $B_2(OH)_2$ or $B(OH)_3$ gas. This can decompose the glass and greatly limit the lifetime of the seal. Thus many of the new investigations have emphasized "low" or "no" boron glass compositions.

High-temperature sealing is also required for some modern designs of tubular-type cells. An example is the Rolls-Royce Integrated Planar (IP)-SOFC design which is a segmented cells-in-series concept in which the cells are deposited onto a flat, porous ceramic support tube. This support tube, which contains multiple channels, acts as the fuel gas delivery tube. In the IP-SOFC design, cells are only deposited on the flat surfaces of the tube, hence the edges of the tubes need to be sealed to prevent mixing of the fuel and air. For these seals, Rolls-Royce Fuel Cell Systems Ltd. (RRFCS) currently uses a glass-ceramic. In addition to sealing the tube surfaces, glass-ceramics are also used in certain parts of the stack to create gas-tight joints between the porous support tubes and adjoining manifolds. For both locations, long-term durability and mechanical stability under thermal cycling are required. The details of the glass-ceramic compositions are proprietary to RRFCS [7].

Long-term physical and chemical stability is very important for seals in both SOFC and high-temperature electrolyzer devices. In addition to reaction with steam and vaporization, glass-ceramics will react with cell components. A series of 13 representative compositions in the AO-SiO_2-Al_2O_3-B_2O_3 (A = Ba, Ca, Mg) systems was evaluated by Lahl et al. [8] where the boron content was kept to less than 5%. MgO-containing glasses were found to have significantly lower reactivity with component materials than those containing BaO or CaO. Investigation of MgO-containing glasses with different Al_2O_3 ratios showed that the detrimental formation of $Mg_2Al_4Si_5O_{13}$ phases can be suppressed by using low Al_2O_3 concentrations and appropriate grain sizes and nucleating agents.

Eichler et al. [9] performed metal-to-metal joining experiment using four glass-ceramic compositions in the BaO-MgO-SrO-Al_2O_3-B_2O_3-SiO_2 system as the "solder." They found that many compositions containing MgO reacted with the interconnect alloy they used, Cr-5Fe-1Y_2O_3, to form the reaction phase $MgCr_2O_4$ spinel. Many joints showed enhanced leak rates after thermal cycling. It was thought the leakage was a result of glass delamination (cracking) at the metal interface probably caused by the presence of the low CTE ($6.9 \times 10^{-6}K^{-1}$) $MgCr_2O_4$ spinel. Barium aluminosilicate sealing glass that is bonded to a chromia-forming alloy can react and deplete the protective chromia on the air side. Example reactions noted by Yang [10] are shown in Reactions (5.1) and (5.2):

$$2Cr_2O_3 + 4\,BaO + 3O_2 \rightarrow 4BaCrO_4 \tag{5.1}$$

$$CrO_2\,(OH_2)\,(g) + BaO(s) \rightarrow BaCrO_4\,(s) + H_2O\,(g) \tag{5.2}$$

The formation of barium chromate often leads to the physical separation of the sealing glass and the metal alloy due to barium chromate's high thermal expansion. Along interfacial regions where oxygen or air access is blocked, chromium or chromia can react with barium-calcium-aluminosilicate (BCAS) glass-ceramic to form a chromium-rich solid solution and a series of pores.

Based on the published data, it appears to be difficult to select a single glass-ceramic composition in the $BaO-MgO-Al_2O_3-B_2O_3-SiO_2$ system that would provide ideal performance in all the desired aspects as a sealant for SOFCs or high-temperature electrolyzers for long periods of time. Therefore, other glass-ceramic compositions and other sealing options continue to be evaluated.

5.2.3 Compressive Seals

A compressive seal places a compliant, high-temperature material between the sealing surfaces. The material is then compressed using external forces to the fuel cell stack (e.g., using a load frame and springs or hydraulics to provide a constant pressure). The addition of a load frame adds complexity and cost to the SOFC. The best compressive seals allow the surfaces to slide past each other while maintaining a good—hermetic, if possible—gas seal. Using this seal, matching CTE values is not as important as with a solid (e.g., glass-based seal). High-temperature compressive seals using various metals have been evaluated [11, 12]. These seals needed some improvement with mechanical behavior and gas tightness. Ceramic powders are oxidation resistant, and some have been used to form a compressive seal in SOFCs. But, ceramic oxide powders typically form a "leaky" seal.

Some success has been found using mica; however, problems with through-seal leakage, interface reactions, and crystallization have been encountered [13]. Compressive seals using metal/mica sandwich composites have been developed as a proposed improvement to rigid glass-ceramics [14]. The metal layer (e.g., FeCrAlY-alloy) is presumably added to increase the overall CTE and plasticity of the seal. This approach did show reduced leak rates after five thermal cycles. A compressive hybrid mica seal is under development at Pacific Northwest National Laboratory (PNNL). Sheets of Phlogopite $[KMg_3(AlSi_3O_{10})(OH)_2]$ (high-temperature mica) have been layered between thin glass or silver layers to form a seal between various metal components or metal to zirconia [15].

5.2.4 Metal Seals

The list of high-temperature metals useful for sealing that will not corrode in air is very short. Candidate metals include platinum (m.p. 1774°C), gold (m.p. 1063°C), and silver (m.p. 961°C). Gold and platinum are prohibitively expensive, while silver is only 1/100 the price of gold or platinum; therefore, silver has been considered as an SOFC sealing material by several developers. Possible drawbacks are that silver has a high vapor pressure, and it has a high thermal expansion (CTE $\cong 20 \times 10^{-6}K^{-1}$).

Stevenson [16] reported that silver was unstable, developed large amounts of porosity, when exposed to a dual atmosphere of H_2—3% H_2O on one side and air on the other for 100 h at 700°C. He found silver was stable when exposed to air or an air environment for 100 h at 700°C in an identical configuration.

PNNL has been working on an Ag-CuO [17, 18] "fluxless" seal to "air" braze cells to metal frames [19]. Braze pastes are formulated by mixing silver and copper powders (4 mol%) with an organic binder. The copper oxidizes *in situ* to CuO during the brazing operation. Hermetic seals were made with bond strength being maintained over 40 thermal cycles. The braze wets the YSZ well, with no reaction zone found at the interface (some CuO precipitates near the interface). If well dispersed, the copper oxide particles could be functioning as a strengthening agent, dispersed oxide, in silver at high temperatures. It also may be possible to form a silver-ceramic composite in which silver forms a three-dimensional network and thereby reduces the high CTE of silver [20].

5.2.5 CERAMIC-COMPOSITE SEALS

Ceramic-composite seals are being investigated by Sandia National Laboratory [21] and NexTech Materials, Ltd. [22] Sandia's stated composite approach is to produce a deformable seal based on using a glass above its Tg with control of the viscosity and CTE modified by using ceramic powder additives. NexTech report using $BaO-CaO-Al_2O_3-SiO_2$ candidate glass compositions (e.g., 15 BaO 25 CaO 7.5 Al_2O_3 45 SiO_2) together with powders of mica, talc, alumina, or zirconia fibers as an additive to increase the viscosity. If these compositions were to contain sufficient glass that the seal was fully densified upon melting, then this approach would produce seals very similar to the dense, machinable mica-glass ceramics SOFC seals that were investigated in Japan during the 1990s. These composite seals consisted of microcrystalline mica (fluoro-phlogopite, $KMg_3AlSi_3O_{11}F_2$) in a $SiO_2-B_2O_3-Al_2O_3-K_2O-MgO-F$ glass matrix. However, the mica-glass composites were reported to react with the $La_{0.8} Ca_{0.22} CrO_3$ separator plate at high temperatures [23].

Sandia National Laboratory scientists are experimenting with a glass-metal seal using $ZnO-CaO-SrO-Al_2O_3-B_2O_3-SiO_2$ glasses (glass-ceramic) in conjunction with adding powdered metallic nickel (5 to 30 vol%) [24]. Adding nickel adjusts the composite CTE to higher values, and the nickel appears to have low reactivity with the glass-ceramic.

5.2.6 COMPLIANT SEALS

Compliant seals might prove to be successful if they can be executed at high temperatures. Compliant seals might include: flexible metal "bellows," viscous glass that includes self-healing glass seals, or perhaps wet seals (material system unknown). These require flexible seal designs, stable glasses that have the appropriate viscosity range or noncrystallized glasses, and some concepts may require an applied pressure.

PNNL [25] describes preliminary joining experiments using flexible foils made from an alumina-forming ferritic steel, DuraFoil [26]. This foil was brazed to Haynes 214 alloy* on the bottom side using BNi-2 braze tape**. The foil was also brazed to YSZ (component of a cell) on the top side using Ag-4 mol% CuO. The foil was in

---

* Haynes 214 has excellent oxidation resistant to temperatures over 1000°C, but has high CTE (15.7 × $10^{-6}K^{-1}$).
** Wall Colmonoy Inc., Madison Heights, Michigan.

the form of a compliant/flexible edge seal; i.e., a single segment of a bellow. Good
strength results were reported. In practice, this seal could be used around round gas
manifold holes which are incorporated into both the ceramic cell and the metal alloy
separator plate. The foils are most applicable to provide thermal expansion relief to
round holes and round structures. This is because side expansion in a square or rectangular structure cannot be accommodated by flexing of the foil.

5.3 HYDRODYNAMICS OF LEAKING SEALS

An examination of the equations describing leakage within an SOFC stack provides
valuable insight to the leakage behavior of the system. This detailed understanding
may in turn enable creation of improved seal materials and application methods, as
well as stack designs and assembly techniques. The most basic seal configuration
to consider is the case where no sealant is used. In an ideal mathematical world, all
of the cell and interconnect components comprising a stack would fit together per-
fectly, with no gaps or spaces between them. In reality, the dimensional tolerances
and surface roughness inherent in all manufactured fuel cell components attest that
a gap-free assembly is not an achievable condition.

5.3.1 THEORY OF SEAL LEAKS

A typical planar SOFC stack has a seal around the entire perimeter, on one electrode or the other. There are some internally manifolded, radial outflow designs (e.g.,
HEXIS, TMI, and CFCL) where a smaller circular seal is used around the internal
manifold ports; however, the conventional planar SOFC stack seal can be approximated as occupying a rectangular volume as shown in Figure 5.1. The SOFC seal
flow regime is characterized by high temperature gases flowing in small channels
at very low flow rates (we hope). High-temperature gases have both low density and
high viscosity. The Reynolds number, a dimensionless group representing the ratio
of inertial to viscous forces in the flow, is calculated as:

$$\text{Re} = \frac{\rho v d}{\mu} \qquad (5.3)$$

where ρ is the gas mass density, v is the local velocity, d is the channel characteristic
dimension, and μ is the gas kinematic viscosity. Since the three terms in the numera-
tor are all small—low density, low velocity, small channels—and the viscosity is
high, the seal leakage flow Reynolds number is nearly always much less than 1. Such
flows are termed "laminar flows" and can be treated with closed-form analytical
solutions, while high Reynolds number flows are termed "transitional" or "turbulent"
and in most cases must be approximated numerically.

The volumetric flow rate Q (leakage rate) of a fluid in laminar flow through a high
aspect ratio $(h/w \ll 1)$ rectangular duct (i.e., the seal volume) of width, w, height, h,
and path length, z, in the flow direction is given by:

$$Q = \frac{w h^3}{12 \mu z} \Delta P \qquad (5.4)$$

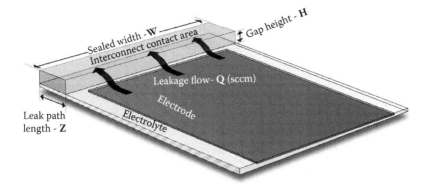

FIGURE 5.3 Schematic of typical planar-type seal, illustrating parameters used in Ceramatec's model.

where ΔP is the pressure difference across the seal. This result emphasizes the primary importance of tightly fitting parts, given the cubic dependence of leak rate on the height of the gap being sealed. A small value of the ratio of w/z is also desirable. This ratio will always be rather large, however, since as z approaches $w/2$, the active area of a square cell goes to zero. This is illustrated schematically in Figure 5.3.

Generally, the absolute magnitude of Q is not as important as the ratio leak rate to the total flow rate (Q_{leak}/Q_{total}). The leakage rate given by Equation (5.4) is the volume flow rate at the temperature and pressure of the leakage flow, and must be corrected to standard conditions for comparison with reactant feed rates. The total required flow rate of fuel or air to the stack is proportional to the stack current, which increases with the electrochemically active area and is inversely proportional to the cell area specific resistance (R'').

$$Q_{total} \propto \frac{A_{active}}{R''} \tag{5.5}$$

For a square cell of dimension, X, with a sealing margin of constant width, z, the active area is given by:

$$A_{active} = (X - 2z)^2 \tag{5.6}$$

while, w, the total seal perimeter per cell varies as $4X$. The resulting variation of leak rate to total flow rate can then be seen to decrease with increasing cell size and decreasing cell resistance.

$$\frac{Q_{leak}}{Q_{total}} \propto \frac{Ch^3}{\mu} \frac{\frac{4X}{z}}{\frac{(X-2z)^2}{R''}} \tag{5.7}$$

Not only does the larger cell have a smaller ratio of seal perimeter to active area, but also the active area fraction of total area is greater as the cell size increases while fixing the seal margin width.

$$\frac{A_{active}}{A_{total}} = \frac{(X - 2z)^2}{X^2} \tag{5.8}$$

The results so far show that whatever the seal material performance, it can be improved by a tightly fit assembly by reducing the cell resistance, increasing the seal margin width, and increasing the cell size itself. Reducing the cell differential pressure also helps reduce the driving pressure for leakage. Equation (5.4) also applies to the flow of reactants in the interconnect channels. The flow path length, z, is replaced by the cell dimension, X, and the flow width, w, by $(X - 2z)$. The pressure required for a given flow Q_{total}, however, still varies as the channel height cubed (h^3). This effect should be balanced against other factors such as stack size, weight, and material costs that favor shallower gas flow channels.

Closed-form expressions for calculating the constant in Equation (5.4) have been found, and the expression generalized for a number of channel geometries [27].

$$Q = \frac{A_c d_h^2}{2f \, \text{Re}} \frac{1}{\mu} \frac{\Delta P}{z} \tag{5.9}$$

Here the A_c is the channel cross-section flow area, and the factor, $f\text{Re}$, is a numerical constant computed and tabulated for various channel geometries. The characteristic dimension has been replaced by a hydraulic diameter defined as four times the flow area divided by the channel perimeter.

$$d_h = \frac{4A}{P} \tag{5.10}$$

For a capillary of circular cross-section, $f\text{Re}$ has a value of 16, and d_{hyd} is equal to the capillary diameter. Equation (5.9) then reduces to Poiseuille's equation.

$$Q = \frac{\pi d^4}{128\mu} \frac{\Delta P}{z} \tag{5.11}$$

Where the seal volume can be filled with a seal material leaving no fluid path through connecting voids or gaps, a gas tight or hermetic seal may be achieved. In other cases, such as where the stack must be easily disassembled or where large thermal expansion differences exist, a porous seal may be employed. The seal volume is filled with particles, felt, or foam leaving a volume fraction, ε, of interconnected void space allowing leakage flows. The total cross-sectional area available for leakage has been reduced by a factor of ε, but the reduction of leak rate is far greater.

Consider a bundle of capillary passages having a cross-sectional open area fraction of ε. The number of passages per unit area, n'', is then given by:

$$n'' = \frac{4\varepsilon}{\pi d_p^2} \tag{5.12}$$

where d_p is the circular equivalent pore diameter. The leakage flow rate through the seal volume is now:

$$Q = whn''Q_p \qquad (5.13)$$

where Q_p is the leak rate through a single-capillary channel of diameter d_p. Substituting Equations (5.11) and (5.12) into Equation (5.13) yields,

$$Q = \frac{wh\varepsilon d_p^2 \Delta P}{32\mu z} \qquad (5.14)$$

Realistically, the flow path through the seal volume will not consist of uniform cylindrical capillaries aligned normal to the seal. The actual leakage paths will be longer, less direct (convoluted), and of nonuniform cross-section. To account for these effects, the effective path length is increased by a tortuosity factor, τ, typically having a value in the range of 2 to 3.

$$Q = \frac{wh\varepsilon d_p^2 \Delta P}{32\mu\tau z} \qquad (5.15)$$

Additionally, Equation (5.15) has assumed the values of d_h and fRe computed for a circular capillary. If it were possible that the geometry of all whn'' passages could be defined and characterized, appropriate values of d_h and fRe could be computed and incorporated into Equation (5.15). However, the utility and significance of Equation (5.15) is in providing the functional relationships of how microstructure characteristics such as porosity, pore diameter, and tortuosity affect seal performance. That is one can expect that reducing seal porosity by half would also cut the leak rate by half, while reducing the pore diameter by a factor of ten would reduce the leak rate 100-fold. In practice, experimental measurements of leak rates are required. Then, rather than trying to extract the full parameter set for Equation (5.15)—i.e., ε, τ, d_p, fRe, etc., from the pore cross-section and path shapes—a single value, k, the Darcy's law permeability, can be used to characterize the seal effectiveness.

$$Q = \frac{kwh\Delta P}{\mu z} \qquad (5.16)$$

Often the seal height or gap is variable and unknown. In this case, kh, the product of permeability, k, and seal height, h, can be replaced by k', a permeability per unit of seal perimeter length. For a given stack geometry, the seal volume basis can also be defined to include the seal path length, z, and perimeter, w, to define a permeability per cell or stack as is convenient.

5.3.2 ANALYSIS OF SEAL LEAKS

The above equations were used to compute the seal leak rate through all four sides of a 10×10 cm cell having a 1-cm seal margin and a seal height of 50 μm (0.002"). An average seal differential pressure of 750 Pa (3" H_2O column) was assumed. The computed unsealed, or open gap, leak rate was 1.8 standard cm³/min per cell unit active area (sccm/cm²), which amounts to a large fraction of the total feed rate, Q_{total}. If the seal volume is filled with a material having a 1-μm pore diameter, a 40% void fraction, and a tortuosity of 2, the leak rate is reduced more than 300-fold, to 0.005 sccm/cm². This is an acceptably low leak rate in many situations. Of course with greater pressure differentials, larger pore diameters or fissures in the seal, leak rates can increase dramatically.

Another important leakage mechanism is a concentration-driven diffusive flux in contrast with the pressure-driven hydrodynamic flux considered previously. The flux of species A relative to the average molar velocity of all components is given by Fick's first law.

$$J_A = -cD_{AB} \frac{\partial x_A}{\partial z} \tag{5.17}$$

The flux, J_A, has units of mol/area-time, the concentration, c, in mol/volume, diffusivity, D_{AB}, in length²/time, while the mole fraction, x_A, is dimensionless naturally. Concentration can be calculated using from the ideal gas law.

$$c \equiv \frac{n}{V} = \frac{P}{RT} \tag{5.18}$$

The leak rate through a porous seal volume can be computed as flux times area and converting from moles to volume with the ideal gas law. The seal void fraction and tortuosity have the same effect as in the hydrodynamic leakage calculations.

$$Q_{diffusive} = \frac{wh\varepsilon}{\tau} \frac{RTJ_A}{P} \tag{5.19}$$

Substituting Equations (5.17) and (5.18) gives,

$$Q_{diffusive} = \frac{wh\varepsilon}{\tau} \frac{D_{AB}(x_{A1} - x_{A2})}{z} \tag{5.20}$$

The negative sign in Equation (5.17) has been dropped, as it is understood that the flux is in the direction of decreasing concentration. Note that the diffusive leak rate equation has no dependence on pore diameter in contrast to the d_p^2 dependence in the pressure-driven leakage equations. However, if the pore diameter can be made much smaller than the mean free path in the gas, the diffusivity undergoes a bulk to Knudsen transition and some small reduction in leak rate is predicted. However, generally porous seal materials will have pore diameters considerably greater than the mean free path. The species mole fractions, x_{A1} and x_{A2}, refer to the high and low concentrations of species A, respectively. Diffusion of oxygen into the fuel stream and fuel into the oxygen stream are of most interest to the fuel cell performance and efficiency. Any oxygen that diffuses into the fuel stream will react with fuel, while any fuel species that diffuses into the oxidant stream will also be consumed by reaction. The result is that x_{A2} is zero in these cases, and the leak rate expression simplifies to:

$$Q_{diffusive} = \frac{wh\varepsilon}{\tau} \frac{D_{AB}x_{A1}}{z} \tag{5.21}$$

TABLE 5.1
Mass Diffusivities of Fuel Cell
Feed Stream Components

Binary Diffusion Pairs	Diffusivity, cm²/sec 800°C, 1 bar
H_2-H_2O	7.4
CH_4-H_2O	2.5
H_2O-H_2O	2.6
CO-H_2O	2.4
N_2-H_2O	2.4
CO_2-H_2O	1.9
O_2-N_2	2.0

Diffusivity varies as T^n where $n = 1.5$–2, and inversely with total pressure and square root of molecular weight [28]. Therefore, high temperature, low molecular weight, and low-pressure gases will have high values of diffusivity resulting in high-diffusive losses. Binary diffusion coefficients at 800°C for fuel gases in water vapor and for air are listed in Table 5.1. Water vapor, at 18 g/mol is representative of the fuel stream average molecular weight, calculated at 16.8 g/mol assuming a reformer Steam/CH_4 ratio of 2, and a mid-cell fuel utilization. Note that the diffusivity of hydrogen in steam, a low molecular weigh mixture, is triple that of any other pair shown in Table 5.1. Computation of multicomponent diffusivities is more involved, but the diffusivity of a reformate mixture is close to that of steam-methane or steam self-diffusion [29]. Diffusive leakage will be much higher with stacks running pure hydrogen fuel streams, and can be reduced by using reformate or diluting the stream with an inert gas such as nitrogen. Dilution with a higher molecular weight inert gas not only reduces the diffusivity, but also decreases the driving mole fraction, x_{AI}. An example calculation using the diffusivity estimation method presented in Treybal is shown in Figure 5.4.

Estimates of diffusive leakage using the same geometry used for pressure-driven leakage, and the diffusivity values in Table 5.1 show leak rates of 0.04 sccm/cm² for pure hydrogen, or 0.008 sccm/cm² for reformate or nitrogen diluted hydrogen streams. Thus, diffusive leak rates of 2 to 10 times that of hydrodynamic may be expected.

It is possible to design an experiment to estimate the relative losses due to pressure-driven and concentration-driven leakage. In this approach, a conventional stack performance model [30, 31] is modified to incorporate leakage effects from both pressure- and concentration-driven sources. The pressure driven leakage is assumed to be relatively constant as reactant stream compositions change due to leakage, feed composition (e.g., nitrogen or steam dilution), and stack operation (consuming fuel and generating steam). The model contains leakage parameters, which are initially unknown, but subsequently determined by a nonlinear regression of the model against experimental data. A single parameter can be used to characterize the pressure leakage response and another the diffusive leakage. The various experimental operating

Mass Diffusivity Estimate

Diffusivity Estimation Method for Binary Gas Mixtures from Treybal, *Mass-Transfer Operations*, 3rd Edition

Model for diffusivity estimation: H_2-H_2O. 800C

Ma:= 2   Mb:= 18   Molecular weight of hydrogen and water (g/mole)

T := 1073.0   Stack temperature (K)

Pt:= $1.013.10^5$ Total pressure (Pa)

Look up r and ε from Table 2.2 in Treybal

rA := 0.2827   rA = 0.28   Molecular separation at collision (nm)

rB := 0.2641   rB := 0.26

κ := $1.381.10^{-23}$ Boltzmann Constant

εA := 59.7·κ $\dfrac{εA}{κ} = 59.7$ ε energy of molecular attraction (ε/κ in K)

εB := 809.1·κ $\dfrac{εB}{κ} = 809.1$

$rAB := \dfrac{rA + rB}{2}$ rAB = 0.273   AB collision separation (nm)

$εAB := κ·\sqrt{\dfrac{εA}{κ}·\dfrac{εB}{κ}}$ $\dfrac{εAB}{κ} = 220$ AB attraction energy (ε/κ in K) from Treybal Table 2.2

$\dfrac{κ·T}{εAB} = 4.882$ f := 0.42   Collision function f from Trebal Figure 2.5

$$Dab := \frac{10^{-4}·\left(1.084 - 0.249·\sqrt{\frac{1}{Ma} + \frac{1}{Mb}}\right)·T^{\frac{3}{2}}·\sqrt{\frac{1}{Ma} + \frac{1}{Mb}}}{Pt·rAB^2·f} \frac{m^2}{sec} \qquad Dab = 7.4\frac{cm^2}{sec}$$

Dh2-h2o, mass diffusivity of hydrogen in steam

FIGURE 5.4 "MathCAD" worksheet of diffusivity calculation (following Treybal).

conditions can be obtained by systematically varying stack operating conditions, such as fuel flow rate, operating voltage, temperature, and fuel inlet composition. The nonlinear regression analysis finds the optimal set of leakage parameter values such that stack performance model, with leakage included, predictions best match the experimental data over the full range of experimental operating conditions.

Results from an example calculation are shown in Figure 5.5. The curves are generated using the leakage parameters fit to the data from one stack. Stack current was measured over a range of fuel flow rates, operating voltages, temperatures, and with and without nitrogen addition to the fuel stream. The complete data set was used in the nonlinear regression to determine a fixed leakage component (pressure driven)

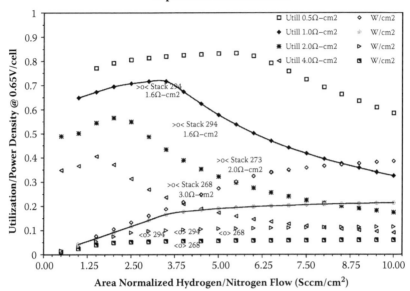

FIGURE 5.5 Comparison of measured stack test data to that predicted by Ceramatec's Leak Model.

and a concentration dependent leakage component (diffusive). Operating points from other stacks of similar construction but widely varying performance were superimposed on the model-generated curves and were found to be in general agreement.

The model shows that as the fuel flow is reduced at constant operating voltage, the utilization increases at first, then begins to decrease. In the absence of leakage, the stack utilization and efficiency would continue to climb as fuel flow rates are reduced, independent of stack resistance. With leakage effects considered, stacks with lower area-specific resistance (ASR, units of Ω-cm^2) can achieve higher values of fuel utilization. This is in effect a system of competing rates, with the desired electrochemical reaction in competition for fuel with the undesired leakage losses.

In summary, understanding the leakage mechanisms gives insight into designing better seal materials and seal configurations. Both pressure and concentration driven leakage mechanisms are important in high-temperature fuel cell systems. Fitting of experimental results to a stack leakage model can provide a useful characterization of the overall leakage effects, which can be used in turn to select optimal stack operating conditions.

5.4 TESTING OF SEAL PROPERTIES AND BEHAVIOR

Physical isolation of reactant and product gases from direct mixing is an obvious requirement of the seal material. However, as mentioned in Section 5.1, the leak requirement is more stringent for an SOEC than an SOFC. In general, around 1%

fuel consumption due to leak is considered acceptable for a fuel cell, provided the leak is not localized resulting in detrimental hot spots. In a fuel cell, there is no significant pressure differential between the air and fuel chambers, whereas the need for collecting the product in an electrolyzer cell may lead to considerable back pressure that the seal needs to accommodate.

Mechanical and chemical stability in relevant atmospheres as well as chemical compatibility with mating components are critical. It is also essential that the seal remain stable in dual atmosphere conditions where the seal separates the fuel and air chambers. Certain locations within a stack demand an electrically insulating seal material while others may not. For cost considerations, not only the seal material needs to be low cost and requires an inexpensive application technique, it should not demand very high mechanical tolerances in the mating components. In other words, the seal material must accommodate conventional manufacturing dimensional tolerances of the mating surfaces during assembly and initial heating up to affect the seal.

The specific property requirements of a seal material largely depend on the seal concept. For example, the CTE match is more critical for a rigid seal than for a compressive seal while the latter needs to withstand the load applied to affect the seal. Among the glass or glass composite seals, the glass transition temperature needs to be above or below the stack operating temperature depending on whether one wants a rigid or a compliant seal. Thus, a single set of specifications is unlikely to cover the range of options that may be available based on the approach to sealing. In general, knowledge of thermal expansion, chemical stability and compatibility with the stack components, and stability of necessary properties during operating life and potential thermal cycles is essential. Certain properties near the stack operating temperature are useful to measure in order to modify them. Researchers at Sandia National Laboratories [32] have measured the viscosity of glass and glass-ceramic composites using an optical technique of photographing the deformation of the seal material under a load. Using such measurements, the viscosity, contact angle, and seal spreading rate are measured.

While necessary, the property measurements alone do not provide all the necessary information about the functionality of the seal material. However, material screening and evaluation using stack tests are not practical. In this section experimental techniques to evaluate the seal material in addition to property measurements are discussed. While discussions focus on glass or glass-ceramic composite seal materials, many of the techniques apply to other types of seal materials.

5.4.1 WETTING

The wetting behavior is particularly critical for seals with glassy or flowable components. High-temperature optical techniques such as those used in Sandia National Laboratories are an elegant way of monitoring the wetting behavior at various temperatures. This provides an accurate determination of the seal affinity to bond to the mating surfaces. A simpler approach is to apply the seal over a material of interest and evaluate the interfacial wettability. An example of the interface microstructure between scandia-stabilized zirconia (ScSZ) and a glass-ceramic composite is shown in Figure 5.6. While the wetting behavior of the vitreous phase was good,

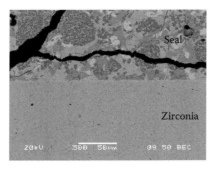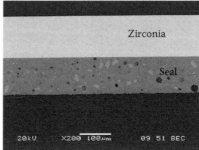

FIGURE 5.6 Evaluation of seal wetting and cracking behavior on a zirconia substrate.

comparison of milled and unmilled crystalline blend showed that an intimate mix of the two phases is necessary to achieve crack-free seals (see Figure 5.6). A similar wetting characterization was done by making a sandwich of stainless steel (CTE = ~10 ppm/°C) and a mixed conducting composition, La and Co doped $SrFeO_3$ (CTE = ~19 ppm/°C). As can be seen in Figure 5.7, the composite seal showed good wettability on both the ferrite and stainless steel surfaces. It should be noted that some areas of the seal showed cracks in the ferrite–stainless couple likely due to the CTE mismatch between the two mating materials.

5.4.2 STABILITY

In addition to matching thermal expansion, the stability of the seal materials at the stack operating temperature is a critical requirement. Both phase stability and dimensional stability of the components of the seal material need to be evaluated. In particular, glass-based seals tend to devitrify resulting in continually changing phase composition and their thermomechanical behavior. Aging study of the selected seal material may reveal potential changes that may affect its stability. The seal material aged for various lengths of time at the stack operating temperature can be analyzed

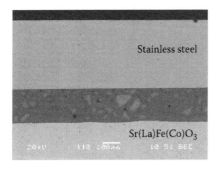

FIGURE 5.7 Wetting behavior of composite seal on $SrFeO_3$ and ferritic stainless steel.

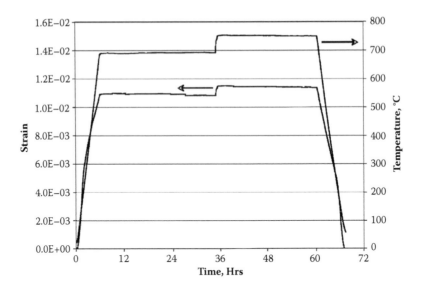

FIGURE 5.8 High temperature stability of a seal component.

for phase stability. Similarly, a bar sample can be aged and its TEC can be measured. Alternatively, the expansion behavior can also be continuously monitored as shown in Figure 5.8. A rod of a seal constituent is heated in a dilatometer to 700°C and held for 24 h and then taken to 750°C and held for another 24 h. No dimensional change at 700 and 750°C is noted during the nearly 50 h of total hold time. Such short-term tests will help screen a variety of material to identify potential candidate materials. The material, however, needs to be evaluated at higher temperatures, in the 800 to 900°C range, to expand its applicability. It should be recognized that an operating stack may experience a temperature difference of 150 to 200°C from air inlet to air outlet.

5.4.2.1 Chemical Reactivity and Stability

The chemical compatibility of the seal material with respect to the sealing surfaces is essential to achieve a good seal. In general, some diffusion of species between the materials should be expected. This is particularly likely between metal components, such as an interconnect, and a glass-based seal. Chromium from the stainless steel interconnect is known to migrate into alkaline earth oxide based glasses. While such migration can in fact improve the wetting characteristics of the seal, strength reduction of the seal has also been reported [33]. However, the extent of migration over time may determine whether the metal interface must be treated with a more compatible material. It is common to treat the metal surface with materials such as alumina or zirconia to mitigate diffusion of species, provided the surface treatment gives a stable intermediate layer.

A high-temperature seal for a solid oxide cell needs to be hermetic and be able to withstand thermal cycles. However, there is sparse literature available regarding seal function tests of high-temperature fuel cells. The hermeticity of the seal material can be tested using two techniques: (1) pressure test for hydrodynamic leak, and (2) oxygen sensor test for a concentration-driven leak.

5.4.2.2 Pressure-Leakage Test

This is the most common test method employed to qualify the leak characteristics of a new seal material. The test method involves applying the seal between two ceramic discs or between a ceramic and a metal disc, pressurizing the cavity formed by the seal and monitoring the pressure decay as a function of time.[22] Alternatively, a metal tube and a ceramic disc can also be used [34]. Typically, the cavity is pressurized to about 2 psi and the leak rate is determined by the pressure decay as a function of time. These tests can be done at room temperature or elevated temperatures. Similar test arrangement has also been used to test a plastically deformable brazed metal seal between fuel cell anode material and Haynes 214 washer [35]. The cavity is pressurized to measure the rupture strength of the seal material.

A thin zirconia membrane is sealed to a stainless cup of about one inch diameter (Figure 5.9). The stainless cup has a feed tube welded to it. The inside cavity is evacuated to about 5 psi and sealed off. One can calculate the leak rate by monitoring the pressure decay as a function of time (see Section 5.2.1).

The arrangement is then placed in a furnace, heated to 800°C, and the pressure decay test is repeated. The setup can be subjected to thermal cycles with pressure decay tests at room temperature and at 800°C. A good hermetic seal, however, will not show any pressure decay at a low pressure differential. The results of one series of tests using a glass composite material that has a glass transition temperature (T_g) of about 700°C, at ~5 psi are shown in Figure 5.10. The combination of pressure and thermal cycle tests showed that the seal is capable of withstanding the pressure

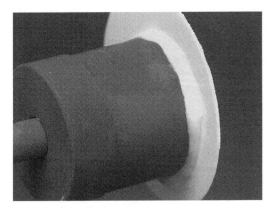

FIGURE 5.9 Photograph of the pressure test assembly after five thermal cycles.

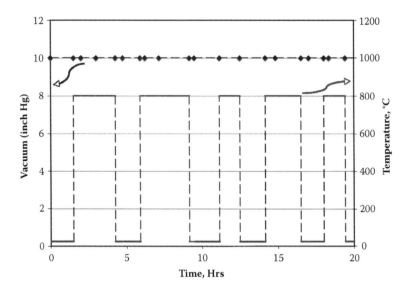

FIGURE 5.10 Leak characteristics with thermal cycles.

differential expected in a fuel cell stack. This arrangement can also be used to verify the seal hermeticity at various temperatures and various lengths of time. At a higher pressure of 20 psi, the seal failed; however, when heated to 900°C (sealing temperature) and retested at 800°C, the seal appeared to be hermetic. Thus, this arrangement can also be used to determine the healing conditions when the seal fails.

The above test provides a basis for evaluating a seal material's capability at the desired operating temperature. However, in realistic stack conditions, a seal material is under a shear stress. A double tube arrangement can be used to study the seal behavior. A disc can be sealed on both sides, and both tube enclosures can be pressurized to the same level. Such condition will eliminate the flexing of the membrane causing the seal to delaminate at a fairly low pressure when tested above T_g. In fact, a repeat test of the above seal with a double-tube arrangement showed that the seal could withstand 20 psi pressure before a small leak developed.

A test method to evaluate the shear stress capability of a seal material is reported [36]. An electrolyte-anode-electrolyte trilayer was glass sealed to two metal interconnect plates as shown in Figure 5.11. Shear testing was done in two different modes, constant loading rate and constant displacement rate, to determine the shear modulus and viscosity.

5.4.2.3 Pressure-Sensor Tests for Leaks

The pressure test is a good indication of leaks when there is a pressure difference across the seal. However, there are situations when even at low-pressure difference across the seal a concentration-driven leak may cause direct mixing of gases through the seal affecting the device efficiency. As mixing hydrogen and oxygen through

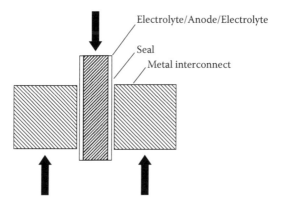

FIGURE 5.11 Shear test sample schematic.

diffusion is more precisely measurable using an oxygen sensor technique, small fuel cells (~1.5 inch diameter button cells) are used to quantify the leak characteristics of the seal. Two oxygen ion-conducting membranes were selected for this test: Sc-doped zirconia (180 μm thick) and Sr- and Mg-doped lanthanum gallate (300 μm thick). For both cells, the hydrogen electrode was nickel-ceria. For the air electrode Sr-doped lanthanum manganite was used for the zirconia cell and Sr-doped lanthanum cobaltite for the gallate cell. In addition to the primary air and hydrogen electrodes, the same materials were used as the reference electrodes on the air and fuel sides of the cell. Platinum mesh and wires were attached as current collectors. Each cell was joined to a zirconia tube using a glass composite material at 950°C. Photographs of a zirconia electrolyte button cell are shown in Figure 5.12.

FIGURE 5.12 Photograph of a zirconia button cell sealed to a zirconia tube.

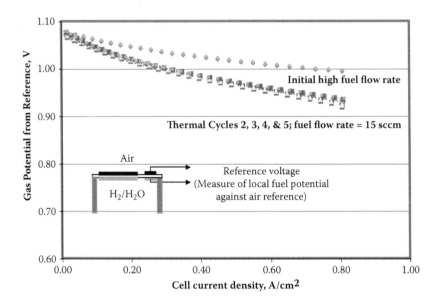

FIGURE 5.13 Reference electrode response in a zirconia membrane cell over five thermal cycles.

Using air and hydrogen reference electrodes, one can measure the sensor (Nernst) voltage generated by the difference in oxygen partial pressures. In addition, by connecting the fuel cell to a load, one can transport oxygen from air to the hydrogen side in a continuous and controlled manner. This will change the local H_2/H_2O ratio in the fuel and, in turn, the sensor voltage measured using the reference electrodes. Hydrogen bubbled through water at room temperature was used as the fuel (H_2/H_2O = ~97%). The current through the cell was varied using a potentiostat set at 50 mA increments and the reference voltage was monitored as a function of cell current. Initial test at 800°C was done with a very high flow rate of hydrogen. After the cells were cooled down to room temperature and inspected, they were retested at 800°C with 15 sccm of hydrogen bubbled through water. As expected, the sensor voltage was lower at the low flow rate corresponding to a lower H_2/H_2O ratio resulting from complete saturation of water vapor. Four thermal cycles were done using the low hydrogen flow rate and both cells reproduced the sensor response in each of the cycles. This clearly demonstrated that the seal did not allow concentration-driven mixing of fuel and air even through multiple thermal cycles. The sensor response data are plotted in Figure 5.13.

5.5 SUMMARY AND CONCLUSIONS

Over the years, a large number of materials and approaches have been investigated for sealing of SOFC stacks. The best of these approaches have been successful in sealing the cells to "acceptable" leakage rates. Glass-ceramic and composite seals were described that have been successfully developed and utilized at Ceramatec, Inc. The following seal properties and behavior have been found to be important: wetting of components, phase and dimensional stability, and low chemical reactivity.

However, there can be significant problems with all types of seals during long-term service. This is due to corrosion effects, long-term phase changes such as devit-rification or cracking associated with multiple thermal cycles. In plane as well as vertical temperature gradients during heatup, cooldown, and dynamic load conditions can also be a major source of seal failure. Addressing these issues requires modeling as well as functional testing using actual stack components.

Work is ongoing to improve the understanding of both pressure- and concentration-driven leakage mechanisms. Understanding the leakage mechanisms can help design better seal materials and seal configurations. Fitting of experimental results to a stack leakage model provided a useful characterization of the overall leakage effects. This information can be used to select optimal operating conditions for solid oxide cell stacks.

REFERENCES

1. J.E. O'Brien, J.S. Herring, P.A. Lessing, and C.M. Stoots, "High Temperature Steam Electrolysis from Advanced Nuclear Reactors using Solid Oxide Fuel Cells," presented at the First International Conference on Fuel Cell Science, Engineering, and Technology, Rochester, NY, April 21–23, 2003.
2. J.S. Herring, J.E. O'Brien, C. Stoots, and P.A. Lessing, "Hydrogen Production from Nuclear Energy via High Temperature Electrolysis," paper 4322, 2004 International Congress on Advances in Nuclear Power Plants (CAPP '04), Pittsburgh, Pa, June 13–17, 2004.
3. R.N. Singh, "High Temperature Seals for Solid Oxide Fuel Cells," 28th International Conference on Advanced Ceramics and Composites, eds., E. Lara-Curzio and M.J. Readey, Cocoa Beach, FL, 25 (3), pp. 299–307 (2004).
4. J.W. Fergus, "Sealants for solid oxide fuel cells," *Journal of Power Sources*, 147, pp. 46–57 (2005).
5. P.A. Lessing, "A review of sealing technologies applicable to solid oxide electrolysis cells," *Journal of Materials Science*, 42, pp. 3465–3476 (2007).
6. Z.G. Yang, K.S. Weil, K.D. Meinhardt, J.W. Stevenson, D.M. Paxton, G.-G. Xia, and D.-S. Kim, "Chemical compatibility of barium-calcium-aluminosilicate base sealing glasses with heat resistant alloys," in *Joining of Advanced and Speciality Materials V*, pp. 116-124, J.E. Indacochea, J.N. DuPont, T.J. Lienert, W. Tillmann, N. Sobczak, W.F. Gale, and M. Singh, eds., ASM International, Materials Park, OH, 2002.
7. Personal Communication, Mr. Stephen H. Pike, Rolls-Royce Fuel Cell Systems, Ltd., Charnwood Building, Holywell Park, Ashby Road, Loughborough LE11 3GR, UK, Phone: +44 1509 225439.
8. N. Lahl et al. "Aluminosilicate Glass Ceramics as Sealant in SOFC Stacks," *Solid Oxide Fuel Cells Vol. VI*, Electrochemical Society Proceedings Vol. 99-19, pp. 1057–1066.
9. K. Eichler et al., "Degradation Effects at Sealing Glasses for the SOFC," *Proc. 4th European SOFC Forum*, pp. 899–906 (2000).
10. Z. Yang et al., "Chemical Compatibility of Barium-Calcium-Aluminosilicate-Based Sealing Glasses with the Ferritic Stainless Steel Interconnect in SOFCs," *Journal of the Electrochemical Society*, 150(8), pp. A1095–A1101 (2003).

11. M. Bram et al., "Basic investigations on metallic and composite gaskets for an application in SOFC stacks," in *Proceedings of the Fifth European Solid Oxide Fuel Cell Forum*, J. Huijsmans (ed.), 1-5 July 2002, Lucerne, Switzerland, 2202, pp. 847–854.

12. S. Reckers et al., "Leakage investigations of compressive metallic seals in SOFC stack," in *Proceedings of the Fifth European Solid Oxide Fuel Cell Forum*, J. Huijsmans (ed.), 1-5 July 2002, Lucerne, Switzerland, 2002, pp. 847–854.

13. Y.-S. Chou and J.W. Stevenson, "Phlogopite Mica-Based Compressive Seals for Solid Oxide Fuel Cells: Effect of Mica Thickness," *Journal of Power Sources*, 124, pp. 473–478 (2003).

14. M. Bram et al., "Deformation Behavior and Leakage Tests of Alternate Sealing Materials for SOFC Stacks," *Journal of Power Sources*, 138, pp. 111–119 (2004).

15. Y.-S. Chou and J.W. Stevensen, "Novel Silver/Mica Multilayer Compressive Seals for Solid-Oxide Fuel Cells: The Effect of Thermal Cycling and Material Degradation on Leak Behavior," *Journal of Power Sources*, 135, pp. 72–78 (2004).

16. J. Stevenson, "SOFC Seals: Materials Status," PNNL Presentation at SECA Core Technology Program—SOFC Seal Meeting, July 8, 2003, Sandia National Laboratory, Albuquerque, NM.

17. Z.B. Zhao et al., "Equilibrium Phase Diagrams in the Systems PbO-Ag and CuO-Ag," *J. American Ceramic Soc.*, 76[10], pp. 2663–2664 (1993).

18. A.M. Meier et al., "A Comparison of the Wettability of Copper-Copper Oxide and Silver-Copper Oxide on Polycrystalline Alumina," *J. Materials Science*, 30[19] pp. 4781–4786 (1995).

19. K.S. Weil, C.A. Coyle, J.S. Hardy, J.Y. Kim, and G-G Xia, "Alternative Planar SOFC Sealing Concepts," *Fuel Cells Bulletin*, May 2004, p. 14.

20. P. Kofstad and R. Bredesen, "High Temperature Corrosion in SOFC Environments," *Solid State Ionics*, 52, pp. 69–75 (1992).

21. R. Loehman et al., "Development of Reliable Methods for Sealing Solid Oxide Fuel Cell Stacks," SECA Core Technology Program Review, January 27–28, 2005, Tampa, FL.

22. M.M. Seabaugh and B. Emley, "Textured Composite Seals for Low Temperature SOFCs," SECA Core Technology Program, SOFC Seal Meeting, July 7, 2003.

23. T. Yamamoto et al., "Application of Mica Glass-Ceramics as Gas-Sealing Materials for SOFC," in *Proceedings of the Fourth International Symposium on Solid Oxide Fuel Cells (SOFC-IV)*, Vol. 95-1, pp. 245–253, The Electrochemical Society (1995).

24. R. Loehman et al., "Development of Reliable Methods for Sealing Solid Oxide Fuel Cell Stacks," SECA Core Technology Program Review, January 27–28, 2005, Tampa, FL.

25. K.S. Weil, C.A. Coyle, J.S. Hardy, J.Y. Kim, and G-G Xia, "Alternative Planar SOFC Sealing Concepts," *Fuel Cells Bulletin*, May 2004, p. 15.

26. DuraFoil composition: 22%Cr, 7%Al, 0.1%La+Ce, Bal. Fe. Manufactured by Engineered Materials Solutions Inc., Attleboro, MA.

27. R.K. Shah and A.L. London, *Laminar Flow Forced Convection in Ducts*, Academic Press, New York, 1978.

28. R.E. Treybal, *Mass-Transfer Operations Third Edition*, pp. 31–34, McGraw-Hill, New York, 1980.

29. R.C. Reid, J.M. Prausnitz, and B.E. Poling, *The Properties of Gasses & Liquids Fourth Edition*, pp. 596-597, McGraw-Hill, New York, 1986.

30. J. Hartvigsen, S. Elangovan, and A. Khandkar, *Science and Technology of Zirconia V*, pp. 682–693, S. Badwal, M. Bannister, and R. Hannink, eds., Technomic, Lancaster–Basel, 1993.

31. S.K. Pradhan, S.V. Mazumder, J.J. Hartvigsen, and M. Hollist, "Effects of Electrical Feedbacks on Planar Solid Oxide Fuel Cells," *Transactions of the ASME*, 166(4), pp. 154–166, May 2007.

32. E. Corral, B. Gauntt, and R. Loehman, "Controlling Seal Materials Properties for Reliable Seal Performance Using Glass-Ceramic Composites," 30th International Conference on Advanced Ceramics and Composites, Daytona Beach, Fl. January 21–26, 2007.

33. Y-S Chou, J.W. Stevenson, X. Li, G. Yang, and P. Singh, "High Temperature Glass Seal," *SECA Annual Review*, San Antonio, Texas August 7–9, 2007.

34. Y-S. Chou and J.W. Stevenson, "Compressive Mica Seals for Solid Oxide Fuel Cells," *Journal of Materials Engineering and Performance*, 15(4), pp. 414–421, August 2006.

35. K.S. Weil, J.S. Hardy, and B.J. Koeppel, "New Sealing Concept for Planar Solid Oxide Fuel Cells," *Journal of Materials Engineering and Performance*, 15(4), pp. 427–432, August 2006.

36. J. Malzbender, J. Mönch, R.W. Steinbrech, T. Koppitz, S.M. Gross, and J. Remmel, "Symmetric Shear Rest of Glass-Ceramic Sealants at SOFC Operation Temperature," *J. Mater. Sci.*, 42, pp. 6297–6301, 2007.

6 Processing

Olivera Kesler and Paolo Marcazzan

CONTENTS

6.1 INFLUENCE OF PROCESSING ON SOFC MICROSTRUCTURE, PROPERTY, AND PERFORMANCE

In developing new materials for targeted applications, the influence of processing methods on the properties and performance of the materials must be taken into account, in addition to the influence of the material composition. By altering the microstructure of the resulting material, a change in the processing method used to prepare it can drastically alter its properties. The importance of processing is reflected, as one example, in the designations of different heat treatments as part of the designations of different alloys along with information on the elemental compositions.

Similarly, in the development of solid oxide fuel cells (SOFCs), it is well recognized that the microstructures of the component layers of the fuel cells have a tremendous influence on the properties of the components and on the performance of the fuel cells, beyond the influence of the component material compositions alone. For example, large electrochemically active surface areas are required to obtain a high performance from fuel cell electrodes, while a dense, defect-free electrolyte layer is needed to achieve high efficiency of fuel utilization and to prevent crossover and combustion of fuel.

The existence in the scientific literature of contradictory studies regarding the properties and performance of fuel cell layers with different material compositions highlights the important influence of the microstructure on the determination of material properties. For example, in studies carrying out a direct comparison of the conductivities of 10 mol% Sm-doped ceria and 10 mol% Gd-doped ceria, some have found that the Gd doping leads to higher conductivity over the 500 to 700°C temperature range [1], while others have found that the Sm doping leads to higher conductivity over the same temperature range [2]. These contradictory results indicate that other factors besides the material composition play a significant role in the material properties. Specifically, the microstructure, and especially the grain structure in ionically conducting solids [2], plays a substantial role in the material properties. In turn, the processing methods used to prepare the materials play an extremely important role in determining the features of the microstructures produced such as grain size, particle size, total porosity, and pore size distribution. Those microstructural features directly influence properties such as electrical conductivity and permeability, which translate into different performance characteristics of the fuel cell such as the activation, ohmic, and concentration polarizations.

The microstructure of a fuel cell can also determine its long-term durability and performance stability, in addition to its short-term electrochemical performance. For example, very fine, high surface area microstructures in an electrode can result in high short-term performance but rapid degradation as the electrodes sinter during operation, while a coarser initial microstructure can lead to lower initial performance but to higher performance stability and hence higher performance in a longer time frame compared to a finer microstructure.

As a result of these processing-microstructure-property-performance inter-relationships, it is essential to optimize not only the material compositions to be utilized in the fuel cell components, but also the processing methods used to produce those components. Such optimization must be performed considering both short- and long-term performance and durability, and also additional considerations of practical concern such as process and material cost, scalability for mass production, and compatibility with other processes and components in the fuel cell.

6.1.1 MICROSTRUCTURAL, PROPERTY, AND PERFORMANCE REQUIREMENTS OF SOFC COMPONENTS

Each of the components of an SOFC stack: anodes, cathodes, electrolytes, and interconnects must be thermally, chemically, mechanically, and dimensionally stable at the operating conditions and compatible with the other layers with which they come into contact in terms of thermal expansion and chemical inter-reaction. They must also have compatible processing characteristics. In addition to those requirements, the individual layers have additional microstructural, property, and processing target requirements, as summarized in Table 6.1.

In addition to the general requirements for each layer, there are specific requirements generated by the choice of fuels or the operating conditions such as temperature or extent of thermal cycling. Anodes that directly utilize hydrocarbon fuels have more stringent chemical requirements to avoid carbon deposition than anodes in fuel cells operating on either pure hydrogen gas or natural gas reformate [3–5]. In addition, when sulfur is present in the fuel, the anode must also be tolerant to high levels of fuel impurities such as H_2S [6, 7]. Because oxygen reduction at the cathode generally has slower reaction kinetics than hydrogen oxidation at the anode, microstructural requirements for high reaction surface area are particularly important at the cathode, or for anodes directly utilizing fuels other than hydrogen, for which the electrochemical oxidation kinetics are also slower than for hydrogen.

These requirements have led to the investigation of processing methods that allow higher fuel cell performances to be achieved from a given set of material compositions, by adjusting the microstructures and processing methods. For example, many studies have been performed with the aim of optimizing the particle size distributions in each of the anode, electrolyte, and cathode layers to maximize electrochemical performance and minimize diffusion losses. Composite electrodes have been utilized to extend the reaction surface area further into the electrode volume, and graded microstructures and compositions have been utilized to simultaneously provide high reaction surface area and high diffusivity where they are most needed, and to minimize thermal stresses. Each of these approaches is discussed in more detail in the following sections.

The use of two-phase composites or graded or multilayered structures introduces additional processing requirements beyond those generally needed to prepare single-phase materials. For example, the influence of sintering temperature varies with the melting temperature of the materials being processed, and allowable processing temperatures depend also on whether individual components can react with each other during the fabrication of the fuel cell. These factors ultimately determine which processing methods will be feasible for an SOFC utilizing a given material set. The ultimate choice of processing route from among the feasible methods then becomes a matter of satisfying the practical requirements of commercial production while producing high performance cells.

TABLE 6.1

Microstructural, Property, and Processing Requirements of SOFC Component Layers

	Anode	Electrolyte	Cathode	Interconnect
Microstructure	Porous, many triple-phase boundaries, stable to sintering.	Dense, thin, free of cracks and pinholes.	Porous, many triple-phase boundaries, stable to sintering.	Dense separation between cells, porous or channeled gas transport paths.
Electrical	Electronically and preferably ionically conductive.	Ionically but not electronically conductive.	Electronically and preferably ionically conductive.	Electronically but not ionically conductive; conductive oxide layer.
Chemical	Stable in fuel atmosphere; preferably also stable in air for redox tolerance. Catalytic for oxidation and reforming but not for carbon deposition.	Stable in both oxidizing and reducing environment. Minimal reduction and resulting electronic conductivity in reducing conditions.	Stable in air environments. Catalytic for oxygen reduction. Resistant to performance loss caused by chromium deposition.	Stable in both air and fuel environments. Resistant to rapid oxidation. Minimal chromium evaporation.
Thermal Expansion	Compatible with other layers, especially electrolyte.	Compatible with other layers, especially structural support layer.	Compatible with other layers, especially electrolyte.	Compatible with other layers, especially electrolyte.
Chemical Compatibility	Minimal reactivity with electrolyte and interconnect.	Minimal reactivity with anode and cathode.	Minimal reactivity with electrolyte and interconnect.	Minimal reactivity with anode and cathode.
Processing Methods	Compatible with subsequently produced layers.			

6.1.2 Composite Electrodes

To meet the requirements for electronic conductivity in both the SOFC anode and cathode, a metallic electronic conductor, usually nickel, is typically used in the anode, and a conductive perovskite, such as lanthanum strontium manganite (LSM), is typically used in the cathode. Because the electrochemical reactions in fuel cell electrodes can only occur at surfaces where electronic and ionically conductive phases and the gas phase are in contact with each other (Figure 6.1), it is common

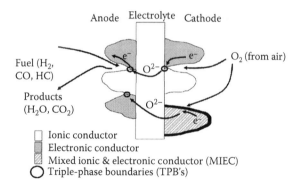

Anode Electrolyte Cathode

Fuel (H$_2$, CO, HC)

Products (H$_2$O, CO$_2$)

O$_2$ (from air)

e$^-$ O^{2-} e$^-$

O^{2-}

e$^-$

☐ Ionic conductor
▨ Electronic conductor
▧ Mixed ionic & electronic conductor (MIEC)
○ Triple-phase boundaries (TPB's)

FIGURE 6.1 Triple-phase boundaries (TPBs) in SOFC electrodes at which electrochemical reactions take place. Cathode mixed conductor materials have larger potentially electrochemically reactive surface areas (entire particle surfaces rather than only the TPBs).

practice to include a second solid phase, usually an ionically conductive or mixed ionically and electronically conductive (MIEC) ceramic, in both the anode and cathode. This material is often either doped zirconia (yttria-stabilized zirconia, YSZ, or scandia-stabilized zirconia, ScSZ), doped ceria (samaria-doped ceria, SDC, or gadolinia-doped ceria, GDC), or doped lanthanum gallate (lanthanum strontium gallium magnesium oxide, LSGM). The addition of a ceramic phase in the anode also slows agglomeration or sintering of the nickel or other metallic phases during operation and process firing. Ceramic composite anodes have also been utilized to improve the anode stability to reduction-oxidation cycling [8] or to prevent carbon deposition in the anodes when utilizing dry methane or higher hydrocarbon fuels [9].

A comparison of single-phase LSM cathodes with LSM-YSZ composite cathodes was carried out by Murray et al. [10], showing that the composite cathodes had a larger triple-phase boundary (TPB) length than the single-phase cathodes, leading to a corresponding increase in the electrochemical performance. The two cathode types were found by impedance spectroscopy to have the same rate-limiting reaction mechanism, thereby confirming that the improved performance was due to an increase in the electrochemically active surface area, and not due to a change in the rate-limiting reaction mechanism step.

Similarly, composite LSM-YSZ cathodes produced by atmospheric plasma spraying (APS) were found to have a lower cathode polarization resistance, R$_p$ = 0.4 Ωcm^2 at 750°C [11], than single-phase lanthanum strontium cobaltite (LSC) cathodes produced by APS, cathode R$_p$ = 0.6 Ωcm^2 at 800°C [12], even though the latter material has a higher electronic and ionic conductivity than that of LSM. The increase in the TPB length produced by the use of a composite cathode therefore has the potential to produce a greater benefit than the use of an MIEC cathode material alone.

Single-phase perovskite MIECs such as Sr-doped lanthanum cobaltite (LSC), lanthanum ferrite (LSF), lanthanum cobalt ferrite (LSCF) and samarium cobaltite (SSC), and Ca-doped lanthanum ferrite (LCF) [13] are sometimes used alone in SOFC cathodes, as depicted in the lower right-hand corner of Figure 6.1, but combining an

MIEC with an additional ionically conductive phase, such as GDC or SDC, typically extends the electrochemically active region still further due to the higher ionic conductivity of GDC and SDC compared to that of the perovskites. The optimal composition of a two-phase composite depends in part on the operation temperature, due to the larger dependence of ionic conductivity on temperature compared to electronic conductivity. A two-phase composite of LSCF-GDC therefore has an increasingly large optimal GDC content as the operating temperature is reduced [14]. A minimum cathode R_p for temperatures above approximately 650°C has been found for 70–30 wt% LSCF-GDC composite cathodes, while at lower temperatures, a 50-50 wt% LSCF-SDC composite cathode was found to have a lower R_p [15].

Another example of composite SOFC cathodes was produced by glycine-nitrate combustion to form a cathode perovskite with nominal composition of $Gd_{0.6}Sr_{0.4}Fe_{0.8}Co_{0.2}O_{3-\delta}$ [16]. It was found by energy dispersive x-ray spectroscopy (EDX) that the process instead produced a composite of two perovskites, a cubic phase of composition $Gd_{0.76}Sr_{0.24}Fe_{0.75}Co_{0.22}O_{3-\delta}$, and an orthorhombic phase of composition $Gd_{0.49}Sr_{0.51}Fe_{0.79}Co_{0.24}O_{3-\delta}$. The separate phases were then also produced individually, and it was found that the composite cathode had a significantly lower area-specific polarization resistance (0.025 Ωcm^2 at 800°C) than the orthorhombic or cubic phases alone (0.41 and 3.33 Ωcm^2, respectively, at 800°C). The microstructure of the composite cathode was found to have the two phases present in different particle size distributions, which likely led to a combination of high surface area and sufficient porosity that could not be produced by either of the two phases alone. The microstructure of the composite cathode is shown in Figure 6.2.

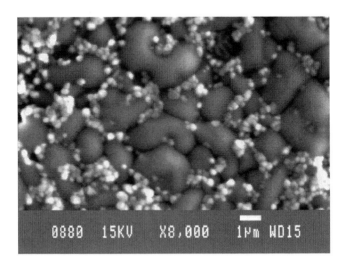

FIGURE 6.2 Microstructure of composite cathode showing a two-phase mixture of $Gd_{0.76}Sr_{0.24}Fe_{0.75}Co_{0.22}O_{3-\delta}$ and $Gd_{0.49}Sr_{0.51}Fe_{0.79}Co_{0.24}O_{3-\delta}$, which resulted in a low polarization resistance compared to a cathode made from either of the two phases alone [16]. Reproduced by permission of ECS-The Electrochemical Society.

Because the individual components of composite anodes and cathodes typically have different melting temperatures and reactivities with neighboring fuel cell component materials, the processing requirements of each component in a composite electrode must be considered in designing a fabrication process for the fuel cells. For example, in APS LSM-YSZ cathodes [11], the higher melting temperature of YSZ and lower stability of LSM in reducing environments constrains the process parameters used in producing the cathode to sufficiently melt the YSZ phase while not decomposing the LSM phase during deposition.

6.1.3 EFFECTS OF PARTICLE SIZE AND MICROSTRUCTURE ON PERFORMANCE

In addition to the use of composite anodes and cathodes, another commonly used approach to increase the total reaction surface area in SOFC electrodes is to manipulate the particle size distribution of the feedstock materials used to produce the electrodes to create a finer structure in the resulting electrode after consolidation. Various powder production and processing methods have been examined to manipulate the feedstock particle size distribution for the fabrication of SOFCs and their effects on fuel cell performance have also been studied. The effects of other process parameters, such as sintering temperature, on the final microstructural size features in the electrodes have also been examined extensively.

Introduction of nano-scale (~40 nm crystallites) SDC particles into a Ni-ScSZ structural anode support, even in quantities as small as 2 wt%, has been found to improve the fuel cell performance at 850°C by approximately 33% compared to the performance with the coarser structural support layer alone [17]. Impregnation of a $La_{0.75}Sr_{0.25}Cr_{0.5}Mn_{0.5}O_3$ (LSCM) anode with GDC after firing was found to allow the retention of finer particle sizes in the final anode microstructure, fired at 950°C after infiltration, compared with the cathode microstructural features, which were sintered at a higher temperature of 1200°C (0.3 vs. 1.6 µm, respectively) [18]. This infiltration technique was therefore considered a promising way to produce higher-surface area electrodes than can otherwise be produced when co-firing with the SOFC structural support layer is necessary.

In addition to particle size, the particle morphology used in the formation of SOFC components can also affect the fuel cell performance. One study produced NiO, SDC, and NiO-SDC composite particles of similar size ranges by spray pyrolysis (0.6 to 0.7 µm, 0.8 to 1.2 µm, and approximately 1 µm, respectively) [19]. The authors found that anodes produced from the composite powders resulted in fuel cells that exhibited higher peak power densities in air at 700°C compared to fuel cells produced from ball-milled mixtures of the separately produced powders (0.45 vs. 0.31 W/cm²). The authors attributed this performance difference to the morphology of the composite powders produced. The SDC phase, by partially coating the nickel phase, was hypothesized to be more effective at preventing agglomeration of the nickel particles during heating to the operating temperature than in the mixture of separately produced powders, as shown schematically in Figure 6.3 [19]. Scanning electron microscopy (SEM) of the tested cells confirmed that the anodes made from the composite powders had a higher surface area than the anodes made from the mechanically mixed powders after testing.

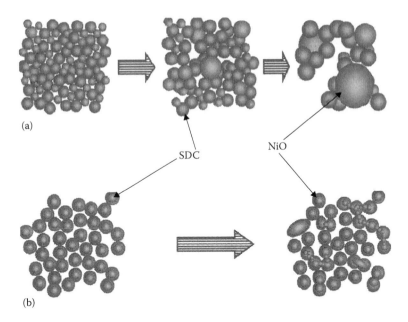

FIGURE 6.3 (a) Morphology of ball-milled NiO and SDC powders produced separately and (b) NiO-SDC composite powders coproduced by spray pyrolysis, showing a possible mechanism by which the composite powders retain a higher surface area resulting in a higher performance [19]. Reprinted from [19] with permission from Elsevier.

Many research efforts have also been focused on the improvement of cathode performance through the control of the microstructure, to increase the electrochemically active surface area for oxygen reduction. One such study investigated the influence of the LSM particle size in the starting powder mixture on the performance of LSM-YSZ composite cathodes [20]. SOFC cathodes made from LSM powder with a smaller particle size (81 nm) were found to have a comparable performance at 600°C (R_P = 0.46 Ωcm^2) to that of cathodes made from coarser powder (210 nm, R_P = 0.47 Ωcm^2), but only when different sintering temperatures were used for the two cathodes: 1100°C for the former and 1200°C for the latter. It was found that the smaller particle size LSM allowed a comparable performance to be achieved at a lower sintering temperature compared to the coarser particles, thus simplifying the processing conditions.

An important influence of both particle size and powder processing method was found by Hagiwara et al. [21] in the preparation of LSM-ScSZ composite cathodes by mechano-chemical mixing of powders in an attrition particle composing machine. The authors prepared ScSZ-LSM mechano-chemically combined powders by combining fine 0.5 μm ScSZ powder with LSM over a range of particle sizes with d_{50} ranging from 0.5 to 5 μm. The fine ScSZ powder was found to coat the LSM powder, and the resulting fuel cell cathodes exhibited a lower polarization resistance than both a pure LSM cathode and LSM-ScSZ composite cathodes made by mixing of small LSM particles without the mechano-chemical treatment, although traditionally processed composite cathodes with larger LSM particles outperformed the mechano-chemically produced cathodes. The processing technique was found to provide a useful mechanism for providing more control over the microstructure and morphology of the cathodes than is normally achieved using mixtures of particles. It also allows for higher TPB lengths to be achieved with lower ScSZ contents than with traditional mixing, thus providing a mechanism for lowering the material cost by reducing the amount of expensive ScSZ needed for a given level of performance [21].

To facilitate the introduction of silver into an LSCF cathode for use in an intermediate-temperature (IT) SOFC, a similar approach was utilized for Ag-LSCF cathodes by Simner et al. [22]. A mechano-fusion powder manufacturing process was applied to the fabrication of LSCF- or SSC-coated silver powder to prevent excessive silver agglomeration. The fuel cell made with those materials was found to exhibit unacceptably low performance, 0.35 W/cm^2 at 650°C, and unacceptably high degradation rates (4.5 to 7.5% per 1000 h) at higher temperatures of 700 and 750°C, respectively. Long periods of operation were required to reach the highest performances, which was attributed to the time required for the silver to electromigrate from within the coated particles and form more interfaces in the material. They proposed modifying the process to achieve such a higher-surface area structure prior to cell conditioning to both increase the performance and reduce the degradation rates [22].

The influence of cathode particle sizes on performance was also investigated by researchers at Forschungzentrum Jülich (FZJ), who found that coarse, nonground LSM powder ($d_{90} = 26$ μm) used to prepare a cathode current collector layer resulted in higher SOFC performance than finer, ground LSM powder ($d_{90} = 2.5$ μm), while the effect of calcining or of grinding the YSZ powder, or both, did not significantly influence the fuel cell performance, although the differences in particle sizes between the three YSZ processing methods was quite small ($d_{90} = 1.04$, 0.92, and 2.52 μm for calcined and ground, ground but not calcined, and neither ground nor calcined powders, respectively) [23].

Due to the important relationship between particle size of starting powders and resulting electrode microstructure and corresponding performance, much work has been performed to modify the particle size and morphology of the starting powders used in SOFC processing. Additional methods have been investigated to better control the microstructure and properties of fuel cell components, which are discussed in more detail in Section 6.2.

6.1.4 Multi-Layered and Graded Components

As discussed in Sections 6.1.2 and 6.1.3, the use of composite electrodes increases the TPB length compared to single-phase electrodes, and decreases the extent of metal sintering in cermet anodes, while controlling the particle size and morphology of starting feedstock materials provides a mechanism for controlling the resulting electrode microstructures. Combing these two concepts, it has become common practice to prepare electrodes from composite materials in which the concentrations are varied across the electrode thickness, while different particle sizes and porosities are introduced across the electrode thickness to better control both the electrochemical activity and mass transport properties of the layers simultaneously. Similarly, two-layer electrolytes have also become quite common to obtain higher performance and efficiency at operating temperatures below 700°C.

6.1.4.1　Bilayered Electrodes

SOFC electrodes are commonly produced in two layers: an anode or cathode functional layer (AFL or CFL), and a current collector layer that can also serve as a mechanical or structural support layer or gas diffusion layer. The support layer is often an anode composite plate for planar SOFCs and a cathode composite tube for tubular SOFCs. Typically the functional layers are produced with a higher surface area and finer microstructure to maximize the electrochemical activity of the layer nearest the electrolyte where the reaction takes place. A coarser structure is generally used near the electrode surface in contact with the current collector or interconnect to allow more rapid diffusion of reactant gases to, and product gases from, the reaction sites. A typical microstructure of an SOFC cross-section showing both an anode support layer and an AFL is shown in Figure 6.4 [24].

The introduction of such a layer can dramatically improve the fuel cell performance. For example, in the SOFC with bilayered anode shown in Figure 6.4, the area-specific polarization resistance for a full cell was reduced to 0.48 Ωcm^2 at 800°C from a value of 1.07 Ωcm^2 with no anode functional layer [24]. Use of an immiscible metal oxide phase (SnO_2) as a sacrificial pore former phase has also been demonstrated as a method to introduce different amounts of porosity in a bilayered anode support, and high electrochemical performance was reported for a cell produced from that anode support (0.54 W/cm^2 at 650°C) [25]. Use of a separate CFL and current collector layer to improve cathode performance has also been frequently reported (see for example reference [23]).

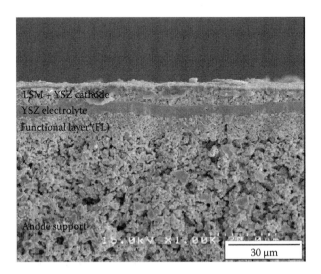

FIGURE 6.4 SEM cross-sectional micrograph of an SOFC, showing an anode support layer, anode functional layer, electrolyte, and cathode [24]. Reprinted from [24] with permission from Elsevier.

6.1.4.2   Multilayered and Graded Electrodes

In addition to bilayered electrodes with a functional layer and a support layer, electrodes have also been produced with multilayered or graded structures in which the composition, microstructure, or both are varied either continuously or in a series of steps across the electrode thickness to improve the cell performance compared to that of a single- or bilayered electrode. For example, triple-layer electrodes commonly utilize a functional layer with high surface area and small particle size, a second functional layer (e.g., reference [26]) or diffusion layer with high porosity and coarse structure, and a current collector layer with coarse porosity and only the electronically conductive phase (e.g., reference [27]) to improve the contact with the interconnect.

Graded electrodes with multilayered transitions from a functional layer to a current collector layer have been prepared with between four and ten layers and gradual transitions in composition and microstructure between layers to improve electrode performance. For example, Ruiz-Morales et al. [28] have prepared composite anodes graded in four layers of varying YSZ composition to improve the oxide ion diffusivity by introducing disordered defect transitions in a ceramic composite anode. They utilized Mn- and Ga-doped lanthanum strontium titanate (LSTMG)-YSZ anodes prepared in four layers, and found comparable performance in wet hydrogen to that of more traditional Ni-YSZ anode composites, while also obtaining high activity for methane oxidation at high temperatures (950°C). Such ceramic-based anodes are more resistant to carbon deposition during the utilization of hydrocarbon fuels, but typically suffer from lower electrochemical activities, so the utilization of a graded anode structure represents a strategy for making ceramic composite anodes more viable for hydrocarbon utilization. The authors used multiple depositions of the anode layers followed by prefiring steps after each deposition step at 300°C, followed by a cofiring step at 1200°C [28].

Graded cathodes have also been studied to improve the electrochemical performance, with many studies focusing on cathode processing, microstructure, and performance due to the slower reaction kinetics compared to that of hydrogen oxidation. Barthel and Rambert [29] prepared graded LSM-YSZ cathodes by vacuum plasma spraying (VPS), and reported lower polarization resistances of the graded cathodes compared to single-layer composite cathodes.

Multilayered cathodes prepared by infiltration and screen printing with layer compositions of LSM+SDC/LSC+LSM+SDC/LSC+SDC/LSC resulted in very low polarization resistances of 0.06 Ωcm^2 at 750°C [30]. Hart et al. [31] prepared cathodes with 10 discrete screen printing steps to prepare cathodes with compositions ranging from an LSM-YSZ active layer near the electrolyte interface to LSM alone to LSC as a current collector layer in three different gradient profiles, as well as bilayered LSM-LSC cathodes for comparison. They found that all three of the graded cathodes exhibited an approximately 10 times lower polarization resistance compared with the bilayered LSM-LSC cathode, although each of the graded compositions contained YSZ in some proportion in the active layers, so a direct separation of the effects of grading compared to the effects of extending the TPB length with the ionically conducting YSZ phase cannot be quantified from the results. The authors also noted that a composition of electronic conductor of 20% is too low for the layer adjacent to the electrolyte, resulting in a significantly larger series resistance compared to the graded layers that started with a higher LSM content, due to incomplete percolation of the electronically conducting phase at low concentrations.

Although the electrode performance can be improved by the introduction of compositional and microstructural gradients into the anode or cathode, the processing required to produce such graded layers also increases in complexity as the number of discrete layers increases, particularly when separate deposition and firing steps are required for each increment in layer composition or structure.

6.1.4.3   Bilayered Electrolytes

In addition to bilayered anode and cathode functional layer and current collector/support layer combinations, bilayered electrolyte structures are commonly fabricated, particularly for low-temperature operation below 700°C, by a variety of processing methods. Bilayered electrolytes are used for several purposes:

1. To prevent inter-reactions between YSZ electrolytes and perovskite cathode materials [32, 33].
2. To introduce a barrier layer when utilizing doped ceria electrolytes (SDC, GDC, or lanthanum-doped ceria, LDC) to prevent the reduction of Ce^{4+} to Ce^{3+}. Reduction of cerium cations results in unwanted electronic conductivity that lowers fuel efficiency [34], and mechanical degradation that results from the volume expansion of cerium ions upon reduction [35].
3. To prevent low-temperature LSGM electrolytes from inter-reacting with Ni in the anode and forming a nonconductive barrier layer [36].

Inter-reactions between Ni and LSGM electrolytes have been prevented by the introduction of barrier layers consisting of LDC fabricated by uniaxially pressing LDC and LSGM powders and cofiring them [36], of SDC screen printed on an LSGM electrolyte support [37], or of LDC interlayers screen printed on LSGM electrolyte supports with $Nd_{2-x}La_xNiO_{4+\delta}$ ($x = 0, 0.8, 2$) cathodes [38]. In reference [36], a high OCV is reported, indicating that reduction of the LDC is prevented by the LSGM, while low series resistances indicate that no major inter-reactions between Ni and LSGM have occurred. In reference [37], very low series and polarization resistances at 800°C (~0.10 to 0.265 Ωcm^2 for a variety of anode compositions) suggested that inter-reaction between Ni and LSGM had been successfully prevented. In reference [38], inter-reactions between Nd and LSGM were found, while heavy substitution of the A-site Nd cation with La reduced the extent of inter-reaction and improved performance, in both cases with no barrier layer present. The introduction of an LDC barrier layer also prevented inter-reaction when screen printed between the electrolyte and cathode. Therefore, the authors propose utilizing the new cathode material studied as a lower-cost, less-reactive alternative to Sr- and Co-containing perovskites. LDC-LSGM bilayer electrolytes were also fabricated for use in cells with oxide anodes for use in methane and H_2/H_2S mixtures, with high open-circuit voltages (up to 1.2 V) [39, 40].

To prevent inter-reactions between perovskite cathodes and YSZ electrolytes, thin barrier layers of GDC have been applied by spin coating a sol onto a screen-printed YSZ electrolyte on a tape cast anode substrate and firing the spin-coated layer at low temperature prior to deposition of the cathode by spray painting [41]. The GDC interlayers succeeded in preventing the inter-reaction between the cathode and electrolyte layers, but required additional processing steps, thus making the manufacturing process more complex. However, the low temperature of firing needed for the interlayer when it is produced from sol precursors avoids high-temperature inter-reactions between the ceria and zirconia phases, thereby avoiding the formation of a ceria-zirconia solid solution with conductivity up to 2 orders of magnitude lower than those of the separate phases [42]. Slurry dip coating of co-precipitated powders has also been utilized to apply GDC and SDC interlayers to cells with ScSZ electrolyte supports [43]. The interlayers were fired at low temperatures to consolidate them, thus minimizing inter-reactions with the zirconia electrolyte, while also providing a barrier to inter-reaction between the SSC cathodes and ScSZ electrolytes. The GDC barriers were found to result in a higher conductivity and lower polarization resistance than the SDC barriers.

In another study of processing bilayered electrolyte structures, a YSZ-SDC bilayer was screen printed in two steps on an anode support with a screen-printed AFL, and co-fired [44]. The overall cell performance was found to be quite high (1.18 W/cm^2 peak power density at 650°C), but evidence of Zr diffusion through the SDC layer was found, raising questions about the long-term stability of the material. Similarly, a 2 µm ScSZ + 4 µm SDC bilayer electrolyte was screen printed in two steps and co-fired. The thin electrolyte combined with electronic blocking by the ScSZ led to a combination of high OCV (>1 V at 750°C) and fairly high power density (545 mW/cm^2 peak power density at 650°C). However, the performance was limited to an extent by inter-reaction between the two electrolyte layers during cofiring at high temperatures [45]. Pulsed-laser deposition (PLD) has also been used to produce bilayered ScSZ-SDC electrolytes with 2 and 20 µm thick ScSZ and SDC layers, respectively, resulting in a peak power density of 0.16 W/cm^2 at 600°C [46]. Lamination and co-firing of multiple tape-cast layers has been used to produce triple-layer electrolytes with $BaCe_{0.8}Y_{0.2}O_{3-\delta}$ (BCY) between two GDC layers [47]. The authors reported an increase in OCV compared to that of cells produced with only doped ceria as electrolytes, although still lower than theoretical OCV values over the temperature range studied of 500 to 700°C.

The results summarized here illustrate the importance of identifying a processing method that produces not only the targeted high density and low thickness for each electrolyte layer, but which is also compatible with the combination of materials used in a bilayer electrolyte to avoid inter-reactions between the layers.

6.2   PROCESSING METHODS FOR SOFC COMPONENTS

The combination of various SOFC component performance, microstructural, and property requirements has led to a variety of structures, such as the composite, graded, and multilayered electrodes and electrolytes described above. The need to produce such complex structures and to process combinations of materials with varying melting temperatures and stabilities requires multiple processing methods to be used for the fabrication of each SOFC. An overview of commonly used SOFC fabrication techniques is presented in the following sections.

6.2.1 SUPPORT LAYERS

Both tubular and planar SOFCs are typically fabricated using one of the cell layers as the structural support layer with a fairly large thickness, on the order of millimeters or hundreds of micrometers, with the other components present as thinner layers of 10s of micrometers for the electrodes and 5 to 40 micrometers for the electrolyte.

6.2.1.1 Tubular Cells

Cathode-Supported Tubes—Extrusion

The most extensively commercialized SOFC stack design, particularly for large, high-power stacks of 100 kW or greater, consists of tubular cells, typically supported on the cathode layer. The cells used in such large stacks are typically approximately 2 cm in diameter and 1 m in length. A schematic diagram of a section of a cathode-supported tubular SOFC developed by Siemens Power Generation is shown in Figure 6.5 [48].

A more recent development in high power density large-scale tubular SOFCs is that of flat tubes, which consist of a tube with two flat, parallel sides, and two rounded sides, with cross-connected current paths connecting the two flat faces of the tubes through the interior to minimize the length of the current path, as shown schematically in Figure 6.6 [48].

Tubular SOFC cathode supports with diameters or distance between flat faces on the order of 1 to 2 cm are commonly prepared by extrusion. Extrusion is a wet-ceramic process used to prepare tubes, and one which facilitates the formation

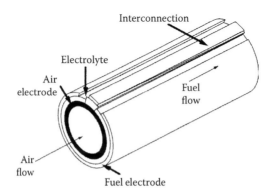

FIGURE 6.5 Tubular cathode-supported solid oxide fuel cell developed by Siemens Power Generation [48]. Reprinted from [48] with permission from Elsevier.

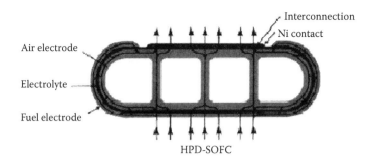

FIGURE 6.6 High power density (HPD) SOFC, consisting of a flattened tube with two flat faces. The vertical struts between the two flat faces provide shorter paths for the electronic current collection, eliminating the need for all of the electronic current to travel around the circumference of the cathode, as in the standard tubular cell design shown in Figure 6.5 [48]. Reprinted from [48] with permission from Elsevier.

of tubes having one end closed, which is a desirable feature of tubular SOFCs to facilitate the separation of the fuel and air supplies. A schematic diagram of the extrusion process is shown in Figure 6.7.

Slurry containing the cathode powder is placed into a cylindrically shaped die with an open end, as shown in Figure 6.7 (a), and then a mandril is inserted into the tube that extrudes the slurry into the shape of a hollow tube with a closed end on one side and an open end on the other side, as shown in Figure 6.7 (b). The electrolyte, anode, and interconnect layers are then deposited onto the cathode support tube, using combinations of processes described in Section 6.2.2.

Anode-Supported Tubes—Electrophoretic Deposition

Although cathode-supported tubular SOFCs in large-scale stacks are the type of SOFC stack most widely commercialized, recent alternative tubular cell designs have been developed with anode-supported designs for smaller-power applications. Cells in these stacks have diameters on the order of several millimeters rather than centimeters,

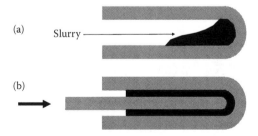

FIGURE 6.7 Extrusion process for fabricating tubular SOFC support layers. (a) Open-ended die with cathode slurry in it and (b) Mandril insertion into the die, extruding the cathode slurry into a closed-ended hollow tube.

resulting in both high volumetric power density and increased resistance to thermal shock caused by rapid thermal cycling. Electrophoretic deposition (EPD) has been used to deposit anode tubes on graphite rods, as shown schematically in Figure 6.8 [49].

In the EPD process, a DC electric field is used to deposit charged particles from a colloidal suspension onto an oppositely charged substrate, as illustrated in Figure 6.8. The graphite rod used for the deposition substrate is later burned out prior to cell operation, leaving a hollow tube. The other fuel cell layers can be deposited by a similar process onto the anode support tube.

Electrolyte-Supported Tubes—Extrusion

Millimeter-scale tubular SOFCs (2.4 to 2.5 mm inner diameter and 5 to 20 mm outer diameter with 2 mm wall thickness) have been produced with an electrolyte tube as the support layer. LSGM [50, 51] and YSZ [51] electrolyte support tubes have been fabricated by extrusion and sintering. Fairly high sintering temperatures are required for LSGM support tubes (1520°C) [50], and while sintering yields of such cells were initially found to be low due to breakage during firing, the yield can be significantly improved by design of furnace support tubes to separate the tubes and keep them immobile during firing [51], which increases the firing yield (avoidance of mechanical damage during firing) to the 60 to 98% range.

6.2.1.2 Planar Cells

Planar SOFCs have received increasing attention recently as an alternative to tubular cells due to their higher power densities, short current paths, and corresponding

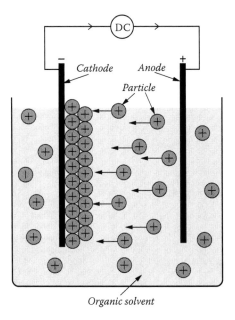

FIGURE 6.8 Electrophoretic deposition used to deposit tubular anode-supported SOFCs [49]. Reprinted with permission from Blackwell Publishing.

higher efficiencies, and simpler processing methods and lower cost. The shorter current paths also allow lower-temperature operation compared to tubular cells (below vs. above 850°C for large-area cells and tubes), which allows lower-cost interconnect materials to be utilized, such as stainless steel alloys rather than the doped LaCrO$_3$ ceramic interconnects that are commonly used for higher-temperature tubular cells. Planar cells have been fabricated that are anode-supported (ASC) [52, 53], electrolyte-supported (ESC) [37, 39, 40, 54], and interconnect- or metallic current-collector-supported (ISC) [11, 46, 55–57], with the structural support layer typically fabricated first, and the non-support layers deposited subsequently in separate processing steps onto the support layer.

Tape Casting

The most common processing method used to fabricate the support layers of planar cells is tape casting. The process involves dispensing a slurry with the desired powder composition from a vessel onto a moving carrier film; for example, a mylar film supported on a flat glass plate. The dispensed slurry passes under a doctor blade to form a flattened layer of green, i.e., unfired ceramic tape, as shown schematically in Figure 6.9. The height of the doctor blade and slurry viscosity can both be adjusted to control the tape thickness. The viscosity is adjusted through the use of organic additives in the slurry, which serve as binders, plasticizers, and dispersants [24, 26, 58–60]. Organic pore formers are often added to increase the porosity of the anode after firing by burning out of the organic phase in air during subsequent processing. The concentration, composition, and shape of the pore former particles used in turn greatly influence the final anode microstructure after firing [58]. A tape with pore former can also be used as a template for further processing steps after firing, such as infiltration steps that are used to introduce a second phase, or a smaller particle size of a similar phase, into the anode [17]. Multiple tapes can also be cast and then laminated together and co-fired to produce a multilayered structure, either within a single component, or with multiple components in the fuel cell [60].

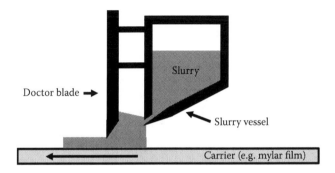

FIGURE 6.9 Schematic diagram of the tape-casting process for producing a ceramic green film.

Uniaxial Pressing

Uniaxial pressing is also commonly used to fabricate SOFC support layers in laboratory-scale research, although typically not for large-scale SOFC fabrication due to the labor-intensiveness of the process and the difficulty in fabricating large-area components. In addition to pressing and firing of the support layer with subsequent deposition of other layers [52], fabrication of multiple layers has also been accomplished by co-pressing and co-firing of the powder layers [18, 26, 36, 61].

6.2.2 NONSTRUCTURAL LAYERS

Once the structural support layers have been fabricated by extrusion or EPD for tubular cells or by tape casting or powder pressing for planar cells, the subsequent cell layers must be deposited to complete the cell. A wide variety of fabrication methods have been utilized for this purpose, with the choice of method or methods depending on the cell geometry (tubular or planar, and overall size) materials to be deposited and support layer material, both in terms of compatibility of the process with the layer to be deposited and with the previously deposited layers, and desired microstructure of the layer being deposited. In general, the methods can be classified into two very broad categories: wet-ceramic techniques and direct-deposition techniques.

The wet-ceramic techniques utilize one of a variety of methods to deposit a wet layer typically consisting of ceramic powders suspended in a liquid carrier, along with additional organic additives to control the properties of the suspension or slurry. After the layer is formed in the wet state or green state (the unfired state in ceramics processing terminology), a drying step is performed, followed by one or several heating steps to densify and solidify the deposited layers. The consolidation steps usually involve high-temperature sintering either of the layers individually after their deposition or of multiple layers together, known as co-firing. Occasionally low-temperature firing steps can be utilized to consolidate green bodies made from very fine and intimately mixed particles, typically on the nano-scale.

Direct-deposition techniques generally allow a coating to be formed without a liquid suspension carrier fluid, so that the component is deposited already in a solid state. Even after some of the direct-deposition techniques, high-temperature firing is sometimes required to densify the coatings that have been deposited or to control the phase in cases where the as-deposited coating is amorphous, while in other methods, no further consolidation is required to densify the coatings or to increase their bonding strength.

Both wet-ceramic techniques and direct-deposition techniques require preparation of the feedstock, which can consist of dry powders, suspensions of powders in liquid, or solution precursors for the desired phases, such as nitrates of the cations from which the oxides are formed. Section 6.1.3 presented some processing methods utilized to prepare the powder precursors for use in SOFC fabrication. The component fabrication methods are presented here. An overview of the major wet-ceramic and direct-deposition techniques utilized to deposit the thinner fuel cell components onto the thicker structural support layer are presented below.

6.2.2.1 Wet-Ceramic Processing Methods

Screen Printing

By far the most commonly utilized method for depositing nonstructural layers onto planar SOFC supports is screen printing. The technique is widely used both in commercial production of planar SOFCs, for example, at Versa Power Systems, in Calgary, Alberta, Canada, and in research and development scale SOFCs [17–23, 27, 33, 36–40, 52–54, 59–66]. The screen printing and subsequent firing method is a fairly simple method for depositing a green layer of ceramic slurry, and has been demonstrated to produce electrodes with low polarization resistances. For example, a $La_{1.7}Sr_{0.3}CuO_{4-\delta}$ cathode screen printed on an SDC electrolyte resulted in a cathode area-specific polarization resistance of 0.16 Ωcm^2 at 700°C [63], while a $Ba_{0.5}Sr_{0.5}Co_{0.8}Fe_{0.2}O_{3-\delta}$ (BSCF) cathode screen printed on an electrolyte disc exhibited an area-specific polarization resistance of 0.035 Ωcm^2 at 700°C [64]. A composite LSCF-SDC cathode screen printed onto an SDC electrolyte exhibited a cathode polarization resistance of 0.23 Ωcm^2 at 700°C [65], and a NiO-SDC anode with SSC-SDC cathode exhibited a full-cell polarization resistance of 0.23 Ωcm^2 at 550°C at open-circuit voltage (OCV) conditions [53]. As seen from the examples, very high performances, or low polarization resistances, can be achieved by screen-printed electrodes on electrolyte or electrode supports, in particular when utilizing mixed conducting oxides at either the anode or cathode side of the cell or both.

Screen printing steps are typically followed by sintering steps to consolidate the deposited slurry layer. Screen-printed electrolytes are typically fired at high temperatures of approximately 1400°C, while screen-printed cathodes are typically fired at lower temperatures, ranging from 900 to 1200°C, either in a separate cathode firing step, or a first co-firing step containing all of the cell layers fired only once, or *in situ* after mounting the cell in a test station and forming a seal with the surroundings.

Painting, Colloidal Spray Deposition, Spin Coating, and Slurry Coating

Painting is a simple technique in which a slurry of the powder to be deposited is applied to the support layer with a brush. Although simple, it is difficult to scale up for mass production and not readily reproducible, since the painting technique (pressure applied, speed of coating stroke, brush properties) influences the overall coating properties. Painting has been used to produce barium lanthanum cobalt oxide (BLC) cathodes, YSZ-LSCM anode layers (LSCM = lanthanum strontium chromium manganese oxide) [8], $Sm_{2-x}Sr_xNiO_4$ cathodes on SDC electrolytes [67], SSC cathodes on ScSZ pellets coated with GDC [43], and BSCF-LSC composite cathodes [68], the latter achieving a cathode ASR_p of 0.21 Ωcm^2 at 600°C. BSCF has also been sprayed in colloidal form onto an SDC electrolyte, with a very low cathode ASR_p of 0.055 to 0.071 Ωcm^2 reported at 600°C at OCV [69]. Spin coating of ceramic slurries has also been used to achieve thin and dense coatings of YSZ electrolyte layers as thin as 1 μm after firing [70]. Slurry coating is sometimes used to deposit the anode layer on top of the cathode support tube and electrolyte layer for cathode-supported cells [48]. Painting, dip coating, and other wet-ceramic techniques have also been used to deposit electrode layers onto electrolyte-supported tubular SOFCs [51].

Sol-Gel Methods

Because screen printing, painting, and colloidal spray deposition with standard powders require high-temperature firing steps to consolidate the deposited slurries, the possibility for inter-reactions between adjacent layers limits the material combinations that can be used or the extent of consolidation that can be achieved when lower firing temperatures are chosen to consolidate the materials. To lower the firing temperatures necessary to process the SOFC component layers, solution-gelation, or sol-gel methods, have been investigated to obtain nanometer-scale green parts that can be fully densified or fully bonded and consolidated at significantly lower temperatures compared to the micron-size particles more commonly used in screen printing and painting methods.

The sol-gel technique utilizes solutions as precursors of the cations of interest, often in nitrate or alkoxide form. The precursors are dissolved in solution, where they form a well-dispersed (with mixing on an atomic scale), nanometer-scale colloidal system, called a sol, upon undergoing a hydration reaction. The colloid can then either be dried and the solid extracted and ground to a powder [71], which is then processed using a traditional wet-ceramic technique to incorporate it into a fuel cell, or it can be densified by partial evaporation of the suspension fluid to produce a high-viscosity slurry that can then be used for other wet-ceramic deposition techniques such as screen printing [72]. The sol can also be deposited directly onto a substrate through a technique such as dip coating [24, 73, 74] or spin coating. The coating would then be fired at a low temperature to consolidate the green layer into a well-bonded ceramic layer. The nanometer-size particles allow fully dense bilayer electrolyte coatings to be obtained by sol-gel methods with a firing step as low as 600°C [75]. Additional organic additives can be incorporated into the sol to increase the dispersion of the particles and reduce the crystallite size (e.g., [73]), or solid particles can be incorporated into a sol to form a composite sol [74] to increase the intimacy of mixing between the two phases.

Electrophoretic Deposition (EPD)

As shown schematically in Figure 6.8, EPD is an electrochemical deposition method in which a direct current (DC) electric field is used to deposit thin layers of particles from a charged colloidal suspension onto a charged substrate. That substrate can then be removed, as in the case of the anode-supported tubular micro-SOFCs discussed in Section 6.2.1.1 [49]. EPD has also been utilized to deposit YSZ coatings less than 10 μm thick on LSM substrates that were subsequently sintered at high temperatures [76], and to deposit LSGM and GDC coatings to form dense electrolyte layers upon high-temperature firing in a reducing atmosphere [77]. The bath composition, current density, and applied voltage during the deposition process can be varied to alter the deposition rate. YSZ films can also be deposited as electrolyte layers by EPD in alternating and repeated deposition and sintering cycles [78]. It has been found that five cycles of deposition and sintering were needed to obtain films dense enough to achieve OCVs greater than 1 V during testing. Three minutes of deposition were required to obtain 2 μm thick YSZ layers, although some pinholes were still present in the thin electrolyte coatings [78]. EPD has also been used, among other techniques, to deposit electrode layers onto extruded tubular micro-SOFCs [51]. EPD is also currently being investigated as a low-cost alternative to electrochemical vapor deposition (EVD) as a process to deposit electrolytes on cathode-supported tubular cells [48].

Sintering

All of the wet-ceramic processing methods described above deposit layers in green, or unconsolidated, form using slurries of various viscosities with a suspension liquid as the carrier. Subsequent to the screen printing, painting, colloidal spray deposition, EPD, sol-gel dip coating, or spin coating methods described above, sintering (firing) steps are needed to densify the layers or to consolidate them to achieve good contact and bonding between the particles. The firing temperature must be high enough to fully densify the electrolyte—typically 1400°C for most electrolytes. Because such temperatures are high enough for inter-reactions to occur between some electrolyte-cathode pairs, the firing steps are often performed in at least two separate sintering steps, with one high-temperature step used to co-fire or co-sinter the anode and electrolyte together, and a second, lower-temperature, sintering step used to consolidate the cathode [53, 60], which can be performed *ex situ* in a separate processing step, or *in situ* as part of the fuel cell testing procedure [44].

The choice of sintering temperature for each of the consolidation steps requires careful consideration of the tradeoffs between using a temperature that is sufficiently high to achieve a good bond between the layers being joined, such as a cathode and electrolyte layer, while remaining sufficiently low to avoid problems created during high-temperature firing. Examples of such problems include unwanted inter-reactions that produce nonconductive phases between adjacent layers, such as strontium or lanthanum zirconates between YSZ electrolytes and perovskite cathodes doped with Sr. Other examples are unnecessary coarsening of the microstructure and the corresponding decrease in electrochemically active surface area of the cell, and decomposition of certain perovskite materials that are not fully stable at high-sintering temperatures, such as $La_{1-x}Ca_xFeO_{3-\delta}$ [13]. Many studies have been performed on the topic of determining an optimal sintering temperature to achieve a desirable tradeoff among those factors [13, 30, 63, 65–67, 79–81].

Co-firing of an entire cell involves a tradeoff between fully densifying the electrolyte and preventing adjacent electrode layers from inter-reacting or over-densifying. One study of the effect of full-cell sintering temperature for a Ni-ScSZ anode functional layer (AFL), ScSZ electrolyte, SDC electrolyte interlayer, and LSC cathode co-fired on a Ni-YSZ support found that firing at a lower temperature of 1200°C resulted in a lower performance than firing at 1300°C, with the latter resulting in a power density of 0.54 W/cm² at 700°C [59]. Many other studies have been performed specifically to select a good sintering temperature for the cathode firing step subsequent to the anode-electrolyte co-firing. For example, a study of $Sm_{2-x}Sr_xNiO_4$ as a cathode painted on a GDC pellet and sintered at various temperatures demonstrated a tradeoff between ensuring contact between particles and high electrochemically active surface area [67]. The researchers found that sintering the cathode at 900°C resulted in poor contact between particles and a correspondingly high R_p, while microstructures sintered at 1100°C were too fully sintered, leaving little porosity, and also resulting in a high polarization resistance. On the other hand, firing at

1000°C resulted in the lowest R_p of the three sintering temperatures studied (3.06 Ωcm^2 at 700°C at OCV), with an acceptable tradeoff between particle size and connectivity [67]. Figure 6.10 shows micrographs of the cathodes produced at the three firing temperatures and of a test cell [67].

A study of the influence of sintering temperature on the cathode properties for SSC-SDC cathodes examined sintering temperatures of 950, 1050, and 1150°C and their effect on cell performance and microstructure. The cell with the cathode sintered at 1050°C had the highest performance of the three cathode sintering temperatures (0.75 W/cm^2 peak power density at 600°C) and lower anode + cathode R_p (0.102 Ωcm^2 at 600°C), with both the lower and higher sintering temperature leading to a lower peak power density and higher electrode polarization resistance [79]. SEM observations confirmed that the cathodes sintered at 950°C had poor contact between the cathode and electrolyte, while those sintered at 1150°C were excessively densified. The 1050°C sintered cathodes exhibited a good balance between sufficient porosity and contact between the cathode and electrolyte.

Similar microstructural evolution and polarization resistance trends were observed for $La_{2-x}Sr_xCuO_{4-\delta}$ cathodes on GDC electrolytes, with 900°C found to be the optimal sintering temperature, 1000°C resulting in an overly densified microstructure and higher R_p, and 800°C sintering temperature resulting in a cathode–electrolyte interface with poor contact, and also a higher R_p than that resulting from sintering at 900°C [63].

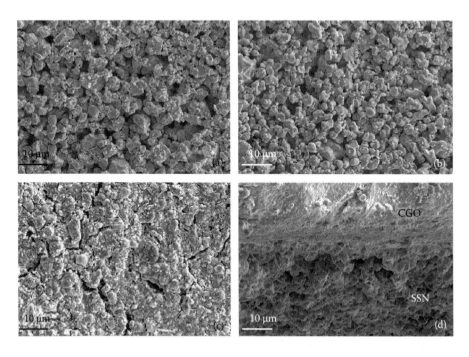

FIGURE 6.10 Microstructures of $Sm_{2-x}Sr_xNiO_4$ cathodes sintered at three firing temperatures: (a) 900°C, (b) 1000°C, and (c) 1100°C, and (d) of a test cell [67]. Reprinted from [67] with permission from Elsevier.

A similar trend was found for LSC-SSC composite cathodes, with a sintering temperature of 1000°C recommended by the authors, and a temperature of 1200°C observed to cause excessive sintering and lower cathode performance [81].

A similar study for $Nd_{0.6}Sr_{0.4}Co_{1-y}Mn_yO_{3-\delta}$ cathodes materials showed high interfacial polarization resistance at 800°C for samples fired at 1000°C due to poor contact, which was lowered by sintering at 1100°C. At 1200°C, inter-reaction phases were observed between the cathode and LSGM electrolyte, so the 1100°C sintering temperature was recommended for that material system [66]. A similar trend was observed for LSM + SDC/LSC + LSM + SDC/LSC + SDC/LSC multilayered cathodes prepared by infiltration and screen printing. A very low R_p of 0.06 Ωcm^2 at 750°C was observed when the cathode was fired at 1000°C, but higher R_p values were found at both higher and lower sintering temperatures (0.39 Ωcm^2 for sintering at 1050°C and 0.33 Ωcm^2 for sintering at 950°C) [30].

As seen from the examples presented, the choice of sintering temperature is an important part of determining the optimal processing conditions for fabricating SOFCs. The electrolyte sintering temperature must be sufficiently high to fully densify the electrolyte layers to ensure gas tightness, which typically requires sintering temperatures of approximately 1400°C. On the other hand, electrolyte and cathode materials commonly inter-react with each other to form nonconductive phases at temperatures of approximately 1200°C or higher, so often the cathodes are fired in a separate sintering step from the electrolyte and anode. The choice of cathode firing temperature presents a tradeoff between a sufficiently high temperature to ensure good contact between cathode and electrolyte, and therefore low interfacial polarization resistance, and sufficiently low temperature to avoid over-densifying the cathode and excessively reducing its porosity, thereby degrading performance, especially at high current densities. The optimal sintering temperature depends on the material system and its melting temperatures, as well as on the presence or absence of sintering aids that allow a reduction of sintering temperature [82]. If no sintering aid is used, then in general materials with higher melting temperatures will require higher sintering temperatures to ensure that good contacts are formed between the electrodes and electrolyte, and that sufficient densification of the electrolyte, but not excessive densification of the electrodes, occurs.

Infiltration

Because of the tradeoff that exists between good electrical contact and high surface area in the selection of a sintering temperature, a recent development in electrode processing has been to prepare electrodes in a series of steps. The first step involves forming a sintered porous preform using a firing temperature that achieves good contact between the porous electrode structure and the adjacent electrolyte layer. The porous preform is then infiltrated with either a slurry of nano-particles or solution precursors for the phase to be added to the electrode, typically with nano-scale particles resulting from the solution. The infiltrated preform is then fired at a low temperature to consolidate the nano-structured added phase, and the infiltration and low-temperature firing steps are often repeated, sometimes many times, to obtain sufficient reaction sites in the electrode.

An infiltration method as described above was used to obtain a high-surface area electrode structure for LSM-YSZ cathodes, and the resulting polarization resistance achieved was quite low for that material set: 0.4 to 0.5 Ωcm^2 at 700°C [83]. The same research group has also developed an infiltration procedure to prepare anodes for direct oxidation of hydrocarbon fuels that use copper as the electronically conductive anode phase rather than nickel, to avoid catalyzing unwanted carbon deposition. Because both copper and copper oxide have lower melting temperatures than the sintering temperature required to densify the electrolyte of an SOFC, anodes made with Cu cannot be co-fired with the electrolyte, as is commonly done for anode-supported cells with Ni-based anodes. Therefore, an infiltration method was utilized to introduce Cu cations from a nitrate solution into a porous preform after first sintering the anode preform with pore formers, leaving a skeleton structure ready for subsequent infiltration of a second phase [84]. Figure 6.11 shows a schematic diagram of the processing method used for the introduction of the direct-oxidation anode materials into a porous preform prepared by sintering a tape made with pore formers [84].

A similar infiltration method was used to form anodes with Cu-Co alloys as the electronically conducting phase, with Co added to enhance the anode catalytic activity without catalyzing carbon deposition [85], in contrast with Ni-Cu alloys, which were found to catalyze carbon deposition even when small quantities of Ni were present [86]. The authors of the studies [84–86] have reported that the percolation of the infiltrated phases is incomplete following the processing of the cells, such that subsequent carbon deposition can actually serve to connect previously isolated islands of the metallic phase, thus increasing the electronic conductivity and decreasing R_s in the short term. Since carbon deposition was observed by the same authors to cause severe

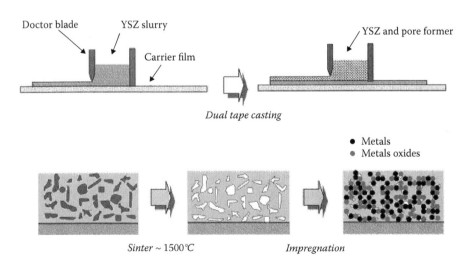

FIGURE 6.11 Diagram of the processing technique used to prepare Cu-CeO$_2$-YSZ anodes for direct oxidation of hydrocarbon fuels by preparing a porous preform of YSZ and then infiltrating it with cerium nitrates to form ceria and then with copper nitrates to form metallic copper [84]. Reprinted from [84] with permission from Elsevier.

and irreversible degradation in Ni-based anodes [84], it is likely that the short-term benefit of deposited carbon compensating for incomplete percolation of the metallic phase in the anode would be counteracted by the longer-term microstructural degradation caused by carbon deposition on nickel. The results of the above infiltration studies suggest that while infiltration is currently the most commonly utilized method to fabricate anodes containing copper or Cu-Co as the metallic phase, incomplete connectivity of the infiltrated phase could limit the performance in the early stages, and possibly also make the electrodes more susceptible to subsequent degradation caused by coarsening of the metallic phase and further de-percolation of the conductive path as a result. Multiple impregnation steps have been utilized to address this issue, to build up a thicker and more connected coating of the electronically conductive material, requiring a separate firing step after each infiltration [87].

The impregnation technique has also been utilized to make anodes for standard hydrogen fueled SOFCs, based on vacuum impregnation of a porous YSZ layer on a dense YSZ support with a solution of cerium and nickel nitrates. The presence of Ce was found to increase the Ni dispersion on the YSZ compared to anodes without added ceria, thereby decreasing Ni agglomeration and enhancing the anode properties [88]. Infiltration has also been utilized to introduce a nano-sized GDC gel precursor into a tape cast Ni-ScSZ anode [17], impregnate an LSM-YSZ cathode support with GDC to enhance the activity of the cathode [18], infiltrate a porous ScSZ layer with nano-particulate LSM [89], and to infiltrate an LSM-YSZ porous preform with SSC nano-particles, reducing R_p by a factor of 3 compared to the polarization resistance without the infiltrated SSC particles [90]. SDC has been impregnated into screen-printed LSM, thereby reducing the polarization resistance to 0.23 from 5.4 Ωcm^2 at 700°C [91], and infiltration has also been used to introduce both anode and cathode materials (LSF cathode and $La_{0.3}Sr_{0.7}TiO_3$ anode, respectively), into porous YSZ layers on both sides of a YSZ electrolyte support within the same cell [92]. Vacuum infiltration has also been utilized to deposit electrodes onto electrolyte support tubes for tubular micro-SOFCs [51].

As seen from the examples presented, infiltration provides the flexibility to add materials as catalysts in smaller quantities compared to the materials used for the bulk structural or percolated conductive paths. Specifically, the technique allows low-melting temperature materials such as Cu to be used in electrodes, and facilitates the use of smaller amounts of highly catalytic materials that are more expensive than the bulk materials and which inter-react when fired at high temperatures, while using a cheaper bulk material with fewer inter-reaction issues and using only low-firing temperatures for the material that is introduced to the component by infiltration. The process also has been demonstrated to improve performance compared to cells that have not been infiltrated, due to the fine particle size that is made possible by eliminating the need for the formation of a structural bond when firing the infiltrated phase. However, the technique typically requires multiple impregnation and firing steps, and often many steps are needed to build up a percolated network of the infiltrated phase. This addition of multiple firing steps increases the processing time and cost and makes mass production less convenient by adding processing complexity, but expands the material set that is accessible by facilitating the use of only lower-temperature firing steps throughout, thus increasing the flexibility of catalyst and electrode design.

6.2.2.2 Direct-Deposition Techniques

Because the high-temperature sintering step is the stage of a multistep wet-ceramic process at which the largest amount of degradation occurs in terms of unwanted layer inter-reactions, microstructural coarsening, and possible oxidation of a metal support layer for metallic interconnect-supported cells, there is a need for direct-deposition techniques that do not require a high-temperature consolidation step after deposition of the layers in cases where high-temperature firing steps are most likely to have negative consequences. Examples of such cases are cathode-supported tubular cells, which cannot be sintered at usual electrolyte densification temperatures due to unwanted inter-reactions between the LSM in the cathode and YSZ electrolyte, metal-supported cells where high-temperature firing steps would be likely to oxidize the metal support or else require that a reducing atmosphere be maintained during sintering, and cells with low-melting temperature components, such as Cu, in the anodes that would melt or nano-structured components that would coarsen excessively during a high-temperature sintering step. Direct-deposition techniques have been developed for many of these applications. An overview of some of the direct-deposition processes that have been applied to SOFCs is presented below.

Electrodeposition

In conjunction with the infiltration methods described above, electrodeposition of cobalt has been investigated as a method of introducing Co into direct oxidation anodes after infiltration with Ce and Cu nitrates and calcining at 450°C, subsequent to the infiltrations to produce CeO_2 and metallic Cu. Wax was utilized to protect the cathode of the cell during the electrodeposition of the Co phase. No carbon deposition was detected during the testing in methane of the anodes deposited this way, but the processing method utilized is complex and does not result in extremely high performance (~2 Ωcm^2 cell R^p at 900°C at high current densities) [87].

Electrochemical Vapor Deposition and Electrochemical Liquid Deposition

Electrochemical vapor deposition (EVD) is currently utilized to produce the electrolyte on cathode-supported tubular cells because the cathode and electrolyte cannot be co-fired at temperatures sufficiently high to densify an electrolyte produced by a wet-ceramic process. High-density electrolyte films can be produced by EVD without a need for post-deposition sintering, as shown in the cross-sectional micrograph of a cathode-supported tubular cell with EVD-deposited electrolyte shown in Figure 6.12 [48]. EVD is sometimes also used for the anode layer of tubular cells because those layers typically also cannot be fired to a very high temperature once they are deposited on the cathode-electrolyte bilayer tubes. The kinetics of the EVD process for depositing electrolyte layers have been studied [93, 94] in developing the technique for use with large-scale tubular cells. Electrochemical liquid deposition (ELD) has also been used for the deposition of electrolyte layers based on ceria [95]. Although EVD and polarized electrochemical vapor deposition (PEVD) allow for relatively fast deposition rates (60 μm/h) compared to other thin-film deposition techniques (e.g., 1 to 5 μm/h in RF sputtering), the EVD technique still suffers from drawbacks such as the high temperature required (~1200°C) for reasonable

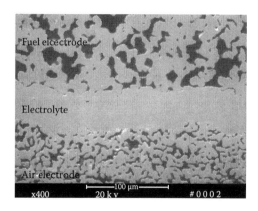

FIGURE 6.12 Siemens Westinghouse SOFC cross-sectional micrograph, showing a dense YSZ electrolyte deposited by EVD [48]. Reprinted from [48] with permission from Elsevier.

deposition rates. Therefore, its use has been mostly limited to the deposition of electrolyte thin films [96]. Although fully dense electrolytes can be deposited by EVD, the technique requires high equipment and processing costs, which has resulted in the consideration of alternative methods to deposit dense electrolyte layers onto tubular cathode supports, such as plasma spraying and EPD.

Chemical Vapor Deposition

Chemical vapor deposition (CVD) is especially targeted at the fabrication of electrolyte thin films by the gas phase reaction of metal halide precursors with a heated substrate [97]. Major drawbacks are the low deposition rates, high deposition temperatures, high equipment costs, and corrosive products [98]. Variants of the technique that allow overcoming some of these obstacles have been developed [98].

Metal-organic CVD (MOCVD) uses volatile metal-organic precursors such as β-diketonates [97] and can be performed at lower temperatures and at atmospheric pressure, but the complexity of the multisource feed system for the single components generally prevents its application in the deposition of multilayer structures.

In the absence of suitable volatile precursors, aerosol-assisted CVD (AACVD) has been used, whereby a precursor solution is atomized and the micron-sized particles deposited onto a heated substrate to grow into thin films [98]. LSM thin layers (4 to 5 μm) have been deposited in this way [99–100].

Combustion CVD (CCVD) has also been developed and employed in the deposition of functionally and porosity graded LSM-GDC/LSM-LSC-GDC/LSC-GDC cathodes on a YSZ electrolyte with a NiO-GDC anode also deposited by CCVD. Reasonable performances of 0.48 W/cm^2 peak power density and 0.11 Ωcm^2 electrode R_p at 800°C were recorded for the cell [101]. CCVD has also been utilized to deposit anode layers with SnO$_2$ as a sacrificial pore former layer [25] and to deposit an SSC-SDC cathode layer, resulting in a low R_p of 0.195 Ωcm^2 at 600°C at OCV [102].

Plasma Spraying

A technique that is under study for the production of SOFC anode, cathode, and electrolyte layers is plasma spraying (PS). PS processing involves the feeding of powder, either suspended in a carrier gas or in a suspension liquid such as water or ethanol, or the feeding of solution precursors, into a plasma jet. The high temperatures of the plasma jet (~10,000 K) melt the particles, and the molten droplets are propelled toward a substrate, where they solidify rapidly to form a coating. A diagram of the process is shown in Figure 6.13, with both radial and axial injection of the feedstock into the plasma jet illustrated schematically.

PS in general is of interest due to the rapid deposition rates, the ability to introduce spatially varying compositions and microstructures in a continuous manner, and the ability to deposit SOFC layers directly onto metallic substrates without a need to sinter them afterward, thus reducing the risk of metal interconnect oxidation during fabrication of the cell, and simplifying the process. The technique is already easily scaled up for mass production and for large cell areas, and is adaptable to a fully automated manufacturing process. Because atmospheric plasma spraying (APS) is approximately one order of magnitude cheaper than vacuum plasma spraying (VPS) [103], a growing interest in applying APS methods for the production

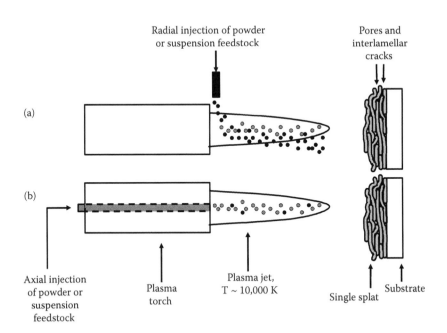

FIGURE 6.13 Schematic diagram of plasma spray deposition processing. (a) Radial injection of feedstock into plasma torch, showing partial entrainment of feed particles into the plasma jet and (b) axial injection of feedstock into plasma torch, showing more complete entrainment of feed particles into the plasma jet.

of SOFC components has arisen in the past approximately 10 years [104–106]. An alternative PS technique uses a liquid slurry feedstock that is atomized and then fed into the plasma. The resulting coating splat size can be controlled by adjusting the size of the liquid droplets and the concentration of the slurry [107]. A combination of liquid slurry and powder feedstocks has been used to produce materials with a range from thin, dense coatings for electrolytes to highly porous electrodes. Other process modifications have been developed such as radio frequency direct current (RF/DC) hybrid PS, which combines the long residence times and high temperatures of RF spraying with the high velocity of DC PS to obtain denser coatings than from either RF or DC spraying alone [108]. Each of these variants of PS processing has different sets of advantages and challenges that make them potentially more suitable for individual layers. A scaled-up manufacturing process could readily make use of combinations of different PS variants, with each layer being deposited potentially by a different torch.

The potential benefits of plasma spraying as an SOFC processing route have generated considerable interest in the process. In the manufacture of tubular SOFCs, APS is already widely used for the deposition of the interconnect layers on tubular cells, and has also been used for the deposition of individual electrode and electrolyte materials, with increasing interest in utilizing APS rather than EVD for electrolyte deposition due to the high cost of the EVD process [48, 51, 104].

In planar SOFCs, individual cathode, anode, and electrolyte layers have been deposited by PS [109–111], as well as coatings on interconnect materials and full cells [108, 110, 112]. In addition to the interconnect layers themselves in tubular SOFCs, dense protective layers with good adhesion have also been deposited to protect planar SOFC interconnects from oxidation [110], and diffusion barriers to inhibit inter-diffusion between the interconnects and anodes have been produced by PS [113].

In the manufacture of planar SOFC components, functionally graded LSM-YSZ cathodes have been deposited by VPS [29], and composite LSM-YSZ cathodes have been deposited by APS, with high performances and low polarization resistances obtained relative to the catalytic activity of the materials set, 0.4 Ωcm^2 at 750°C, indicating that high surface area electrode structures can be deposited by PS, without the adhesion-surface area tradeoff that occurs when sintering cathodes [11]. Cathode microstructures with controlled porosity levels between 15 and 40% have also been fabricated by PS, demonstrating that the technique can produce a wide range of microstructures, suitable for use in graded electrodes or in bilayered electrode structures [111].

Extensive work has been reported on the deposition of individual cell layers and of full anode-electrolyte-cathode fuel cells on metallic interconnect substrates, much of it by VPS, with no sintering or other post-deposition heat treatments required [112]. However, so far relatively thick YSZ electrolytes, approximately 25 to 35 μm, have been needed to provide sufficient gas tightness [108, 114], so further optimization of the process is required to produce thinner, gas-tight electrolytes. Peak power densities of ~300 mW/cm^2 have been reported at 750°C for APS single cells [114], with four-cell stacks exhibiting power densities of approximately 200 mW/cm^2 at 800°C [55].

Full cells deposited by APS have achieved cathode polarization resistances of ~0.6 Ωcm^2 at 800°C, with the anode and electrolyte also fabricated by APS [12].

LSGM electrolytes have also been deposited by APS, but the coatings were found to be amorphous, with post-deposition heat treatments required to crystallize the coatings to obtain adequate conductivity, although the heat treatments were at a much lower temperature than those required for sintering (~800°C) [115]. Interconnect-supported APS anodes and electrolytes with screen-printed cathodes have also been fabricated, with peak power densities of 500 mW/cm^2 at 800°C observed without diffusion barriers between interconnect and anode [116], and 800 mW/cm^2 reported with diffusion barriers based on LaCrO$_3$ [113]. In some cases, increased OCV after some operating time was observed due to healing of microcracks in the as-sprayed YSZ electrolytes, suggesting that there may be some tolerance for defects in the original microstructure [113]. Full cells with suspension plasma sprayed (SPS) electrolytes and anodes and screen-printed cathodes have recently been manufactured on metallic interconnect supports and tested at 600°C, demonstrating a power density of 0.16 W/cm^2. A scanning electron micrograph of a cross-section of the SPS/screen-printed cell is shown in Figure 6.14 [46]. Cu-ceria-based anodes can also be fabricated by PS, thereby simplifying the processing for anodes made from that material set [117]. These results indicate that SPS can be utilized to provide finer microstructures for use in SOFCs that operate at lower temperatures in addition to the higher-temperature SOFCs previously produced by the technique.

Spray Pyrolysis

Spray pyrolysis has been utilized to produce both powders of SOFC materials [19] and also SOFC component layers [118–121]. Thin (1 μm) coatings have been fabricated by spray pyrolysis [121], with zero gas permeability achieved for some conditions of substrate surface roughness and temperature during deposition, but

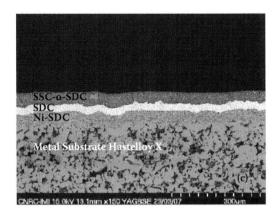

FIGURE 6.14 Cross-sectional scanning electron micrograph of a metal-supported SOFC, with anode and electrolyte produced by SPS and cathode produced by screen printing. Electrolyte is approximately 30 μm thick [46]. Reprinted from [46] with permission from Elsevier.

post-deposition heat treatment was required [121]. A micrograph of an SDC electrolyte and NiO-SDC anode bilayer coating prepared by spray pyrolysis is shown in Figure 6.15 (a), with a higher-magnification view of the electrolyte layer shown in Figure 6.15 (b) [121].

Pulsed-Laser Deposition or Laser Ablation

Pulsed-laser deposition (PLD, or laser ablation) is a vacuum technique based on laser removal of material from a surface, and its subsequent collection and deposition onto a substrate, typically at temperatures of ~700°C. Deposition rates are typically approximately 1 μm/hr, although deposition rates as high as 600 μm/h have been reported [122]. While the technique is not easily scaled up, it offers distinct advantages in producing miniaturized SOFCs due to the low processing temperature and the potential for automation, as well as for its capability to produce controlled nano-structures that have found use in fundamental studies [122]. Electro-chemical impedance spectroscopy (EIS) studies on LSM and LSC cathodes deposited on YSZ (and on LSC cathodes on GDC) by PLD have determined that the charge transfer and surface reactions, respectively, are rate determining [123, 124]. In another cathode study, $Pr_{0.8}Sr_{0.2}Fe_{0.8}Ni_{0.2}O_{3-\delta}$ (PSFN) cathodes deposited on crystalline YSZ by PLD at room temperature and 500°C showed a combination of amorphous and granular structure, but possessed a polycrystalline, needle-shaped topography resulting in increased porosity when deposited at 700°C, and for which an ASR of 0.5 Ωcm^2 was determined at 850°C [125]. PLD has also been used to produce patterned cathodes for micro-SOFCs [126].

PLD has also been utilized to produce bilayer electrolytes. In one study, a NiO-YSZ (anode support, tape cast)/NiO-SDC (AFL, screen-printed)/ScSZ-SDC (electrolyte bilayer, PLD)/SSC (cathode, screen-printed) cell showed excellent performance (0.5 and 0.9 W/cm² at 550 and 600°C, respectively), with an OCV of 1.04 V at 600°C, indicating that the PLD technique was successful in depositing a sufficiently dense ScSZ electronic blocking layer to suppress electronic conduction normally observed across single-layer SDC electrolytes, and which typically result in lower OCV values (0.87 V, 600°C) [46, 127].

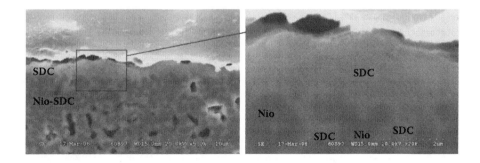

FIGURE 6.15 Coatings produced by spray pyrolysis. (a) NiO-SDC anode layer and SDC electrolyte layer and (b) closeup of electrolyte layer showing SDC layer thickness of approximately 1 μm and fully dense SDC microstructure [121].

Sputtering

The sputtering technique has been used in the preparation of SOFC electrodes, and is generally combined with photolithography in the production of thin-film patterned electrodes that are mainly used in fundamental reactivity and mechanistic studies [120]. Although it is a versatile technique that allows for excellent control of composition and morphology, and relatively low temperatures that help to prevent unwanted reactivity observed at higher temperatures, its major limitations lie in the equipment costs and in the slow deposition rates (~5 µm/h) [120].

Ni-YSZ cermets deposited by RF sputtering (230 nm) were found to have microstructural features consisting of columnar grains 13 to 75 nm long and 9 to 22 nm wide, and showed good adhesion to the YSZ layer on which they were deposited [128]. In a three-layer Ni-YSZ-Ni film deposited on NiO by RF sputtering in another study, the YSZ layer exhibited a columnar structure with some pinholes [129]. Microstructural and electrochemical features of Pt electrodes patterned by lithography on YSZ have also been studied [130, 131].

LSM cathodes have also been deposited on YSZ substrates by RF and DC sputtering methods, and showed good adhesion to the YSZ, good stability, and acceptable electrode activity [132, 133]. All these studies are, however, fundamentally aimed at elucidating mechanistic aspects and establishing correlations between microstructure and reactivity at the electrode. The slow deposition times and high cost generally preclude the use of sputtering for commercial production of SOFCs.

6.3 DESIGN CONSIDERATIONS IN THE SELECTION OF PROCESSING METHODS FOR SOFCS

6.3.1 ELECTROCHEMICAL PERFORMANCE OF FUEL CELLS

One of the first considerations in selecting a processing method for the fabrication of SOFC components is the range of microstructures and layer properties that the technique is capable of producing, and the corresponding contributions to the fuel cell performance that the layers can produce. For example, the minimum thickness of electrolyte layer that a fabrication process can produce without introducing gas leakage due to cracks or pinholes is a key performance parameter to consider in the selection of an electrolyte fabrication process. The electrode layers in a fuel cell can exhibit improved performance from the introduction of certain gradients in the microstructure and composition of the materials used, so selecting a technique that is capable of producing a desired gradient in component properties is important for obtaining the best cell performance possible. Ultimately, if the desired fuel cell component properties cannot be achieved or must be severely compromised when utilizing a particular processing technique, then that technique will not prove useful for the large-scale production of SOFCs.

The choice of techniques that minimize high-temperature firing steps by either lowering the firing temperature, such as sol-gel or infiltration methods, or by eliminating high-temperature firing altogether, such as PS, have the potential to freeze in finer microstructures and nano-structures than those produced by other processing techniques that rely on a high-temperature sintering step to consolidate the electrolyte layer. Such techniques can therefore exhibit higher compatibility with the fabrication requirements of metal-supported SOFCs. On the other hand, other performance challenges can be introduced by those techniques, such as insufficient percolation of the conductive phases for sol-gel or infiltration methods, or insufficient density of the electrolyte for PS methods. Therefore, in selecting a processing method, the balance of performance requirements for each of the cell layers must be considered, and the tradeoffs that can be tolerated in one aspect of performance to compensate for another must be carefully considered.

6.3.2 DURABILITY

In addition to the short-term performance, the changes that a process method is likely to cause in the medium- and long-range performance of the SOFC must also be taken into account in the selection of an appropriate fabrication technique. For example, consider the case of the infiltration method discussed above which results in a nano-structured electrode with high electrochemical surface area and performance. If the number of infiltration steps is insufficient to ensure that the electrode is well above the upper percolation threshold when it is fabricated, it is possible that as the cell operates at high temperatures over extended times, coarsening in the microstructure could cause not only slight decreases in the total conductive path area and corresponding conductivity, but also de-percolation of the conductive phase over time, leading to sharp performance losses. Therefore, it is important in the process of designing a fabrication method that long-term as well as short-term performance of the resulting microstructures and coatings are taken into account.

The choice of cathode firing temperature during wet-ceramic processing is another example of the potential tradeoff that must be considered between initial performance level and durability. Higher cathode sintering temperatures can lead to a coarser starting microstructure and lower electrochemically active surface area compared to cathodes sintered at a lower temperature, thereby leading to a lower performance level. However, a lower sintering temperature, while possibly resulting in a fuel cell with higher peak power density and lower polarization resistance initially due to a higher surface area of reaction sites and finer microstructural features, may also produce a cell whose performance degrades more rapidly over time. Such a trend may result from the cathode microstructure continuing to sinter during the initial stages of operation, rather than retaining the original microstructure, as it would if it were pre-fired to a temperature above the operating temperature. In addition, a lower sintering temperature can result in lower adhesion between the cathode and electrolyte layers, which in turn could result in a lower resistance to thermal cycling and rapid delamination at the cathode–electrolyte interface upon a relatively small number of thermal cycles. Therefore, the long-term stability of the cell performance must be taken into account when selecting an appropriate fabrication process, in addition to the initial performance demands of the application for which the fuel cell is being designed.

6.3.3 COST

In addition to the performance and durability of the fuel cells fabricated, cost is also an extremely important consideration in the decision of whether or not to adopt new energy technologies, including SOFCs. Therefore, any manufacturing process to be utilized must allow the fuel cells to be fabricated at a cost that makes them competitive with other competing technologies, so that fuel cells can ultimately be widely used. Therefore, techniques that are capable of being scaled up for mass production at a reasonable cost and which can lower the cost per kilowatt of a fuel cell stack are more likely to become widely adopted than extremely expensive fabrication methods. There are two main aspects to the cost associated with a given deposition process: the impact of the processing method on the materials that can be used, and their corresponding costs, as well as the equipment and process cost of the manufacturing method itself, which is a strong function of the production volume being manufactured.

6.3.3.1   Material Cost

The choice of manufacturing method can have a significant effect on the materials cost. For example, in comparing a tubular and planar SOFC design, one of the main considerations is the interconnect material used. Tubular cells use expensive ceramic interconnects that are difficult to sinter and process due to the high operating temperatures above 850°C, while planar cells can use planar stainless steel interconnects, in some cases with a suitable coating applied to them, due to the lower operating temperatures in the 700 to 850°C range. Stainless steel interconnects have a lower material cost and are easier to fabricate than ceramic interconnects. In addition, the planar configuration has shorter individual current paths within the cells, so a higher power density can normally be achieved at a lower temperature compared to tubular cells. The higher power density and lower temperature can both contribute to a lower cost per kilowatt, by increasing the power produced per cell (i.e., per given amount of material), as well as facilitating a less expensive choice of materials for each cell due to the less stringent materials requirements of the lower operating temperature.

Lowering the operating temperature further, to approximately 600°C, would allow metal interconnect-supported SOFCs to be used, with relatively inexpensive stainless steel layers as the structural supports, and very thin layers of the more expensive electrochemically active components of the fuel cell. The ability to fabricate metal-supported cells depends on the availability of a manufacturing process that can produce dense electrolyte layers without the high-temperature firing steps that have more traditionally been used to densify a wet-ceramic-produced electrolyte. Therefore, an alternative firing process utilizing a lower temperature, combined with sintering additives in the electrolyte and an inert or reducing atmosphere could be utilized to fabricate metal-supported cells, or a direct-deposition process that requires no firing step at all, such as plasma spraying, could be utilized. By affecting the choice of materials that can be fabricated and therefore the cost of the cell that is produced, the processing method has an extremely important influence on the overall cost of the SOFCs.

6.3.3.2   Equipment and Process Cost and Deposition Time

In addition to the materials costs associated with a given fabrication technique, the cost of the equipment needed to fabricate the cell and the operating costs of the equipment—electricity, natural gas for firing kilns, etc.—and other process costs of the manufacturing (e.g., solvents, gases, and organic phases such as dispersants and pore formers that do not end up in the final SOFCs), as well as labor and factory costs all contribute toward the overall price tag. The production scale also heavily influences the cost per cell or per kilowatt, with the process time influencing the rate at which the cost changes with production scale. For example, a rapid production time will allow more cells to be fabricated per day compared to a slower process, so that scaling up the production volume will require more parallel production lines to be set up for a slower process than for a more rapid one. Some manufacturing processes are also easier to automate, and to scale up to large fuel cell sizes, so that the manufacturing cost per cell would be a strong function of the production volume, but the cost–volume relationship for different manufacturing processes will vary, depending on how readily a given technique can take advantage of the economies of scale when increasing the production rate of the fuel cells. For example, wet-ceramic techniques that require one or more high-temperature firing steps, each lasting approximately one day in length and therefore representing the bottleneck in the manufacturing process, require multiple large furnaces to be utilized when the scale of production is increased dramatically. On the other hand, direct-deposition techniques, such as PS, that do not require high-temperature firing allow for a large increase in the scale of production while utilizing fewer parallel production lines compared to sintering processes, since the rapid deposition process results in more spare capacity for the processing equipment used.

The number of separate and distinct process steps also dramatically influences the cost of the process, in two ways. High process complexity tends to make the manufacturing method more difficult to automate, and more likely to require human intervention in transferring the cells from one process step to another as the nature of the process changes (for example, from a tape caster to a screen printer to a sintering furnace). In addition, the larger the number of distinct processing steps, the lower the yield of the overall process tends to be because the overall yield is the product of the manufacturing yield of each individual step. By reducing the distinct number of firing steps, for example, the chances of a cell cracking during any of the process thermal cycles can be decreased.

6.3.4 COMPATIBILITY WITH OTHER LAYER-PROCESSING METHODS

In designing the overall manufacturing process, the compatibility of the manufacturing steps with each other and with all of the material layers in the cell must be considered. For example, co-firing of anode, electrolyte, and cathode layers can only be accomplished if the sintering temperature is both high enough to densify the electrolyte and low enough to avoid unwanted inter-reactions between adjacent material layers, or only if those layers are modified in such a way as to make them less susceptible to such interactions for a given set of firing conditions, or if the sintering step is eliminated altogether by the use of a direct-deposition process.

6.3.5 MATERIAL COMPOSITION EFFECTS ON CHOICE OF PROCESSING METHOD

The choice of materials to use in the cell has a direct impact on the selection of viable manufacturing processes that can be used. The following considerations are several examples of the factors that must be taken into account.

6.3.5.1 Inter-Reactions between Material Layers

Inter-reaction of zirconia-based electrolytes with cathode perovskites to form nonconductive lanthanum or strontium zirconates occurs at temperatures above approximately 1200°C, limiting the temperatures at which those combinations of materials can be fired together. Similarly, Ni from SOFC anodes inter-reacts with LSGM electrolytes to form a nonconductive phase, so barrier layers are typically introduced to prevent inter-reaction when LSGM electrolytes are used with anodes that contain nickel. Doped zirconia and doped ceria also inter-react with each other, forming a solid solution with a lower conductivity at every mixed composition between the doped ceria and doped zirconia endpoints. Therefore, the process firing sequences must be designed so as to avoid long times at temperatures that result in unwanted inter-reactions.

6.3.5.2 Metal-Supported Cells

The use of metallic interconnects as the structural support of the cell requires consolidating the layers together after the materials are deposited on the metallic structural support. This means that steps must be taken to either reduce the firing temperature, such as adding sintering aids to the electrolyte [82], or to eliminate the high-temperature firing step altogether from the process, such as by utilizing PS that requires no sintering at all [46], or processes that require only low-temperature firing, such as sol-gel processing or infiltration [89].

6.3.5.3 Direct Oxidation Anode Materials

Standard Ni-based cermet anodes have a high catalytic activity for hydrogen oxidation, but also for the formation of solid carbon, which can degrade anode performance. Anodes that utilize mixtures of Cu as a current collector and ceria as a catalyst and ionic conductor, with or without the addition of other catalytic materials, such as Co, can prevent carbon deposition during the utilization of some hydrocarbon fuels [92, 85–87]. However, the fabrication of anodes that contain copper requires new processing methods that are compatible with the low melting temperature of copper and copper oxides, making high-temperature electrolyte firing incompatible with anode firing. Therefore, infiltration [84], electrodeposition [87], PS [117], or other direct-deposition or low-temperature firing methods must be utilized to produce anodes with this material set suitable for direct oxidation of hydrocarbons.

6.3.5.4 Sulfur-Tolerant Anode Materials

Similarly to the case of direct-oxidation anode materials, sulfur-tolerant anode materials based on sulfides [6, 7] or double-perovskite oxides have special requirements for their processing into SOFC layers. For example, nickel sulfide–promoted molybdenum sulfide is tolerant to high sulfur levels [7]. However, it has a low melting temperature [6] that has resulted in the development of cobalt sulfide as a stabilizer of the molybdenum sulfide catalyst [6]. $CoS-MoS_2$ admixed with Ag has an even higher performance in H_2S-containing fuels than in pure H_2 [6]. However, processing methods such as PS, infiltration, or sol-gel techniques that can process low melting temperature materials without high-temperature sintering steps would facilitate more widespread use of such sulfur-tolerant new anode catalyst materials.

Double-perovskite $Sr_2MgMoO_{6-\delta}$ (SMMO) anode materials have also demonstrated a high tolerance to sulfur and to operation in dry methane, but their processing requires sintering in reducing conditions to preserve the n-type electrical conductivity that makes the materials suitable for use in anodes [39, 40]. Therefore, the process must be designed so as to keep the anode material reduced, while not harming the cathode materials that are unstable in reducing atmospheres.

6.3.5.5 Graded or Multilayered Materials

Deposition methods requiring separate firing steps after each deposition are impractical for making graded compositions, and processes with discrete depositions steps, such as painting, screen printing, or tape casting, are not conducive to the production of a continuously graded composition or microstructure rather than step-wise gradients. On the other hand, direct-deposition processes such as CCVD, PS, and spray pyrolysis can be more readily adapted to the deposition of a continuously varying composition or microstructure, thereby increasing the extent of design flexibility in the fuel cells. The manufacturing process selected must be conducive to producing the specific gradient in microstructure and composition desired for maximizing the performance of the fuel cell and increasing the durability and resistance to thermal stresses. Direct-deposition processes can produce more controllable gradients, but discrete-layer deposition processes can also be utilized to fabricate stepped gradients in composition, porosity, or particle size.

6.4 SUMMARY AND CONCLUSIONS

The performance of SOFCs depends critically on the microstructure of each of the component layers, the composition of the materials used, and the resulting physical properties, such as permeability, electronic, and ionic conductivity, and catalytic activity. The microstructures of the materials are affected, in turn, by the processing methods used to fabricate them. Therefore, selecting the appropriate processing technique for a given combination of materials, stack geometry, target operating temperature range, and target composition, is essential for achieving the performance requirements for the stack. A range of processing methods exists for the fabrication of SOFCs, broadly classified into wet-ceramic techniques and direct-deposition techniques. Both tubular and planar cells are typically fabricated by first synthesizing the support layer, and then performing a series of subsequent deposition and consolidation steps, repeated as needed, to build up the functional layers. While tape casting, screen printing, and co-firing remain the most common techniques for the manufacture of planar cells, and extrusion with EVD, slurry coating, and PS remain the most common methods for the fabrication of tubular cells, a wide variety of new or modified process methods are currently under development with the goal of improving the performance and reducing the cost and complexity of SOFC manufacturing. With the introduction of new catalyst materials with carbon or sulfur tolerance or high activity for oxygen reduction, new requirements continue to arise for suitable manufacturing methods that can be used to process the materials compatibly with the remaining SOFC components. The methods overviewed herein have the potential to be adapted for the fabrication of new materials or geometries to continue improving the performance and cost of SOFCs, thereby bringing them closer to commercialization so that their environmental and efficiency benefits can be realized sooner.

REFERENCES

1. Steele BCH. Appraisal of $Ce_{1-y}Gd_yO_{2-y/2}$ electrolytes for IT-SOFC operation at 500°C. *Solid State Ionics* 2000; 129:95–110.
2. Kuharuangrong S. Ionic conductivity of Sm, Gd, Dy, and Er-doped ceria. *J. Power Sources* 2007; 171:506–510.
3. Jiang SP and Chan SH. Development of Ni/Y_2O_3-ZrO_2 cermet anodes for solid oxide fuel cells. *J. Mater. Sci. Tech.* 2004; 9:1109–1118.
4. Shingal SC and Kendall K. *High Temperature Solid Oxide Fuel Cells: Fundamentals, Design, and Applications.* Oxford: Elsevier, 2003.
5. Atkinson A, Barnett S, Gorte RJ, Irvine JTS, McEvoy AJ, Mogensen M, et al. Advanced anodes for high-temperature fuel cells. *Nature Mater.* 2004; 3:17–27.
6. Xu Z, Luo J, and Chuang KT. CoS-promoted MoS_2 catalysts for SOFC using H_2S-containing hydrogen or syngas as fuel. *J. Electrochem. Soc.* 2007; 154:B523–B527.
7. Vorontsov V, An W, Luo JL, Sanger AR, and Chuang KT. Performance and stability of composite nickel and molybdenum sulfide-based anodes for SOFC utilizing H_2S. *J. Power Sources* 2008; 179:9–16.
8. Tao S and Irvine JTS. A redox-stable efficient anode for solid-oxide fuel cells. *Nature Mater.* 2003; 2:320–323.
9. Liu J, Madsen BD, Ji Z, and Barnett SA. A fuel-flexible ceramic-based anode for SOFCs. *Electrochem. Solid State Lett.* 2002; 5:A122–A124.
10. Murray EP, Tsai T, and Barnett SA. Oxygen transfer processes in $(La,Sr)MnO_3/Y_2O_3$–stabilized ZrO_2 cathodes: an impedance spectroscopy study. *Solid State Ionics* 1998; 110:235–243.
11. White BD, Kesler O, and Rose L. Air plasma spray processing and electrochemical characterization of SOFC composite cathodes. *J. Power Sources* 2008; 178:334–343.
12. Zheng R, Zhou XM, Wang SR, Wen TL, and Ding CX. A study of Ni + 7YSZ/8YSZ/$La_{0.6}Sr_{0.4}CoO_{3-\delta}$ ITSOFC fabricated by atmospheric plasma spraying. *J. Power Sources* 2005; 140:217–225.
13. Hung MH, Madhava RMV, and Tsai DS. Microstructures and electrical properties of calcium substituted $LaFeO_3$ as SOFC cathode. *Mater. Chem. Phys.* 2007; 101:297–302.
14. Murray EP, Sever MJ, and Barnett SA. Electrochemical performance of $(La,Sr)(Co,Fe)O_3$-$(Ce,Gd)O_3$ composite cathodes. *Solid State Ionics* 2002; 148:27–34.
15. Wang WG and Mogensen M. High-performance lanthanum-ferrite-based cathode for SOFC. *Solid State Ionics* 2005; 176:457–462.
16. Hansen KK, Søgaard M, and Mogensen M. $Gd_{0.6}Sr_{0.4}Fe_{0.8}Co_{0.2}O_{3-\delta}$: A novel type of SOFC cathode. *Electrochem. Solid State Lett.* 2007; 10:B119–B121.
17. Huang B, Ye XF, Wang SR, Nie HW, Liu RZ, and Wen TL. Performance of $Ni/ScSZ$ cermet anode modified by coating with $Gd_{0.2}Ce_{0.8}O_2$ for a SOFC. *Mater. Res. Bull.* 2007; 42:1705–1714.
18. Chen XJ, Liu QL, Chan SH, Brandon NP, and Khor KA. High performance cathode-supported SOFC with perovskite anode operating in weakly humidified hydrogen and methane. *Electrochem. Commun.* 2007; 9:767–772.

19. Suda S, Itagaki M, Node E, Takahashi S, Kawano M, Yoshida, H et al. Preparation of SOFC anode composites by spray pyrolysis. *J. Eur. Ceram. Soc.* 2006; 26:593–597.

20. Song HS, Kim WH, Hyun SH, and Moon J. Influences of starting particulate materials on microstructural evolution and electrochemical activity of LSM-YSZ composite cathode for SOFC. *J. Electroceram.* 2006; 17:759–764.

21. Hagiwara A, Hobara N, Takizawa K, Sato K, Abe H, and Naito M. Preparation and evaluation of mechanochemically fabricated LSM/ScSZ composite materials for SOFC cathodes. *Solid State Ionics* 2006; 177:2967–2977.

22. Simner SP, Anderson MD, Templeton JW, and Stevenson JW. Silver-perovskite composite SOFC cathodes processed via mechanofusion. *J. Power Sources* 2007; 168:236–239.

23. Mertens J, Haanappel VAC, Tropartz C, Herzhof W, and Buchkremer HP. The electrochemical performance of anode-supported SOFCs with LSM-type cathodes produced by alternative processing routes. *J. Fuel Cell Sci. Technol.* 2006; 3:125–130.

24. Kim SD, Lee JJ, Moon H, Hyun SH, Moon J, Kim J et al. Effects of anode and electrolyte microstructures on performance of solid oxide fuel cells. *J. Power Sources* 2007; 169:265–270.

25. Liu Y and Liu M. Porous SOFC anodes prepared by sublimation of an immiscible metal oxide during sintering. *Electrochem. Solid State Lett.* 2006; 9:B25–B27.

26. Kong J, Sun K, Zhou D, Zhang N, Mu J, and Qiao J. Ni-YSZ gradient anodes for anode-supported SOFCs. *J. Power Sources* 2007; 166:337–342.

27. Song HS, Hyun SH, Moon J, and Song RH. Electrochemical and microstructural characterization of polymeric resin-derived multilayered composite cathode for SOFC. *J. Power Sources* 2005; 145:272–277.

28. Ruiz-Morales JC, Canales-Vázquez J, Savaniu C, Marrero-López C, Zhou W, and Irvine JTS. Disruption of extended defects in solid oxide fuel cell anodes for methane oxidation. *Nature* 2006; 439:568–571.

29. Barthel K and Rambert S. Thermal spraying and performance of graded composite cathodes as SOFC-component. *Mat. Sci. Forum.* 1999; 308–311:800–805.

30. Xu X, Xia C, Xiao G, and Peng D. Fabrication and performance of functionally graded cathodes for IT-SOFCs based on doped ceria electrolytes. *Solid State Ionics* 2005; 176:1513–1520.

31. Hart NT, Brandon NP, Day MJ, and Shemilt JE. Functionally graded cathodes for solid oxide fuel cells. *J. Mat. Sci.* 2001; 36:1077–1085.

32. Basu RN, Tietz F, Wessel E, and Stöver D. Interface reactions during co-firing of solid oxide fuel cell components. *J. Mater. Process. Technol.* 2004; 147:85–89.

33. Bebelis S, Kotsionopoulos N, Mai A, Rutenbeck D, and Tietz F. Electrochemical characterization of mixed conducting and composite SOFC cathodes. *Solid State Ionics* 2006; 177:1843–1848.

34. Zhang X, Robertson M, Deces-Petit C, Qu W, Kesler O, Maric R et al. Internal shorting and fuel loss of a low temperature solid oxide fuel cell with SDC electrolyte. *J. Power Sources* 2007; 164:668–677.

35. Ralph JM, Kilner JA, and Steele BCH. Improving Gd-doped ceria electrolytes for low temperature solid oxide fuel cells. *Mater. Res. Soc. Symp. Proc.* 1999; 575:309–314.

36. Bi Z, Yi B, Wang Z, Dong Y, Wu H, She Y et al. A high-performance anode-supported SOFC with LDC-LSGM bilayer electrolytes. *Electrochem. Solid State Lett.* 2004; 7:A105–A107.

37. Lu XC and Zhu JH. Ni-Fe + SDC composite as anode material for intermediate temperature solid oxide fuel cell. *J. Power Sources* 2007; 165:678–684.

38. Wan J, Goodenough JB, and Zhu JH. $Nd_{2-x}La_xNiO_{4+\delta}$, a mixed ionic/electronic conductor with interstitial oxygen, as a cathode material. *Solid State Ionics* 2007; 178:281–286.

39. Huang YH, Dass RI, Denyszyn JC, and Goodenough JB. Synthesis and characterization of $Sr_2MgMoO_{6-\delta}$. *J. Electrochem. Soc.* 2006; 153:A1266–A1272.

40. Huang YH, Dass RI, Xing ZL, and Goodenough JB. Double perovskites as anode materials for solid-oxide fuel cells. *Science* 2006; 312:254–257.

41. Rose L, Menon M, Kammer K, Kesler O, and Halvor Larsen P. Processing of $Ce_{1-x}Gd_xO_{2-\delta}$ (GDC) thin films from precursors for application in solid oxide fuel cells. *Adv. Mater. Res.* 2007; 15–17:293–298.

42. Tsoga A, Naoumidis A, and Stover D. Total electrical conductivity and defect structure of ZrO_2–CeO_2–Y_2O_3–Gd_2O_3 solid solutions. *Solid State Ionics* 2000; 135:403–409.

43. Nguyen TL, Kato T, Nozaki K, Honda T, Negishi A, Kato K, and Limura Y. Application of $(Sm_{0.5}Sr_{0.5})CoO_3$ as a cathode material to $(Zr,Sc)O_2$ electrolyte with ceria-based interlayers for reduced-temperature operation SOFCs. *J. electrochem. Soc.* 2006; 153:A1310–A1316.

44. Zhang X, Robertson M, Deces-Petit C, Xie Y, Hui R, Yick S et al. NiO-YSZ cermet-supported low temperature solid oxide fuel cells. *J. Power Sources* 2006; 161:301–307.

45. Zhang X, Robertson M, Deces-Petit C, Xie Y, Hui R, Qu W et al. Solid oxide fuel cells with bi-layered electrolyte structure. *J. Power Sources* 2008; 175:800–805.

46. Huang QA, Oberste-Berghaus J, Yang D, Yick S, Wang Z, Wang B et al. Polarization analysis for metal-supported SOFCs from different fabrication processes. *J. Power Sources* 2008; 177:339–347.

47. Tomita A, Teranishi S, Nagao M, Hibino T, and Sano M. Comparative performance of anode-supported SOFCs using a thin $Ce_{0.9}Gd_{0.1}O_{1.95}$ electrolyte with an incorporated $BaCe_{0.8}Y_{0.2}O_{3-\delta}$ layer in hydrogen and methane. *J. Electrochem. Soc.* 2006; 153:A956–A960.

48. Singhal SC. Advances in solid oxide fuel cell technology. *Solid State Ionics* 2000; 135:305–313.

49. Sarkar P, Yamarte L, Rho H, and Johanson L. Anode-supported tubular micro-solid oxide fuel cell. *Int. J. Appl. Ceram. Technol.* 2007; 4:103–108.

50. Du Y and Sammes NM. Fabrication of tubular electrolytes for solid oxide fuel cells using strontium- and magnesium-doped $LaGaO_3$ materials. *J. Eur. Ceram. Soc.* 2001; 21:727–735.

51. Sammes NM and Du Y. Fabrication and characterization of tubular solid oxide fuel cells. *Int. J. Appl. Ceram. Technol.* 2007; 4:89–102.

52. Misono T, Murat, K, Fukui T, Chaichanawong J, Sato K, Abe H et al. Ni-SDC cermet anode fabricated from NiO-SDC composite powder for intermediate temperature SOFC. *J. Power Sources* 2006; 157:754–757.

53. Yin Y, Li S, Xia C, and Meng G. Electrochemical performance of gel-cast NiO-SDC composite anodes in low-temperature SOFCs. *Electrochim. Acta* 2006; 51:2594–2598.

54. Huang K, Wan JH, and Goodenough JB. Increasing power density of LSGM-based solid oxide fuel cells using new anode materials. *J. Electrochem. Soc.* 2001; 148:A788–A794.

55. Lang M, Szabo P, Ilhan Z, Cinque S, Franco T, and Schiller G. Development of solid oxide fuel cells and short stacks for mobile applications. *J. Fuel Cell Sci. Technol.* 2007; 4:384–391.

56. Brandon NP, Blake A, Corcoran D, Cumming D, Duckett A, El-Koury K et al. Development of metal supported solid oxide fuel cells for operation at 500–600°C. *J. Fuel Cell Sci. Technol.* 2004; 1:61–65.

57. Villarreal I, Jacobson C, Leming A, Matus Y, Visco S, and De Jonghe L. Metal-supported solid oxide fuel cells. *Electrochem. Solid State Lett.* 2003; 6:A178–A179.

58. Corbin SF, Lee J, and Qiao X. Influence of green formulation and pyrolyzable particulates on the porous microstructure and sintering characteristics of tape cast ceramics. *J. Am. Cer. Soc.* 2001; 84:41–47.

59. Nguyen TL, Kobayashi K, Honda T, Iimura Y, Kato K, Neghisi, A et al. Preparation and evaluation of doped ceria interlayer on supported stabilized zirconia electrolyte SOFCs by wet ceramic processes. *Solid State Ionics* 2004; 174:163–174.

60. Wang Z, Qian J, Cao J, Wang S, and Wen TJ. A study of multilayer tape casting method for anode-supported planar type solid oxide fuel cells (SOFCs). *J. Alloys Compounds* 2007; 437:264–268.

61. Yin Y, Li S, Xia C, and Meng GJ. Electrochemical performance of IT-SOFCs with a double-layer anode. *J. Power Sources* 2007; 167:90–93.

62. Wang Y, Nie H, Wang S, Wen TL, Guth U, and Valshook V. $A_{2-\alpha}A'_{\alpha}BO_4$-type oxides as cathode materials for IT-SOFCs (A=Pr, Sm; A' = Sr; B= Fe, Co). *Mater. Lett.* 2006; 60:1174–1178.

63. Li Q, Zhao H, Huo L, Sun L, Cheng X, and Grenier JC. Electrode properties of Sr doped La_2CuO_4 as new cathode material for intermediate-temperature SOFCs. *Electrochem. Commun.* 2007; 9:1508–1512.

64. Lee S, Lim Y, Lee EA, Hwang HJ, and Moon JW. $Ba_{0.5}Sr_{0.5}Co_{0.8}Fe_{0.2}O_{3-\delta}$ (BSCF) and $La_{0.6}Ba_{0.4}Co_{0.2}Fe_{0.8}O_{3-\delta}$ (LBCF) cathodes prepared by combined citrate-EDTA method for IT-SOFCs. *J. Power Sources* 2006; 157:848–854.

65. Fu C, Sun K, Zhang N, Chen X, and Zhou D. Electrochemical characteristics of LSCF-SDC composite cathode for intermediate temperature SOFC. *Electrochim. Acta* 2007; 52:4589–4594.

66. Lee KT and Manthiram A. Synthesis and characterization of $Nd_{0.6}Sr_{0.4}Co_{1-y}Mn_yO_{3-\delta}$ ($0\leq y\leq1.0$) cathodes for intermediate temperature solid oxide fuel cells. *J. Power Sources* 2006; 158:1202–1208.

67. Li Q, Fan Y, Zhao H, Sun LP, and Huo LH. Preparation and electrochemical properties of a $Sm_{2-x}Sr_xNiO_4$ cathode for an IT-SOFC. *J. Power Sources* 2007; 167:64–68.

68. Zhou W, Shao Z, Ran R, Zeng P, Gu H, Jin W et al. $Ba_{0.5}Sr_{0.5}Co_{0.8}Fe_{0.2}O_{3-\delta}$ + $LaCoO_3$ composite cathode for $Sm_{0.2}Ce_{0.8}O_{1.9}$-electrolyte based intermediate-temperature solid-oxide fuel cells. *J. Power Sources* 2007; 168:330–337.

69. Shao Z and Halle SM. A high-performance cathode for the next generation of solid oxide fuel cells. *Nature* 2004; 431:170–173.

70. Hui R, Wang Z, Yick S, Maric R, and Ghosh D. Fabrication of ceramic films for solid oxide fuel cells via slurry spin coating technique. *J. Power Sources* 2007; 172:840–844.

71. Subramania A, Saradha T, and Muzhumathi S. Synthesis of nano-crystalline $(Ba_{0.5}Sr_{0.5})$ $Co_{0.8}Fe_{0.2}O_{3-\delta}$ cathode material by a novel sol-gel thermolysis process for IT-SOFCs. *J. Power Sources* 2007; 165:728–732.

72. Liu J, Co AC, Paulson S, and Birss VI. Oxygen reduction at sol-gel derived $La_{0.8}Sr_{0.2}Co_{0.8}Fe_{0.2}O_3$ cathodes. *Solid State Ionics* 2006; 177:377–387.

73. Tang Z, Xie Y, Hawthorne H, and Ghosh D. Sol-gel processing of $Sr_{0.5}Sm_{0.5}CoO_3$ film. *J. Power Sources* 2006; 157:385–388.

74. Xia C, Zhang Y, and Liu M. LSM-GDC composite cathodes derived from a sol-gel process. *Electrochem. Solid State Lett.* 2003; 6:A290–A292.

75. Mehta K, Xu R, and Virkar AV. Two-layer fuel cell electrolyte structure by sol-gel processing. *J. Sol Gel Sci. Tech.* 1998; 11:203–207.

76. Xu Z, Rajaram G, Sankar J, and Pai D. Electrophoretic deposition of YSZ electrolyte coatings for SOFCs. *Fuel Cells Bull.* 2007; March:12–16.

77. Zhitomirsky I and Petric A. Electrophoretic deposition of electrolyte materials for solid oxide fuel cells. *J. Mater. Sci.* 2004; 39:825–831.

78. Ishihara T, Sato K, and Takita Y. Electrophoretic deposition of Y_2O_3-stabilized ZrO_2 electrolyte films in solid oxide fuel cells. *J. Am. Ceram. Soc.* 1996; 79:913–919.

79. Zhang X, Robertson M, Yick S, Deces-Petit C, Qu W, Xie Y et al. $Sm_{0.5}Sr_{0.5}CoO_3$ + $Sm_{0.2}Ce_{0.8}O_{1.9}$ composite cathode for cermet supported thin $Sm_{0.2}Ce_{0.8}O_{1.9}$ electrolyte SOFC operating below 600°C. *J. Power Sources* 2006; 160:1211–1216.

80. Godoi GS and de Souza DPF. Electrical and microstructural characterization of $La_{0.7}Sr_{0.3}MnO_3$ (LSM), $Ce_{0.8}Y_{0.2}O_2$ (CY) and LSM-CY composites. *Mater. Sci. Eng. B* 2007; 140:90–97.

81. Bansal NP and Zhong Z. Combustion synthesis of $Sm_{0.5}Sr_{0.5}CoO_{3-x}$ and $La_{0.6}Sr_{0.4}CoO_{3-x}$ nanopowders for solid oxide fuel cell cathodes. *J. Power Sources* 2006; 158:148–153.

82. Zhang X, Deces-Petit C, Yick S, Robertson M, Kesler O, Maric R et al. A study on sintering aids for $Sm_{0.2}Ce_{0.8}O_{1.9}$ electrolyte. *J. Power Sources* 2006; 162:480–485.

83. Huang Y, Vohs JM, and Gorte RJ. SOFC cathodes prepared by infiltration with various LSM precursors. *Electrochem. Solid State Lett.* 2006; 9:A237–A240.

84. Gorte RJ, Kim H, and Vohs JM. Novel SOFC anodes for the direct electrochemical oxidation of hydrocarbon. *J. Power Sources* 2002; 106:10–15.

85. Lee SI, Ahn K, Vohs JM, and Gorte RJ. Cu-Co bimetallic anodes for direct utilization of methane in SOFCs. *Electrochem. Solid State Lett.* 2005; 8:A48–A51.

86. Kim H, Lu C, Worrell WL, Vohs JM, and Gorte RJ. Cu-Ni cermet anodes for direct oxidation of methane in solid-oxide fuel cells. *J. Electrochem. Soc.* 2002; 149:A247–A250.

87. Gross MD, Vohs JM, and Gorte RJ. A study of thermal stability and methane tolerance of Cu-based SOFC anodes with electrodeposited Co. *Electrochim. Acta* 2007; 52:1951–1957.

88. Qiao J, Sun K, Zhang N, Sun B, Kong J, and Zhou DJ. Ni/YSZ and Ni-CeO$_2$/YSZ anodes prepared by impregnation for solid oxide fuel cells. *J. Power Sources* 2007; 169:253–258.

89. Sholklapper TZ, Radmilovic V, Jacobson CP, Visco SJ, and De Jonghe LC. Synthesis and stability of a nanoparticle-infiltrated solid oxide fuel cell electrode. *Electrochem. Solid State Lett.* 2007; 10:B74–B76.

90. Lu C, Sholklapper TZ, Jacobson CP, Visco SJ, and De Jonghe LC. LSM-YSZ cathodes with reaction-infiltrated nanoparticles. *J. Electrochem. Soc.* 2006; 153:A1115–A1119.

91. Xu X, Jiang Z, Fan X, and Xia C. LSM-SDC electrodes fabricated with an ion-impregnating process for SOFCs with doped ceria electrolytes. *Solid State Ionics* 2006; 177:2113–2117.

92. Gross MD, Vohs JM, and Gorte RJ. A strategy for achieving high performance with SOFC ceramic anodes. *Electrochem. Solid State Lett.* 2007; 10: B65–B69.

93. Tanner C, Jue JF, and Virkar AV. Temperature dependence of the kinetics of electrochemical vapor deposition of CeO_2. *J. Electrochem. Soc.* 1993; 140:1073–1080.

94. Jue JF, Jusko J, and Virkar AV. Electrochemical vapor deposition of CeO_2: Kinetics of deposition of a composite, two-layer electrolyte. *J. Electrochem. Soc.* 1992; 139:2458–2465.

95. Tanner CW and Virkar AV. Electrochemical liquid deposition of ceria. *J. Am. Cer. Soc.* 1994; 77:2209–2212.

96. Haldane MA and Etsell TH. Fabrication of composite SOFC anodes. *Mater. Sci. Eng. B* 2005; 121:120–125.

97. Choy KL. Chemical vapour deposition of coatings. *Progr. Mater. Sci.* 2003; 48:57–170.

98. Meng G, Song H, Xia C, Liu X, and Peng D. Novel CVD techniques for micro- and IT-SOFC fabrication. *Fuel Cells* 2004; 4:48–55.

99. Wang HB, Meng GY, and Peng DK. Aerosol and plasma assisted chemical vapor deposition process for multi-component oxide $La_{0.8}Sr_{0.2}MnO_3$ thin film. *Thin Solid Films* 2000; 368:275–278.

100. Meng G, Song H, Dong Q, and Peng D. Application of novel aerosol-assisted chemical vapor deposition techniques for SOFC thin films. *Solid State Ionics* 2004; 175:29–34.

101. Liu Y, Compson C, and Liu M. Nanostructured and functionally graded cathodes for intermediate-temperature SOFCs. *Fuel Cells Bull.* 2004; 10:12–15.

102. Liu Y and Liu M. Porous electrodes for low-temperature solid oxide fuel cells fabricated by a combustion spray process. *J. Am. Ceram. Soc.* 2004; 87:2139–2142.

103. Fauchais P. Understanding plasma spraying. *J. Phys. D: Appl. Phys.* 2004; 37:R86–R108.

104. Hui R, Wang Z, Kesler O, Rose L, Jankovic J, Yick S et al. Thermal plasma spraying of SOFCs: applications, potential advantages, and challenges. *J. Power Sources* 2007; 170:308–323.

105. Kesler O, Plasma spray processing of planar solid oxide fuel cells. *Mat. Sci. Forum* 2007; 539–543:1385–1390.

106. Henne R. Solid oxide fuel cells: a challenge for plasma deposition processes. *J. Therm. Spray Tech.* 2007; 16:381–403.

107. Karthikeyan J, Berndt CC, Tikkanen J, Reddy S, and Herman H. Plasma spray synthesis of nanomaterial powders and deposits. *Mat. Sci. Eng. A* 1997; 238:275–286.

108. Gitzhofer F, Boulos M, Heberlein J, Henne R, Ishigaki T, and Yoshida T. Integrated fabrication processes for solid-oxide fuel cells using thermal plasma spray technology. *Mater. Res. Soc. Bull.* 2000; July:38–42.

109. Khor KA, Yu LG, Chan SH, and Chen XJ. Densification of plasma sprayed YSZ electrolytes by spark plasma sintering (SPS). *J. Eur. Ceram. Soc.* 2003; 23:1855–1863.

110. Schiller G, Henne R, and Ruckdaeschel R. Vacuum plasma sprayed protective layers for solid oxide fuel cell application. *J. Adv. Mat.* 2000; 32:3–8.

111. Tai L and Lessing PA. Plasma spraying of porous electrodes for a planar solid oxide fuel cell. *J. Am. Ceram. Soc.* 1991; 74:501–504.

112. Schiller G, Henne R, Lang M, and Muller M. Development of solid oxide fuel cells (SOFC) for stationary and mobile applications by applying plasma deposition processes. *Mat. Sci. Forum* 2003; 3:2539–2544.

113. Franco T, HoshiarDin Z, Szabo P, Lang M, and Schiller G. Plasma sprayed diffusion barrier layers based on doped perovskite-type $LaCrO_3$ at substrate-anode interface in solid oxide fuel cells. *J. Fuel Cell Sci. Technol.* 2007; 4:406–412.

114. Lang M, Franco T, Schiller G, and Wagner N. Electrochemical characterization of vacuum plasma sprayed thin-film solid oxide fuel cells (SOFC) for reduced operating temperatures. *J. Appl. Electrochem.* 2002; 32:871–874.

115. Ma XQ, Hui S, Zhang H, Dai J, Roth J, Xiao TD et al. Intermediate temperature SOFC based on fully integrated plasma sprayed components. In Thermal Spray 2003: Advancing the Science & Applying the Technology, C. Moreau and B. Marple (eds.) 2003; Materials Park, Ohio: ASM International.

116. Vassen R, Hathiramani D, Mertens J, Haanappel VAC, and Vinke IC. Manufacturing of high performance solid oxide fuel cells (SOFCs) with atmospheric plasma spraying (APS). *Surf. Coat. Technol.* 2007; 202:499–508.

117. Ben Oved N and Kesler O. A new technique for the rapid manufacturing of direct-oxidation anodes for SOFCs. *Adv. Mat. Res.* 2007; 15–17:287–292.

118. Hsu CS and Hwang BH. Microstructure and properties of the $La_{0.6}Sr_{0.4}Co_{0.2}Fe_{0.8}O_3$ cathodes prepared by electrostatic-assisted ultrasonic spray pyrolysis method. *J. Electrochem. Soc.* 2006; 153:A1478–A1483.

119. Beckel D, Dubach A, Studart AR, and Gauckler LJ. Spray pyrolysis of $La_{0.6}Sr_{0.4}$ $Co_{0.2}Fe_{0.8}O_{3-\delta}$ thin film cathodes. *J. Electroceram.* 2006; 16:221–228.

120. Holtappels P, Vogt U, and Graule T. Ceramic materials for advanced solid oxide fuel cells. *Adv. Eng. Mater.* 2005; 7:292–302.

121. Xie Y, Neagu R, Hsu CS, Zhang X, and Deces-Petit C. Spray pyrolysis deposition of electrolyte and anode for metal-supported solid oxide fuel cell. *J. Electrochem. Soc.* 2008; 155:B407–B410.

122. Pederson LR, Singh P, and Zhou XD. Application of vacuum deposition methods to solid oxide fuel cells. *Vacuum* 2006; 80:1066–1083.

123. Imanishi N, Matsumura T, Sumiya Y, Yoshimura K, Hirano A, Takeda Y et al. Impedance spectroscopy of perovskite air electrodes for SOFC prepared by laser ablation method. *Solid State Ionics* 2004; 174:245–252.

124. Sase M, Suzuki J, Yashiro K, Otake T, Kaimai A, Kawada T et al. Electrode reaction and microstructure of $La_{0.6}Sr_{0.4}CoO_{3-\delta}$ thin films. *Solid State Ionics* 2006; 177:1961–1964.

125. Ruiz de Larramendi I, López Antón R, Ruiz de Larramendi JI, Baliteau S, Mauvy F, Grenier JC et al. Structural and electrical properties of thin films of $Pr_{0.8}Sr_{0.2}Fe_{0.8}Ni_{0.2}O_{3-\delta}$. *J. Power Sources* 2007; 169:35–39.

126. Koep E, Jin C, Haluska M, Das R, Narayan R, Sandhage K et al. Microstructure and electrochemical properties of cathode materials for SOFCs prepared via pulsed laser deposition. *J. Power Sources* 2006; 161:250–255.

127. Yang D, Zhang X, Nikumb S, Decès-Petit C, Hui R, Maric R et al. Low temperature solid oxide fuel cells with pulsed laser deposited bi-layer electrolyte. *J. Power Sources* 2007; 164:182–188.

128. La O GJ, Hertz J, Tuller H, and Shao-Horn Y. Microstructural features of RF-sputtered SOFC anode and electrolyte materials. *J. Electroceram.* 2004; 13:691–695.

129. Nagata A and Okayama H. Characterization of solid oxide fuel cell device having a three-layer film structure grown by RF magnetron sputtering. *Vacuum* 2002; 66: 523–529.

130. Radhakrishnan R, Virkar AV, and Singhal SC. Estimation of charge-transfer resistivity of Pt cathode on YSZ electrolyte using patterned electrodes. *J. Electrochem. Soc.* 2005; 152:A927–A936.

131. Hertz JL and Tuller HL. Electrochemical characterization of thin films for a micro-solid oxide fuel cell. *J. Electroceram.* 2004; 13:663–668.

132. Charpentier P, Fragnaud P, Schleich DM, Lunot C, and Gehain E. Preparation of cathodes for thin film SOFCs. *Ionics* 1997; 3:155–160.

133. Labrincha JA, Li-Jian M, dos Santos MP, Marques FMB, and Frade JR. Evaluation of deposition techniques of cathode materials for solid oxide fuel cells. *Mater. Res. Bull.* 1993; 28:101–109.

Index

A

A-site cations
  deficiencies in lanthanum cobaltite/ferrite
    perovskites, 148
  in LSM-based perovskites, 143
Activation energy, 4, 7, 142
  composition dependence for SM-doped
    ceria, 34
  and doping concentration for yttria-doped
    ceria, 37
  effect of Sm-doping concentration for
    codoped ceria electrolytes, 40
  in LSM perovskites, 141–146
  of O_2 reduction reaction, 141
AISI 430, 194
  SEM cross-sections, 194
Alloy degradation, 193
Alloy design
  deemphasis on electrical conductivity, 189
  for SOFC applications, 187
Alloy groups, key properties comparison, 188
Alloys, for metallic interconnects, 187–190
Alumina, effects on conductivity of zirconia-
  based electrolytes, 16
Anode composition, and electrochemical
  performance, 90–92
Anode reduction condition, and conductivity,
  86–89
Anode sintering condition, and conductivity, 84
Anode-supported tubes, electrophoretic
  deposition methods, 253–254
Anodes, 73–74
  adsorption energies of atomic carbon
    adsorption, 117
  alternate anodes for SOFCs, 115–121, 122
  atomic carbon adsorption, 116
  carbon-tolerant, 115–118
  catalytic activity requirements, 74
  chemical and thermal stability requirements,
    74
  design challenges for charge and mass
    transfer, 74
  ionic conductivity requirements, 74
  materials, 75
  Ni-YSZ cermet anodes, 76–115
  nickel mismatch problems, 75
  primary function, 74
  removing processes for carbon-tolerant
    materials, 116

requirements, 74–75
sulfur-tolerant, 118–121
symbols and abbreviations, 122
tolerance to carbon deposition, 75
tolerance to sulfur poisoning, 75
Area specific resistance (ASR), 20
  in MIEC cathodes, 153
  and seal leaks, 227
Arrhenius plots
  for ceria electrolyte production methods, 48
  conductivities for Sm-doped ceria, 35
  for Sm-doped ceria, 36
  for yttria-stabilized zirconia, 7
Atmospheric plasma spraying (APS), 266
Atomic carbon adsorption, of transition
  metals, 47
Austenitic stainless steel alloys, 188

B

Ball-milled powders, morphology, 246
Barium-aluminosilicate-based seals, 217
Barium-calcium-aluminosilicate (BCAS) glass
  ceramic seals, 218
Barium chromate formation, 218
Barrier layers, 250–251
BCAS base glasses, 196
BCAS seals, 217
Bilayered electrodes, 248
  pulsed laser deposition of, 269
Bilayered electrolytes, 250–251
Bismuth-oxide electrolytes, conductivity, 5
Boron, sealant deficiencies, 217

C

Ca-doped lanthanum ferrite (LCF), 243
Calcia-doped ceria, conductivity of, 21
Calcium doped gadolinium chromite, 182
Calcium doped lanthanum chromite, 180, 184
  conductivity, 182
  TEC for, 185
Carbon coking process, 116
Carbon deposition
  chemical requirements to avoid, 241
  long-term effects of, 262–263
Carbon-induced corrosion, in metallic
  interconnects, 193
Carbon-tolerant anodes, 115–118